T0296510

Finite group theory

During the last 40 years the theory of finite groups has developed dramatically. The finite simple groups have been classified and are becoming better understood. Tools exist to reduce many questions about arbitrary finite groups to similar questions about simple groups. Since the classification there have been numerous applications of this theory in other branches of mathematics.

Finite Group Theory develops the foundations of the theory of finite groups. It can serve as a text for a course on finite groups for students already exposed to a first course in algebra. For the reader with some mathematical sophistication but limited knowledge of finite group theory, the book supplies the basic background necessary to begin to read journal articles in the field. It also provides the specialist in finite group theory with a reference in the foundations of the subject.

The second edition of *Finite Group Theory* has been considerably improved, with a completely rewritten chapter 15 considering the 2-Signalizer Functor Theorem and the addition of an appendix containing solutions to exercises.

M. Aschbacher is Shaler Arthur Hanisch Professor of Mathematics at Caltech. His books include *Finite Group Theory* (Cambridge University Press, 1986), *Sporadic Groups* (Cambridge University Press, 1994), and *3-Transposition Groups* (Cambridge University Press, 1997).

CAMBRIDGE STUDIES IN ADVANCED MATHEMATICS

Editorial Board

B. Bollobas, W. Fulton, A. Katok, F. Kirwan, P. Sarnak

FINITE GROUP THEORY

Second Edition

M. ASCHBACHER

California Institute of Technology

CAMBRIDGE UNIVERSITY PRESS
Cambridge, New York, Melbourne, Madrid, Cape Town, Singapore,
São Paulo, Delhi, Dubai, Tokyo, Mexico City

Cambridge University Press
The Edinburgh Building, Cambridge CB2 8RU, UK

Published in the United States of America by Cambridge University Press, New York

www.cambridge.org
Information on this title: www.cambridge.org/9780521786751

First edition published 1986
Second edition published 2000

A catalogue record for this publication is available from the British Library

Library of Congress Cataloguing in Publication data
Aschbacher, Michael, 1944–
Finite group theory / M. Aschbacher. – 2nd ed.
p.   cm. – (Cambridge studies in advanced mathematics ; 10)
Includes bibliographical references and index.
ISBN 0-521-78145-0 (hb) – ISBN 0-521-78675-4 (pbk.)
1. Finite groups.   I. Title.   II. Series.
QA177 .A82   2000
512'.2 – dc21   99-055693

ISBN 978-0-521-78145-9 Hardback
ISBN 978-0-521-78675-1 Paperback

To Pam

Contents

Preface

Finite Group Theory is intended to serve both as a text and as a basic reference on finite groups. In neither role do I wish the book to be encyclopedic, so I've included only the material I regard as most fundamental. While such judgments are subjective, I've been guided by a few basic principles which I feel are important and should be made explicit.

One unifying notion is that of a group representation. The term representation is used here in a much broader sense than usual. Namely in this book a representation of a group G in a category $\mathscr{C}$ is a homomorphism of G into the automorphism group of some object of $\mathscr{C}$. Among these representations, the permutation representations, the linear representations, and the representations of groups on groups seem to be the most fundamental. As a result much of the book is devoted to these three classes of representations.

The first step in investigating representations of finite groups or finite dimensional groups is to break up the representation into indecomposable or irreducible representations. This process focuses attention on two areas of study: first on the irreducible and indecomposable representations themselves, and second on the recovery of the general representation from its irreducible constituents. Both areas receive attention here.

The irreducible objects in the category of groups are the simple groups. I regard the finite simple groups and their irreducible linear and permutation representations as the center of interest in finite group theory. This point of view above all others has dictated the choice of material. In particular I feel many of the deeper questions about finite groups are best answered through the following process. First reduce the question to a question about some class of irreducible representations of simple groups or almost simple groups. Second appeal to the classification of the finite simple groups to conclude the group is an alternating group, a group of Lie type, or one of the 26 sporadic simple groups. Finally invoke the irreducible representation theory of these groups.

The book serves as a foundation for the proof of the Classification Theorem. Almost all material covered plays a role in the classification, but as it turns out almost all is of interest outside that framework too. The only major result treated here which has not found application outside of simple group theory is the Signalizer Functor Theorem. Signalizer functors are discussed near the end of the book. The last section of the book discusses the classification in general terms.

The first edition of the book included a new proof of the Solvable Signalizer Functor Theorem, based on earlier work of Helmut Bender. Bender's proof was valid only for the prime 2, but it is very short and elegant. I've come to believe that my extension to arbitrary primes in the first edition is so complicated that it obscures the proof, so this edition includes only a proof of the Solvable 2-Signalizer Functor Theorem, which is closer to Bender's original proof. Because of this change, section 36 has also been truncated.

In some sense most of the finite simple groups are classical linear groups. Thus the classical groups serve as the best example of finite simple groups. They are also representative of the groups of Lie type, both classical and exceptional, finite or infinite. A significant fraction of the book is devoted to the classical groups. The discussion is not restricted to groups over finite fields. The classical groups are examined via their representation as the automorphism groups of spaces of forms and their representation as the automorphism groups of buildings. The Lie theoretic point of view enters into the latter representation and into a discussion of Coxeter groups and root systems.

I assume the reader has been exposed to a first course in algebra or its equivalent; Herstein's *Topics in Algebra* would be a representative text for such a course. Occasionally some deeper algebraic results are also needed; in such instances the result is quoted and a reference is given for its proof. Lang's *Algebra* is one reference for such results. The group theory I assume is listed explicitly in section 1. There isn't much; for example Sylow's Theorem is proved in chapter 2.

As indicated earlier, the book is intended to serve both as a text and as a basic reference. Often these objectives are compatible, but when compromise is necessary it is usually in favor of the role as a reference. Proofs are more terse than in most texts. Theorems are usually not motivated or illustrated with examples, but there are exercises. Many of the results in the exercises are interesting in their own right; often there is an appeal to the exercises in the book proper. In this second edition I've added an appendix containing solutions to some of the most difficult and/or important exercises.

If the book is used as a text the instructor will probably wish to expand many proofs in lecture and omit some of the more difficult sections. Here are some suggestions about which sections to skip or postpone.

A good basic course in finite group theory would consist of the first eight chapters, omitting sections 14, 16, and 17 and chapter 7, and adding sections 28, 31, 34, 35, and 37. Time permitting, sections 32, 33, 38, and 39 could be added.

The classical groups and some associated Lie theory are treated in chapter 7, sections 29 and 30, chapter 14, and the latter part of section 47. A different sort of course could be built around this material.

Chapter 9 deals with various concepts in the theory of linear representations which are somewhat less basic than most of those in chapters 4 and 12. Much of the material in chapter 9 is of principal interest for representations over fields of prime characteristic. A course emphasizing representation theory would probably include chapter 9.

Chapter 15 is the most technical and specialized. It is probably only of interest to potential simple groups theorists.

Chapter 16 discusses the finite simple groups and the classification. The latter part of section 47 builds on chapter 14, but the rest of chapter 16 is pretty easy reading. Section 48 consists of a very brief outline of the proof of the finite simple groups makes use of results from earlier in the book and thus motivates those results by exhibiting applications of the results.

Each chapter begins with a short introduction describing the major results in the chapter. Most chapters close with a few remarks. Some remarks acknowledge sources for material covered in the chapter or suggest references for further reading. Similarly, some of the remarks place certain results in context and hence motivate those results. Still others warn that some section in the chapter is technical or specialized and suggests the casual reader skip or postpone the section.

In addition to the introduction and the remarks, there is another good way to decide which results in a chapter are of most interest: those results which bear some sort of descriptive label (e.g. Modular Property of Groups, Frattini Argument) are often of most importance.

1

Preliminary results

I assume familiarity with material from a standard course on elementary algebra. A typical text for such a course is Herstein [He]. A few deeper algebraic results are also needed; they can be found for example in Lang [La]. Section 1 lists the elementary group theoretic results assumed and also contains a list of basic notation. Later sections in chapter 1 introduce some terminology and notation from a few other areas of algebra. Deeper algebraic results are introduced when they are needed.

The last section of chapter 1 contains a brief discussion of group representations. The term representation is used here in a more general sense than usual. Namely a representation of a group G will be understood to be a group homomorphism of G into the group of automorphisms of an object X. Standard use of the term representation requires X to be a vector space.

1 Elementary group theory

Recall that a *binary operation* on a set G is a function from the set product $G \times G$ into G. Multiplicative notation will usually be used. Thus the image of a pair (x, y) under the binary operation will be written xy. The operation is *associative* if $(xy)z = x(yz)$ for all x, y, z in G. The operation is *commutative* if $xy = yx$ for all x, y in G. An *identity* for the operation is an element 1 in G such that $x1 = 1x = x$ for all x in G. An operation possesses at most one identity. Given an operation on G possessing an identity 1, an *inverse* for an element x of G is an element y in G such that $xy = yx = 1$. If our operation is associative and x possesses an inverse then that inverse is unique and is denoted by x^{-1} in multiplicative notation.

A *group* is a set G together with an associative binary operation which possesses an identity and such that each element of G possesses an inverse. The group is *abelian* if its operation is commutative. In the remainder of this section G is a group written multiplicatively.

Let $x \in G$ and n a positive integer. x^n denotes the product of x with itself n times. Associativity insures x^n is a well-defined element of G. Define x^{-n} to be $(x^{-1})^n$ and x^0 to be 1. The usual rules of exponents can be derived from this definition:

(1.1) Let G be a group, $x \in G$, and n and m integers. Then
    (1) $(x^n)(x^m) = x^{n+m} = (x^m)(x^n)$.
    (2) $(x^n)^m = x^{nm}$.

A *subgroup* of G is a nonempty subset H of G such that for each x, $y \in H$, xy and x^{-1} are in H. This insures that the binary operation on G restricts to a binary operation on H which makes H into a group with the same identity as G and the same inverses. I write $H \leq G$ to indicate that H is a subgroup of G.

(1.2) The intersection of any set of subgroups of G is also a subgroup of G.

Let $S \subseteq G$ and define

$$\langle S \rangle = \bigcap_{S \subseteq H \leq G} H$$

By 1.2, $\langle S \rangle$ is a subgroup of G and by construction it is the smallest subgroup of G containing S. The subgroup $\langle S \rangle$ is called the *subgroup of G generated by S*.

(1.3) Let $S \subseteq G$. Then

$$\langle S \rangle = \{(s_1)^{\varepsilon_1} \ldots (s_n)^{\varepsilon_n} : s_i \in S, \quad \varepsilon = +1 \text{ or } -1\}.$$

(1.4) Let $x \in G$. Then $\langle x \rangle = \{x^n : n \in \mathbb{Z}\}$.

Of course 1.4 is a special case of 1.3. A group G is *cyclic* if it is generated by some element x. In that case x is said to be a *generator* of G and by 1.4, G consists of the powers of x.

    The *order* of a group G is the cardinality of the associated set G. Write $|G|$ for the order of a set G or a group G. For $x \in G$, $|x|$ denotes $|\langle x \rangle|$ and is called the *order* of x.

    A *group homomorphism* from a group G into a group H is a function $\alpha: G \to H$ of the set G into the set H which preserves the group operations: that is for all x, y in G, $(xy)\alpha = x\alpha y\alpha$. Notice that I usually write my maps on the right, particularly those that are homomorphisms. The homomorphism α is an *isomorphism* if α is a bijection. In that case α possesses an inverse function $\alpha^{-1}: H \to G$ and it turns out α^{-1} is also a group homomorphism. G is *isomorphic* to H if there exists an isomorphism of G and H. Write $G \cong H$ to indicate that G is isomorphic to H. Isomorphism is an equivalence relation. H is said to be a *homomorphic image* of G if there is a surjective homomorphism of G onto H.

A subgroup H of G is *normal* if $g^{-1}hg \in H$ for each $g \in G$ and $h \in H$. Write $H \lhd G$ to indicate H is a normal subgroup of G. If $\alpha: G \to X$ is a group homomorphism then the *kernel* of α is $\ker(\alpha) = \{g \in G: g\alpha = 1\}$ and it turns out that $\ker(\alpha)$ is a normal subgroup of G. Also write $G\alpha$ for the image $\{g\alpha: g \in G\}$ of G in X. $G\alpha$ is a subgroup of X.

Let $H \leq G$. For $x \in G$ write $Hx = \{hx: h \in H\}$ and $xH = \{xh: h \in H\}$. Hx and xH are *cosets* of H in G. Hx is a right coset and xH a left coset.

To be consistent I'll work with right cosets Hx in this section. G/H denotes the set of all (right) cosets of H in G. G/H is the *coset space* of H in G. Denote by $|G : H|$ the order of the coset space G/H. As the map $h \mapsto hx$ is a bijection of H with Hx, all cosets have the same order, so

(1.5) (Lagrange's Theorem) Let G be a group and $H \leq G$. Then $|G| = |H| \; |G : H|$. In particular if G is finite then $|H|$ divides $|G|$.

If $H \lhd G$ the coset space G/H is made into a group by defining multiplication via

$$(Hx)(Hy) = Hxy \quad x, y \in G$$

Moreover there is a natural surjective homomorphism $\pi: G \to G/H$ defined by $\pi: x \mapsto Hx$. Notice $\ker(\pi) = H$. Conversely if $\alpha: G \to L$ is a surjective homomorphism with $\ker(\alpha) = H$ then the map $\beta: Hx \mapsto x\alpha$ is an isomorphism of G/H with L such that $\pi\beta = \alpha$. The group G/H is called the *factor group* of G by H. Therefore the factor groups of G over its various normal subgroups are, up to isomorphism, precisely the homomorphic images of G.

(1.6) Let $H \lhd G$. Then the map $L \mapsto L/H$ is a bijection between the set of all subgroups of G containing H and the set of all subgroups of G/H. Normal subgroups correspond to normal subgroups under this bijection.

For $x, y \in G$, set $x^y = y^{-1}xy$. For $X \subseteq G$ set $X^y = \{x^y: x \in X\}$. X^y is the *conjugate* of X under y. Write X^G for the set $\{X^g: g \in G\}$ of conjugates of X under G. Define

$$N_G(X) = \{g \in G: X^g = X\}.$$

$N_G(X)$ is the *normalizer in G of X* and is a subgroup of G. Indeed if $X \leq G$ then $N_G(X)$ is the largest subgroup of G in which X is normal. Define

$$C_G(X) = \{g \in G: xg = gx \quad \text{for all} \quad x \in X\}.$$

$C_G(X)$ is the *centralizer in G of X*. $C_G(X)$ is also a subgroup of G.

For $X, Y \subseteq G$ define $XY = \{xy : x \in X, y \in Y\}$. The set XY is the *product* of X with Y.

(1.7) Let $X, Y \leq G$. Then
    (1) XY is a subgroup of G if and only if $XY = YX$.
    (2) If $Y \leq N_G(X)$ then XY is a subgroup of G and $XY/X \cong Y/(Y \cap X)$.
    (3) $|XY| = |X||Y|/|X \cap Y|$.

(1.8) Let H and K be normal subgroups of G with $K \leq H$. Then $G/K/H/K \cong G/H$.

Let $G_1, \ldots, G_n$ be a finite set of groups. The *direct product* $G_1 \times \cdots \times G_n = \prod_{i=1}^{n} G_i$ of the groups $G_1, \ldots, G_n$ is the group defined on the set product $G_1 \times \cdots \times G_n$ by the operation

$$(x_1, \ldots, x_n)(y_1, \ldots, y_n) = (x_1 y_1, \ldots, x_n y_n) \quad x_i, y_i \in G_i$$

(1.9) Let G be a group and $(G_i : 1 \leq i \leq n)$ a family of subgroups of G. Then the following are equivalent:

(1) The map $(x_1, \ldots, x_n) \mapsto x_1 \ldots x_n$ is an isomorphism of G with $G_1 \times \cdots \times G_n$.
(2) $G = \langle G_i : 1 \leq i \leq n \rangle$ and for each i, $1 \leq i \leq n$, $G_i \trianglelefteq G$ and $G_i \cap \langle G_j : j \neq i \rangle = 1$.
(3) $G_i \trianglelefteq G$ for each i, $1 \leq i \leq n$, and each $g \in G$ can be written uniquely as $g = x_1 \ldots x_n$ with $x_i \in G_i$.

If any of the equivalent conditions of 1.9 hold, G will be said to be the *direct product of the subgroups* $(G_i : 1 \leq i \leq n)$.

(1.10) Let $G = \langle g \rangle$ be a cyclic group and $\mathbb{Z}$ the group of integers under addition. Then
    (1) If H is a nontrivial subgroup of $\mathbb{Z}$ then $H = \langle n \rangle$, where n is the least positive integer in H.
    (2) The map $\alpha : \mathbb{Z} \to G$ defined by $m\alpha = g^m$ is a surjective homomorphism with kernel $\langle n \rangle$, where $n = 0$ if g is of infinite order and $n = \min \{m > 0 : g^m = 1\}$ if g has finite order.
    (3) If g has finite order n then $G = \{g^i : 0 \leq i < n\}$ and n is the least positive integer m with $g^m = 1$.
    (4) Up to isomorphism $\mathbb{Z}$ is the unique infinite cyclic group and for each positive integer n, the group $\mathbb{Z}_n$ of integers modulo n is the unique cyclic group of order n.

(5) Let $|g| = n$. Then for each divisor m of n, $\langle g^{n/m} \rangle$ is the unique subgroup of G of order m. In particular subgroups of cyclic groups are cyclic.

(1.11) Each finitely generated abelian group is the direct product of cyclic groups.

Let p be a prime. A *p-group* is a group whose order is a power of p. More generally if π is a set of primes then a π-*group* is a group G of finite order such that $\pi(G) \subseteq \pi$, where $\pi(G)$ denotes the set of prime divisors of $|G|$. p' denotes the set of all primes distinct from p. An element x in a group G is a π-*element* if $\langle x \rangle$ is a π-group. An *involution* is an element of order 2.

(1.12) Let $1 \neq G$ be an abelian p-group. Then G is the direct product of cyclic subgroups $G_i \cong \mathbb{Z}_{p^{e_i}}, 1 \leq i \leq n, e_1 \geq e_2 \geq \cdots \geq e_n > 1$. Moreover the integers n and $(e_i : 1 \leq i \leq n)$ are uniquely determined by G.

The *exponent* of a finite group G is the least common multiple of the orders of the elements of G. An *elementary abelian p-group* is an abelian p-group of exponent p. Notice that by 1.12, G is an elementary abelian p-group of order p^n if and only if G is the direct product of n copies of $\mathbb{Z}_p$. In particular up to isomorphism there is a unique elementary abelian p-group of order p^n, which will be denoted by E_{p^n}. The integer n is the *p-rank* of E_{p^n}. The *p-rank* of a general finite group G is the maximum p-rank of an elementary abelian p-subgroup of G, and is denoted by $m_p(G)$.

(1.13) Each group of exponent 2 is abelian.

If π is a set of primes and G a finite group, write $O_\pi(G)$ for the largest normal π-subgroup of G, and $O^\pi(G)$ for the smallest normal subgroup H of G such that G/H is a π-group. $O_\pi(G)$ and $O^\pi(G)$ are well defined by Exercise 1.1.

Define $Z(G) = C_G(G)$ and call $Z(G)$ the *center* of G. If G is a p-group then define

$$\Omega_n(G) = \langle x \in G : x^{p^n} = 1 \rangle$$

$$\mho^n(G) = \langle x^{p^n} : x \in G \rangle.$$

For $X \leq G$ define $\mathrm{Aut}_G(X) = N_G(X)/C_G(X)$ to be the *automizer* in G of X. Notice that by Exercise 1.3, $\mathrm{Aut}_G(X) \leq \mathrm{Aut}(X)$ and indeed $\mathrm{Aut}_G(X)$ is the group of automorphisms induced on X in G.

A *maximal subgroup* of a group G is a proper subgroup of G which is properly contained in no proper subgroup of G. That is a maximal subgroup is a maximal member of the set of proper subgroups of G, partially ordered by inclusion.

If $\alpha\colon S \to T$ is a function and $R \subseteq S$ then $\alpha|_R$ denotes the restriction of α to R. That is $\alpha|_R\colon R \to T$ is the function from R into T agreeing with α.

Here's a little result that's easy to prove but useful.

(1.14) (Modular Property of Groups) Let A, B, and C be subgroups of a group G with $A \leq C$. Then $AB \cap C = A(B \cap C)$.

If G is a group write $G^{\#}$ for the set $G - \{1\}$ of nonidentity elements of G. On the other hand if R is a ring define $R^{\#} = R - \{0\}$.

Denote by $\mathbb{C}$, $\mathbb{R}$, and $\mathbb{Q}$ the complex numbers, the reals, and the rationals, respectively. Often $\mathbb{Z}$ will denote the integers.

Given a group G, a subgroup H of G, and a collection C of subgroups of G, I'll often write $C \cap H$ for the set of members of C which are subgroups of H.

I'll use the *bar convention*. That is I'll often denote a homomorphic image $G\alpha$ of a group G by $\bar{G}$ (or G^* or $\tilde{G}$) and write $\bar{g}$ (or g^* or $\tilde{g}$) for $g\alpha$. This will be done without comment.

Other notation and terminology are introduced in later chapters. The List of Symbols gives the page number where a notation is first introduced and defined.

2 Categories

It will be convenient to have available some of the elementary concepts and language of categories. For a somewhat more detailed discussion, see chapter 1 of Lang [La].

A *category* $\mathscr{C}$ consists of

(1) A collection $\mathrm{Ob}(\mathscr{C})$ of *objects*.
(2) For each pair A,B of objects, a set $\mathrm{Mor}(A,B)$ of *morphisms* from A to B.
(3) For each triple A, B, C of objects a map

$$\mathrm{Mor}(A, B) \times \mathrm{Mor}(B, C) \to \mathrm{Mor}(A, C)$$

called *composition*. Write fg for the image of the pair (f, g) under the composition map.

Moreover the following three axioms are required to hold:

Cat (1) For each quadruple A, B, C, D of objects, $\mathrm{Mor}(A, B) \cap \mathrm{Mor}(C, D)$ is empty unless $A = C$ and $B = D$.

Cat (2) Composition is associative.

Cat (3) For each object A, $\mathrm{Mor}(A, A)$ possesses an *identity morphism* 1_A such that for all objects B and all f in $\mathrm{Mor}(A, B)$ and g in $\mathrm{Mor}(B, A)$, $1_A f = f$ and $g 1_A = g$.

Almost all categories considered here will be categories of *sets with structure*. That is the objects of the category are sets together with some extra structure, $\text{Mor}(A, B)$ consists of all functions from the set associated to A to the set associated to B which preserve the extra structure, and composition is ordinary composition of functions. The identity morphism 1_A is forced to be the identity map on A. Thus we need to know the identity map preserves structure. We also need to know the composition of maps which preserve structure also preserves structure. These facts will usually be obvious in the examples we consider.

We'll be most interested in the following three categories, which are all categories of sets with structure.

(1) The category of sets and functions: Here the objects are the sets and $\text{Mor}(A, B)$ is the set of all functions from the set A into the set B.
(2) The category of groups and group homomorphisms: The objects are the groups and morphisms are the group homomorphisms.
(3) The category of vector spaces and linear transformations: Fix a field F. The objects are the vector spaces over F and the morphisms are the F-linear transformations.

Let f be a morphism from an object A to an object B. An *inverse* for f in $\mathscr{C}$ is a morphism $g \in \text{Mor}(B, A)$ such that $1_A = fg$ and $1_B = gf$. The morphism f is an *isomorphism* if it possesses an inverse in $\mathscr{C}$. An *automorphism* of A is an isomorphism from A to A. Denote by $\text{Aut}(A)$ the set of all automorphisms of A and observe $\text{Aut}(A)$ forms a group under the composition in $\mathscr{C}$.

If $\alpha\colon A \to B$ is an isomorphism define $\alpha^*\colon \text{Mor}(A, A) \to \text{Mor}(B, B)$ by $\beta \to \alpha^{-1}\beta\alpha$ and observe α^* restricts to a group isomorphism of $\text{Aut}(A)$ with $\text{Aut}(B)$.

Let $(A_i : i \in I)$ be a family of objects in a category $\mathscr{C}$. A *coproduct* of the family is an object C together with morphisms $c_i\colon A_i \to C$, $i \in I$, satisfying the universal property: whenever X is an object and $\alpha_i\colon A_i \to X$ are morphisms, there exists a unique morphism $\alpha\colon C \to X$ with $c_i\alpha = \alpha_i$ for each $i \in I$. As a consequence of the universal property, the coproduct of a family is determined up to isomorphism, if it exists.

The *product* of the family is defined dually. That is to obtain the definition of the product, take the definition of the coproduct and reverse the direction of all arrows.

Exercise 1.2 gives a description of coproducts and products in the three categories listed above.

3 Graphs and geometries

This section contains a brief discussion of two more categories which will make occasional appearances in these notes.

A *graph* $\mathscr{G} = (V, *)$ consists of a set V of *vertices* (or objects or points) together with a symmetric relation $*$ called *adjacency* (or incidence or something else). The ordered pairs in the relation are called the *edges* of the graph. I write $u * v$ to indicate two vertices are related via $*$ and say u is *adjacent* to v. A *path of length n* from u to v is a sequence of vertices $u = u_0, u_1, \ldots, u_n = v$ such that $u_i * u_{i+1}$ for each i. Denote by $d(u, v)$ the minimal length of a path from u to v. If no such path exists set $d(u, v) = \infty$. $d(u, v)$ is the *distance* from u to v.

The relation $\sim$ on V defined by $u \sim v$ if and only if $d(u, v) < \infty$ is an equivalence relation on V. The equivalence classes of this relation are called the *connected components* of the graph. The graph is *connected* if it has just one connected component. Equivalently there is a path between any pair of vertices.

A morphism $\alpha\colon \mathscr{G} \to \mathscr{G}'$ of graphs is a function $\alpha\colon V \to V'$ from the vertex set V of $\mathscr{G}$ to the vertex set V' of $\mathscr{G}'$ which preserves adjacency; that is if u and v are vertices adjacent in $\mathscr{G}$ then $u\alpha$ is adjacent to $v\alpha$ in $\mathscr{G}'$.

So much for graphs; on to geometries. In this book I adopt a notion of geometry due to Tits. Let I be a finite set. A *geometry* over I is a triple $(\Gamma, \tau, *)$ where Γ is a set of objects, $\tau\colon \Gamma \to I$ is a type function, and $*$ is a symmetric incidence relation on Γ such that objects u and v of the same type are incident if and only if $u = v$. $\tau(u)$ is the *type* of the object u. Notice $(\Gamma, *)$ is a graph. I'll usually write Γ for the geometry $(\Gamma, \tau, *)$.

A *morphism* $\alpha\colon \Gamma \to \Gamma'$ of geometries is a function $\alpha\colon \Gamma \to \Gamma'$ of the associated object sets which preserves type and incidence; that is if $u, v \in \Gamma$ with $u * v$ then $\tau(u) = \tau'(u\alpha)$ and $u\alpha *' v\alpha$.

A *flag* of the geometry Γ is a set T of objects such that each pair of objects in T is incident. Notice our one (weak) axiom insures that a flag T possesses at most one object of each type, so that the type function τ induces an injection of T into I. The image $\tau(T)$ is called the *type* of T. The *rank* and *corank* of T are the order of $\tau(T)$ and $I - \tau(T)$, respectively. The *residue* Γ_T of the flag T is $\{v \in \Gamma - T \colon v * t \text{ for all } t \in T\}$ regarded as a geometry over $I - \tau(T)$.

The geometry Γ is *connected* if its graph $(\Gamma, *)$ is connected. Γ is *residually connected* if the residue of every flag of corank at least 2 is connected and the residue of every flag of corank 1 is nonempty.

Here's a way to associate geometries to groups. Let G be a group and $\mathscr{F} = (G_i \colon i \in I)$ a family of subgroups of G. Define $\Gamma(G, \mathscr{F})$ to be the geometry whose set of objects of type i is the coset space G/G_i and with objects $G_i x$ and $G_j y$ incident if $G_i x \cap G_j y$ is nonempty. For $J \subseteq I$ write J' for the complement $I - J$ of J in I and define $G_J = \bigcap_{j \in J} G_j$. Observe that for $x \in G$, $S_{J,x} = \{G_j x \colon j \in J\}$ is a flag of $\Gamma(G, \mathscr{F})$ of type J.

A group H of automorphisms of a geometry Γ is said to be *flag transitive* if H is transitive on flags of type J for each subset J of I.

4 Abstract representations

Let $\mathscr{C}$ be a category. A $\mathscr{C}$-*representation* of a group G is a group homomor-phism $\pi\colon G \to \mathrm{Aut}(X)$ of G into the group $\mathrm{Aut}(X)$ of automorphisms of some object X in $\mathscr{C}$. (Recall the definition of $\mathrm{Aut}(X)$ in section 2.) We will be most concerned with the following three classes of representations.

A *permutation representation* is a representation in the category of sets and functions. The group $\mathrm{Aut}(X)$ of automorphisms of set X is the *symmetric group* $\mathrm{Sym}(X)$ of X. That is $\mathrm{Sym}(X)$ is the group of all permutations of X under composition.

A *linear representation* is a representation in the category of vector spaces and linear transformations. $\mathrm{Aut}(X)$ is the *general linear group* $\mathrm{GL}(X)$ of the vector space X. That is $\mathrm{GL}(X)$ is the group of all invertible linear transforma-tions of X.

Finally we will of course be interested in the category of groups and group homomorphisms. Of particular interest is the representation of G via conjuga-tion on itself (cf. Exercise 1.3).

Two $\mathscr{C}$-representations $\pi_i\colon G \to \mathrm{Aut}(X_i)$, $i = 1, 2$, are said to be *equiva-lent* if there exists an isomorphism $\alpha\colon X_1 \to X_2$ such that $\pi_2 = \pi_1\alpha^*$, where $\alpha^*\colon \mathrm{Aut}(X_1) \to \mathrm{Aut}(X_2)$ is the isomorphism described in section 2. The map α is said to be an *equivalence* of the representations. $\mathscr{C}$-representations $\pi_i\colon G_i \to \mathrm{Aut}(X_i)\, i = 1, 2$, are said to be *quasiequivalent* if there exists a group isomor-phism $\beta\colon G_2 \to G_1$ of groups and a $\mathscr{C}$-isomorphism $\alpha\colon X_1 \to X_2$ such that $\pi_2 = \beta\pi_1\alpha^*$.

Equivalent representations of a group G are the same for our purposes. Quasiequivalent representations are almost the same, differing only by an au-tomorphism of G.

A representation π of G is *faithful* if π is an injection. In that event π induces an isomorphism of G with the subgroup $G\pi$ of $\mathrm{Aut}(X)$, so G may be regarded as a group of automorphisms of X via π.

Let $\pi_i\colon G \to \mathrm{Aut}(X_i)$, $i = 1, 2$, be $\mathscr{C}$-representations. Define a G-*morphism* $\alpha\colon X_1 \to X_2$ to be a morphism α of X_1 to X_2 which commutes with the action of G in the sense that $(g\pi_1)\alpha = \alpha(g\pi_2)$ for each $g \in G$. Write $\mathrm{Mor}_G(X_1, X_2)$ for the set of G-morphisms of X_1 to X_2. Notice that the composition of G-morphisms is a G-morphism and the identity morphism is a G-morphism. Similarly define a G-*isomorphism* to be a G-morphism which is also an iso-morphism. Notice the G-isomorphisms are the equivalences of representations of G.

One focus of this book is the decomposition of a representation π into smaller representations. Under suitable finiteness conditions (which are always present in the representations considered here) this process of decomposition must terminate, at which point we have associated to π certain indecomposable or irreducible representations which cannot be broken down further. It will develop that the indecomposables associated to π are determined up to equivalence. Thus we are reduced to a consideration of indecomposable representations.

In general indecomposables are not irreducible, so an indecomposable representation π can be broken down further, and we can associate to π a set of irreducible constituents. Sometimes these irreducible constituents are determined up to equivalence, and sometimes not. Even when the irreducible constituents are determined, they usually do not determine π. Thus we will also be concerned with the *extension problem*: Given a set S of irreducible representations, which representations have S as their set of irreducible constituents? There is also the problem of determining the irreducible and indecomposable representations of the group.

It is possible to give a categorical definition of indecomposability (cf. Exercise 1.5). There is also a uniform definition of irreducibility for the classes of representations considered most frequently (cf. Exercise 1.6). I have chosen however to relegate these definitions to the exercises and to make the appropriate definitions of indecomposability and irreducibility for each category in the chapter discussing the elementary representation theory of the category. This process begins in the next chapter, which discusses permutation representations.

However one case is of particular interest. A representation of a group G on itself via conjugation (in the category of groups and group homomorphisms) is irreducible if G possesses no nonidentity proper normal subgroups. In this case G is said to be *simple*. To my mind the simple groups and their irreducible linear and permutation representations are the center of interest in finite group theory.

Exercises for chapter 1

1. Let G be a finite group, π a set of primes, Ω the set of normal π-subgroups of G, and Γ the set of normal subgroups X of G with G/X a π-group. Prove
   (1) If $H, K \in \Omega$ then $HK \in \Omega$. Hence $\langle \Omega \rangle$ is the unique maximal member of Ω.
   (2) If $H, K \in \Gamma$ then $H \cap K \in \Gamma$. Hence $\bigcap_{H \in \Gamma} H$ is the unique minimal member of Γ.
2. Let $\mathscr{C}_1$ be the category of sets and functions, $\mathscr{C}_2$ the category of vector spaces and linear transformations, and $\mathscr{C}_3$ the category of groups and homomorphisms. Let $F = (A_i : 1 \le i \le n)$ be a family of objects in $\mathscr{C}_k$. Prove
   (1) Let $k = 1$. Then the coproduct C of F is the disjoint union of the sets A_i with $c_i : A_i \to C$ the inclusion map. The product P of F is

the set product $A_1 \times \cdots \times A_n$ with $p_i: P \to A_i$ the projection map
$p_i: (a_1, \ldots, a_n) \mapsto a_i$.

(2) Let $k = 2$. Then $C = P = \bigoplus_{i=1}^n A_i$ is the direct sum of the subspaces
A_i, with $c_i: A_i \to C$ defined by $a_i c_i = (0, \ldots, a_i, \ldots, 0)$ and $p_i: P \to$
A_i the projection map.

(3) Let $k = 3$. Then the product P of F is the direct product $A_1 \times \cdots \times A_n$.
with $p_i: P \to A_i$ the projection map. (The coproduct turns out to be
the so-called free product of the family.)

3. Let G be a group, $H \trianglelefteq G$, and for $g \in G$ define $g\pi: H \to H$ by
$x(g\pi) = x^g, x \in H$. Let $\mathscr{C}$ be the category of groups and homomorphisms.
Prove π is a $\mathscr{C}$-representation of G with kernel $C_G(H)$. π is the *representation* by *conjugation* of G on H. If $H = G$, the image of G under π
is the *inner automorphism group* of G and is denoted by $\mathrm{Inn}(G)$. Prove
$\mathrm{Inn}(G) \trianglelefteq \mathrm{Aut}(G)$. Define $\mathrm{Out}(G) = \mathrm{Aut}(G)/\mathrm{Inn}(G)$ to be the *outer automorphism group* of G.

4. Let $\mathscr{C}$ be a category, $F = (A_i: 1 \le i \le n)$ a family of objects in $\mathscr{C}$, C and
P the coproduct and product of the family with canonical maps $c_i: A_i \to C$
and $p_i: P \to A_i$, respectively. For $\alpha_i \in \mathrm{Aut}(A_i)$ define $\bar{\alpha}_i$ to be the unique
member of $\mathrm{Mor}(C, C)$ with $\alpha_i c_i = c_i \bar{\alpha}_i$ and $c_j = c_j \bar{\alpha}_i$ for all $i \ne j$. Define
$\bar{\alpha}_i \in \mathrm{Mor}(P, P)$ dually. Prove the map

$$\phi: \prod_{i=1}^n \mathrm{Aut}(A_i) \to \mathrm{Aut}(X)$$

$$(\alpha_1, \ldots, \alpha_n) \mapsto \bar{\alpha}_1 \ldots \bar{\alpha}_n$$

is an injective group homomorphism for $X = C$ and P.

5. Assume the hypothesis and notation of Exercise 1.2, and let $\pi: G \to \mathrm{Aut}(X)$
be a $\mathscr{C}_k$-representation, where $X = C$ if $k = 1$ or 2, and $X = P$ if $k = 3$.
Prove the following are equivalent:

(a) There exist $\mathscr{C}_k$-representations $\pi_i: G \to \mathrm{Aut}(A_i)$, $1 \le i \le n$, such that
$\pi = \psi\phi$, where ϕ is the injection of Exercise 1.4 and

$$\psi: G \to \prod_{i=1}^n \mathrm{Aut}(A_i) \text{ is defined by } g\psi = (g\pi_1, \ldots, g\pi_n).$$

(b) π is decomposable.

(If $k = 1$, transitivity is the same as indecomposability. See chapter 2 for
the definition of transitivity. See chapters 5 and 4 for the definitions of
decomposability when $k = 2$ and 3.)

6. Assume the hypothesis and notation of Exercise 1.2, let X be an object in
$\mathscr{C}_k$, and $p: G \to \mathrm{Aut}(X)$ a $\mathscr{C}_k$-representation. A $\mathscr{C}_k$-*equivalence relation*
on X is an equivalence relation $\sim$ on X such that $\sim$ is preserved by the
operations on X if $k = 2$ or 3 (i.e. if γ is an n-ary operation on X and

$x_i \sim y_i$ then $\gamma(x_1, \ldots, x_n) \sim \gamma(y_1, \ldots, y_n)$). Define $G\pi$ to preserve $\sim$ if $x \sim y$ implies $xg\pi \sim yg\pi$ for each $g \in G$. Prove that (a) and (b) are equivalent:

(a) $G\pi$ preserves no nontrivial $\mathscr{C}_k$-equivalence relation on X.

(b) is an irreducible $\mathscr{C}_k$-representation.

   (See chapters 5 and 4 for the definition of an irreducible $\mathscr{C}_k$-representation when $k = 2$ and 3. A $\mathscr{C}_1$-representation is irreducible if it is primitive, and primitivity is defined in chapter 2.)

7. Let $\pi, \sigma: G \to \mathrm{Aut}(X)$ be faithful $\mathscr{C}$-representations. Prove π is quasi-equivalent to σ if and only if $G\pi$ is conjugate to $G\sigma$ in $\mathrm{Aut}(X)$.

8. Let G be a group and $\mathscr{F} = (G_i : i \in I)$ a family of subgroups of G. Prove
   (1) The geometry $\Gamma = \Gamma(G, \mathscr{F})$ is connected if and only if $G = \langle \mathscr{F} \rangle$.
   (2) For $g \in G$, define $g\pi: \Gamma \to \Gamma$ by $(G_i x)g\pi = G_i xg$. Prove π is a representation of G as a group of automorphisms of Γ.

2

Permutation representations

Section 5 develops the elementary theory of permutation representations. The foundation for this theory is the notion of the transitive permutation representation. The transitive representations play the role of the indecomposables in the theory. It will develop that every transitive permutation representation of a group G is equivalent to a representation by right multiplication on the set of cosets of some subgroup of G. Hence the study of permutation representations of G is equivalent to the study of the subgroup structure of G.

Section 6 is devoted to a proof of Sylow's Theorem. The proof supplies a nice application of the techniques developed in section 5. Sylow's Theorem is one of the most important results in finite group theory. It is the first theorem in the local theory of finite groups. The local theory studies a finite group from the point of view of its p-subgroups and the normalizers of these p-subgroups.

5 Permutation representations

In this section X is a set, G a group, and $\pi: G \to \text{Sym}(X)$ is a permutation representation of G. Recall $\text{Sym}(X)$ is the symmetric group on X; that is $\text{Sym}(X)$ is the group of all permutations of X. Thus $\text{Sym}(X)$ is the automorphism group of X in the category of sets and functions, and π is a representation in that category.

For $x \in X$ and $\alpha \in \text{Sym}(X)$ write $x\alpha$ for the image of x under α. Notice that, by definition of multiplication in $\text{Sym}(X)$:

$$x(\alpha\beta) = (x\alpha)\beta \quad x \in X, \quad \alpha, \beta \in \text{Sym}(X).$$

I'll often suppress the representation π and write xg for $x(g\pi)$, $x \in X$, $g \in G$. One feature of this notation is that:

$$x(gh) = (xg)h \quad x \in X, \quad g, h \in G.$$

The relation $\sim$ on X defined by $x \sim y$ if and only if there exists $g \in G$ with $xg = y$ is an equivalence relation on X. The equivalence class of x under this relation is

$$xG = \{xg: g \in G\}$$

and is called the *orbit* of x under G. As the equivalence classes of an equivalence relation partition a set, X is partitioned by the orbits of G on X.

Let Y be a subset of X. G is said to *act* on Y if Y is a union of orbits of G. Notice G acts on Y precisely when $yg \in Y$ for each $y \in Y$, and each $g \in G$. Further if G acts on Y then $g|_Y$ is a permutation of Y for each $g \in G$, and the restriction map

$$G \to \operatorname{Sym}(Y)$$
$$g \mapsto g|_Y$$

is a permutation representation with kernel

$$G_Y = \{g \in G : yg = y \quad \text{for all} \quad y \in Y\}.$$

Hence $G_Y \trianglelefteq G$ when G acts on Y. Even when G does not act on Y, we can consider

$$G(Y) = \{g \in G : Yg = Y\},$$

where $Yg = \{yg : y \in Y\}$. G_Y and $G(Y)$ are subgroups of G called the *pointwise stabilizer of Y in G* and the *global stabilizer of Y in G*, respectively. $G(Y)$ is the largest subgroup of G acting on Y. Write G^Y for the image of $G(Y)$ under the restriction map on Y. We have seen that:

(5.1) The restriction map of $G(Y)$ on Y is a permutation representation of $G(Y)$ with kernel G_Y and image $G^Y \cong G(Y)/G_Y$, for each subset Y of X.

For $x \in X$ write G_x for $G_{\{x\}}$. Next for $S \subseteq G$ define

$$\operatorname{Fix}(S) = \{x \in X : xs = x \quad \text{for all} \quad s \in S\}.$$

$\operatorname{Fix}(S)$ is the set of *fixed points* of S. Notice $\operatorname{Fix}(S) = \operatorname{Fix}(\langle S \rangle)$. Also

(5.2) If $H \trianglelefteq G$ then G acts on $\operatorname{Fix}(H)$. More generally G permutes the orbits of H of cardinality c, for each c.

For $Y \subseteq X$, I'll sometimes write $C_G(Y)$ and $N_G(Y)$ for G_Y and $G(Y)$, respectively, and I'll sometimes write $C_X(G)$ for $\operatorname{Fix}(G)$. Usually this notation will be used only when X possesses a group structure preserved by G.

(5.3) Let P be the set of all subsets of X. Then $\alpha : G \to \operatorname{Sym}(P)$ is a permutation representation of G on P where $g\alpha : Y \to Yg$ for each $g \in G$ and $Y \subseteq X$.

π is a *transitive permutation representation* if G has just one orbit on X; equivalently for each $x, y \in X$ there exists $g \in G$ with $xg = y$. G will also be said to be *transitive on X*.

Here's one way to generate transitive representations of G:

(5.4) Let $H \leq G$. Then $\alpha: G \to \text{Sym}(G/H)$ is a transitive permutation representation of G on the coset space G/H, where

$$g\alpha: Hx \mapsto Hxg.$$

α is the *representation of G on the cosets of H by right multiplication*. H is the stabilizer of the coset H in this representation.

We'll soon see that every transitive representation of G is equivalent to a representation of G by right multiplication on the cosets of some subgroup. But first here is another way to generate permutation representations of G.

(5.5) The map $\alpha: G \to \text{Sym}(G)$ is a permutation representation of G on itself, where

$$g\alpha: x \mapsto x^g \quad x, g \in G.$$

α is the *representation of G on itself by conjugation*. For $S \subseteq G$, the global stabilizer of S in G is $N_G(S)$, while $C_G(S)$ is the pointwise stabilizer of S.

Notice that 5.5 is essentially a consequence of Exercise 1.3. Recall $N_G(S) = \{g \in G: S^g = S\}$, $S^g = \{s^g: s \in S\}$, and $C_G(S) = \{g \in G: s^g = s \text{ for all } s \in S\}$. By 5.3, G is also represented on the power set of G, and evidently, for $S \subseteq G$, the set $S^G = \{S^g: g \in G\}$ of conjugates of S under G is the orbit of S under G with respect to this representation.

(5.6) Let $Y \subseteq X$ and $g \in G$. Then $G(Y^g) = G(Y)^g$ and $G_{Y^g} = (G_Y)^g$.

(5.7) Assume π is a transitive representation and let $x \in X$ and $H = G_x$. Then

$$\ker(\pi) = \bigcap_{g \in G} H^g = \ker_H(G)$$

is the largest normal subgroup of G contained in H.

(5.8) Assume π is a transitive permutation representation, let $x \in X$, $H = G_x$, and let α be the representation of G on the cosets of H by right multiplication. Define

$$\beta: G/H \to X$$
$$Hg \mapsto xg.$$

Then β is an equivalence of the permutation representations α and π.

Proof. We must show β is a well-defined bijection of G/H with X and that, for each $a, g \in G, (Ha)\beta(g\alpha) = (Ha)g\pi\beta$. Both computations are straightforward.

(5.9) (1) Every transitive permutation representation of G is equivalent to a representation of G by right multiplication on the cosets of some subgroup.

(2) If $\pi': G \to \mathrm{Sym}(X')$ and π are transitive representations, $x \in X$, and $x' \in X'$, then π is equivalent to π' if and only if G_x is conjugate to $G_{x'}$ in G.

Proof. Part (1) follows from 5.8. Assume the hypothesis of (2). If $\beta: X \to X'$ is an equivalence of π and π' then $G_x = G_{x\beta}$ and, by 5.6, $G_{x\beta}$ is conjugate to $G_{x'}$ in G. Conversely if $G_{x'}$ is conjugate to G_x in G, then by 5.6 there is $y \in X$ with $G_y = G_{x'}$ and by 5.8 both π and π' are equivalent to the representation of G on the cosets of $G_{x'}$ and hence equivalent to each other.

Let $(X_i : i \in I)$ be the orbits of G on X and π_i the restriction of π to X_i. By 5.1, π_i is a permutation representation of G on X_i and, as X_i is an orbit of G, π_i is even a transitive representation. $(\pi_i : i \in I)$ is the family of *transitive constituents* of π and we write $\pi = \sum_{i \in I} \pi_i$. Evidently:

(5.10) The transitive constituents $(\pi_i : i \in I)$ of π are uniquely determined by π, and if π' is a permutation representation of G with transitive constituents $(\pi'_j : j \in J)$, then π is equivalent to π' if and only if there is a bijection α of I with J such that $\pi'_{i\alpha}$ is equivalent to π_i for each $i \in I$.

So the study of permutation representations is effectively reduced to the study of transitive representations, and 5.9 says in turn that the transitive permutation representations of a group are determined by its subgroup structure.

The transitive representations play the role of the indecomposable permutation representations. For example see Exercise 1.5.

(5.11) If G is transitive on X then X has cardinality $|G : G_x|$ for each $x \in X$.

Proof. This is a consequence of 5.8.

(5.12) Let $S \subseteq G$. Then S has exactly $|G : N_G(S)|$ conjugates in G.

Proof. We observed earlier that G is transitively represented on the set S^G of conjugates of S via conjugation, while, by 5.5, $N_G(S)$ is the stabilizer of S with respect to this representation, so the lemma follows from 5.11.

Let p be a prime and recall that a *p-group* is a group whose order is a power of p.

(5.13) If G is a p-group then all orbits of G on X have order a power of p.

Proof. This follows from 5.11 and the fact that the index of any subgroup of G divides the order of G.

(5.14) Let G be a p-group and assume X is finite. Then $|X| \equiv |\text{Fix}(G)| \bmod p$.

Proof. As the fixed points of G are its orbits of length 1, 5.14 follows from 5.13.

Here are a couple applications of 5.14:

(5.15) Let G and H be p-groups with $H \neq 1$ and let $\alpha: G \to \text{Aut}(H)$ be a group homomorphism. Then $C_H(G) \neq 1$.

Proof. α is also a permutation representation of G on H. By 5.14, $|H| \equiv |\text{Fix}(G)| \bmod p$. But $\text{Fix}(G) = C_H(G)$ in this representation, while $|H| \equiv 0 \bmod p$ as H is a p-group with $H \neq 1$. So $C_H(G) \neq 1$, as claimed.

(5.16) If G is a p-group with $G \neq 1$, then $Z(G) \neq 1$.

Proof. Apply 5.15 to the representation of G on itself by conjugation and recall $Z(G) = C_G(G)$.

The following technical lemma will be used in the next section to prove Sylow's Theorem:

(5.17) Let X be finite and assume for each $x \in X$ that there exists a p-subgroup $P(x)$ of G such that $\{x\} = \text{Fix}(P(x))$. Then
  (1) G is transitive on X, and
  (2) $|X| \equiv 1 \bmod p$.

Proof. Let $X = Y + Z$ be a partition of X with G acting on Y and Z. Let $Y \neq \varnothing$ and pick $y \in Y$. For $V = Y$ or Z and $H \leq G$ denote by $\text{Fix}_V(H)$ the fixed points of H on V. By hypothesis $\{y\} = \text{Fix}(P(y))$, so $1 = |\text{Fix}_Y(P(y))|$ and $0 = |\text{Fix}_z(P(y))|$. Hence, by 5.14, $|Y| \equiv 1 \bmod p$ and $|Z| \equiv 0 \bmod p$. But if Z is nonempty, then, by symmetry, $|Y| \equiv 1 \bmod p$, a contradiction. Thus

$Y = X$ and, as $|Y| \equiv 1 \bmod p$, (2) holds. Since we could have chosen Y to be an orbit of G on X, (1) holds.

Let Q be a partition of X. Q is *G-invariant* if G permutes the members of Q. Equivalently regard Q as a subset of the power set P of X and represent G on P as in 5.3; then Q is G-invariant if G acts on the subset Q of P with respect to this representation. In particular notice that if Q is G-invariant then there is a natural permutation representation of G on Q. Q is *nontrivial* if $Q \neq \{\{x\}: x \in X\}$ and $Q \neq \{X\}$.

Let G be transitive on X. G is *imprimitive* on X if there exists a nontrivial G-invariant partition Q of X. In this event Q is said to be a *system of imprimitivity* for G on X. G is *primitive* on X if it is transitive and not imprimitive.

(5.18) Let G be transitive on X and $y \in X$.

(1) If Q is a system of imprimitivity for G on X and $y \in Y \in Q$, then G is transitive on Q, the stabilizer H of Y in G is a proper subgroup of G properly containing G_y, Y is an orbit of H on X, $|X| = |Y||Q|$, $|Q| = |G:H|$, and $|Y| = |H:G_y|$.

(2) If $G_Y < H < G$ then $Q = \{Yg: g \in G\}$ is a system of imprimitivity for G on X, where $Y = yH$ and H is the stabilizer of Y in G.

The proof is left as an exercise. As a direct consequence of 5.18 we have:

(5.19) Let G be transitive on X and $x \in X$. Then G is primitive on X if and only if G_x is a maximal subgroup of G.

Let G be finite and transitive on X, let $|X| > 1$, and let $x \in X$. Then there is a sequence $G_x = H_0 \leq H_1 \leq \cdots \leq H_n = G$ with H_i maximal in H_{i+1}. This gives rise to a family of primitive permutation representations: the representations of H_{i+1} on the cosets of H_i. This family of primitive representations can be used to investigate the representation π of G on X.

From this point of view the primitive representations play the role of irreducible permutation representations. See also Exercise 1.6.

I close this section with two useful lemmas. The proofs are left as exercises.

(5.20) Let G be transitive on X, $x \in X$, and $H \leq G$. Then H is transitive on X if and only if $G = G_x H$.

(5.21) Let G be transitive on X, $x \in X$, $H = G_x$ and $K \leq H$. Then $N_G(K)$ is transitive on $\mathrm{Fix}(K)$ if and only if $K^G \cap H = K^H$.

6 Sylow's Theorem

In this section G is a finite group. If n is a positive integer and p a prime, write n_p for the highest power of p dividing n. n_p is the *p-part* of n.

A *Sylow p-subgroup* of G is a subgroup of G of order $|G|_p$. Write $\mathrm{Syl}_p(G)$ for the set of Sylow p-subgroups of G.

In this section we prove:

Sylow's Theorem. Let G be a finite group and p a prime. Then
(1) $\mathrm{Syl}_p(G)$ is nonempty.
(2) G acts transitively on $\mathrm{Syl}_p(G)$ via conjugation.
(3) $|\mathrm{Syl}_p(G)| = |G : N_G(P)| \equiv 1 \bmod p$ for $P \in \mathrm{Syl}_p(G)$.
(4) Every p-subgroup of G is contained in some Sylow p-subgroup of G.

Let Γ be the set of all p-subgroups of G and Ω the set of all maximal p-subgroups of G; that is partially order Γ by inclusion and let Ω be the maximal members of this partially ordered set. It follows from 5.3 and 5.5 that G is represented as a permutation group via conjugation on the power set of G, and it is evident that G acts on Γ, Ω, and $\mathrm{Syl}_p(G)$ with respect to this representation.

Let $R \in \Omega$. Claim R is the unique point of Ω fixed by the subgroup R of G. For if R fixes $Q \in \Omega$ then, by 5.5, $R \le N_G(Q)$, so, by 1.7.2, $RQ \le G$ and, by 1.7.3, $|RQ| = |R||Q|/|R \cap Q|$. Thus $RQ \in \Gamma$, so, as $R \le RQ \ge Q$ and $R, Q \in \Omega$, we conclude $R = RQ = Q$. So the claim is established.

I've shown that for each $R \in \Omega$ there is a p-subgroup $P(R)$ of G such that R is the unique point of Ω fixed by $P(R)$; namely $R = P(R)$. So it follows from 5.17 that:

(i) G is transitive on Ω, and
(ii) $|\Omega| \equiv 1 \bmod p$.

Let $P \in \Omega$ and suppose $|P| = |G|_p$. Then $P \in \mathrm{Syl}_p(G)$, so, as G is transitive on Ω and G acts on $\mathrm{Syl}_p(G)$, we have $\Omega \subseteq \mathrm{Syl}_p(G)$. On the other hand as $|R|$ divides $|G|$ for each $R \in \Omega$ it is clear $\mathrm{Syl}_p(G) \subseteq \Omega$, so $\Omega = \mathrm{Syl}_p(G)$. Thus (i) implies parts (1) and (2) of Sylow's Theorem, while (ii) and 5.12 imply part (3). Evidently each member of Γ is contained in a member of Ω, so, as $\Omega = \mathrm{Syl}_p(G)$, the fourth part of Sylow's Theorem holds.

So to complete the proof of Sylow's Theorem it remains to show $|P| = |G|_p$ for $P \in \Omega$. Assume otherwise and let $M = N_G(P)$. By (i), (ii), and 5.12, $|G : M| \equiv 1 \bmod p$, so $|M|_p = |G|_p$, and hence p divides $|M/P|$. Therefore, by Cauchy's Theorem (Exercise 2.3), there exists a subgroup R/P of M/P of order p. But now $|R| = |R/P||P|$ is a power of p, so $P < R \in \Gamma$, contradicting $P \in \Omega$.

This completes the proof of Sylow's Theorem.

Next a few consequences of Sylow's Theorem.

(6.1) Let $P \in \mathrm{Syl}_p(G)$. Then $P \trianglelefteq G$ if and only if P is the unique Sylow p-subgroup of G.

Proof. This is because G acts transitively on $\mathrm{Syl}_p(G)$ via conjugation with $N_G(P)$ the stabilizer of P in this representation.

Lemma 6.1 and the numerical restrictions in part (3) of Sylow's Theorem can be used to show groups of certain orders have normal Sylow groups. See for example Exercises 2.5 and 2.6.

(6.2) (Frattini Argument) Let $H \trianglelefteq G$ and $P \in \mathrm{Syl}_p(H)$. Then $G = HN_G(P)$.

Proof. Apply 5.20 to the representation of G on $\mathrm{Syl}_p(H)$, using Sylow's Theorem to get H transitive on $\mathrm{Syl}_p(H)$.

Actually Lemma 6.2 is a special case of the following lemma, which has a similar proof, and which I also refer to as a Frattini Argument:

(6.3) (Frattini Argument) Let K be a group, $H \trianglelefteq K$, and $X \subseteq H$. Then $K = HN_K(X)$ if and only if $X^K = X^H$. Indeed H has $|K : HN_K(X)|$ orbits of equal length on X^K, with representatives $(X^y : y \in Y)$, where Y is a set of coset representatives for $HN_K(X)$ in K.

(6.4) Let $H \trianglelefteq G$ and $P \in \mathrm{Syl}_p(G)$. Then $P \cap H \in \mathrm{Syl}_p(H)$.

Exercises for chapter 2

1. Prove Lemma 5.18.
2. Prove Lemmas 5.20 and 5.21.
3. Prove Cauchy's Theorem: Let G be a finite group and p a prime divisor of $|G|$. Then G contains an element of order p. (Hint: Prove p divides $|C_G(x)|$ for some $x \in G^{\#}$. Then proceed by induction on $|G|$.)
4. Let G be a finite group and p a prime. Prove:
   (1) If $G/Z(G)$ is cyclic, then G is abelian.
   (2) If $|G| = p^2$ then $G \cong \mathbb{Z}_{p^2}$ or E_{p^2}.
5. (1) Let $|G| = p^e m$, $p > m$, p prime, $(p, m) = 1$. Prove G has a normal Sylow p-subgroup.
   (2) Let $|G| = pq$, p and q prime. Prove G has a normal Sylow p-subgroup or a normal Sylow q-subgroup.

6. Let $|G| = pq^2$, where p and q are distinct primes. Prove one of the following holds:
   (1) $q > p$ and G has a normal Sylow q-group.
   (2) $p > q$ and G has a normal Sylow p-group.
   (3) $|G| = 12$ and G has a normal Sylow 2-group.
7. Let G act transitively on a set X, $x \in X$, and $P \in \mathrm{Syl}_p(G_x)$. Prove $N_G(P)$ is transitive on $\mathrm{Fix}(P)$.
8. Prove Lemmas 6.3 and 6.4.
9. Prove that if G has just one Sylow p-subgroup for each $p \in \pi(G)$, then G is the direct product of its Sylow p-subgroups.

3

Representations of groups on groups

Chapter 3 investigates representations in the category of groups and homomorphisms, with emphasis on the normal and subnormal subgroups of groups.

In section 7 the concept of an irreducible representation is defined, and the Jordan–Hölder Theorem is established. As a consequence, the composition factors of a finite group are seen to be an invariant of the group, and these composition factors are simple.

The question arises as to how much the structure of a group is controlled by its composition factors. Certainly many nonisomorphic groups can have the same set of composition factors, so control is far from complete. To investigate this question further we must consider extensions of a group G by a group A. Section 10 studies split extensions and introduces semidirect products.

Section 9 investigates solvable and nilpotent groups. For finite groups this amounts to the study of groups all of whose composition factors are, in the first case, of prime order and, in the second, of order p for some fixed prime p.

Commutators, characteristic subgroups, minimal normal subgroups, central products, and wreath products are also studied.

7 Normal series

In this section G and A are groups, and $\pi: A \to \mathrm{Aut}(G)$ is a representation of A in the category of groups and homomorphisms. I'll also say that A acts as a group of automorphisms on G. Observe that π is also a permutation representation, so we can use the terminology, notation, and results from chapter 2.

A *normal series of length n* for G is a series

$$1 = G_0 \trianglelefteq G_1 \trianglelefteq \cdots \trianglelefteq G_n = G.$$

The series is *A-invariant* if A acts on each G_i.

(7.1) Let $(G_i: 0 \le i \le n)$ be an A-invariant normal series and H an A-invariant subgroup of G. Then
  (1) The restriction $\pi|_H: A \to \mathrm{Aut}(H)$ is a representation of A on H.
  (2) $(G_i \cap H: 0 \le i \le n)$ is an A-invariant normal series for H.
  (3) If $H \trianglelefteq G$ then $\pi_{G/H}: A \to \mathrm{Aut}(G/H)$ is a representation, where $a\pi_{G/H}: Hg \mapsto H(ga)$ for $a \in A$, $g \in G$.

(4) If $H \trianglelefteq G$ then $(G_i H/H : 0 \le i \le n)$ is an A-invariant normal series for G/H.

(5) If X is an A-invariant subgroup of G and $H \trianglelefteq G$, then $X \cap H$ is an A-invariant normal subgroup of X and XH/H is an A-invariant subgroup of G/H which is A-isomorphic to $X/(X \cap H)$.

A subgroup H of G is *subnormal* in G if there exists a series $H = G_0 \trianglelefteq G_1 \trianglelefteq \cdots \trianglelefteq G_n = G$. Write $H \trianglelefteq \trianglelefteq G$ to indicate H is subnormal in G.

(7.2) If $X = G_0 \trianglelefteq G_1 \trianglelefteq \cdots \trianglelefteq G_n = G$, then either $X = G$ or $\langle X^G \rangle \le G_{n-1} \ne G$, so $\langle X^G \rangle \ne G$.

(7.3) Let X and H be A-invariant subgroups of G with $X \trianglelefteq \trianglelefteq G$. Then
  (1) There exists an A-invariant series
$$X = G_0 \trianglelefteq G_1 \trianglelefteq \cdots \trianglelefteq G_n = G.$$
  (2) $X \cap H \trianglelefteq \trianglelefteq H$.
  (3) If $H \trianglelefteq G$ then $XH/H \trianglelefteq \trianglelefteq G/H$.

Proof. Part (1) follows from 7.2 and induction on the length n of a subnormal series $X = G_0 \trianglelefteq G_1 \trianglelefteq \cdots \trianglelefteq G_n = G$, since $\langle X^G \rangle$ is A-invariant. Parts (2) and (3) are straightforward.

The family of *factors* of a normal series $(G_i : 0 \le i \le n)$ is the family of factor groups $(G_{i+1}/G_i : 0 \le i \le n)$. Partially order the set of normal series for G via
$$(H_i : 0 \le i \le m) \le (G_i : 0 \le i \le n) \quad \text{if} \quad H_i = G_{j(i)}$$
for each $0 \le i \le m$, and some $j(i)$.

The representation π is said to be *irreducible* if G and 1 are the only A-invariant normal subgroups of G. We also say that G is A-*simple*. An A-*composition series* for G is a normal series $(G_i : 0 \le i \le n)$ maximal subject to being A-invariant and to $G_i \ne G_{i+1}$ for $0 \le i \le n$. Of particular importance is the case $A = 1$. G is said to be *simple* if it is 1-simple. Similarly the *composition series* for G are its 1-composition series.

(7.4) If G is finite then G possesses an A-composition series.

(7.5) An A-invariant normal series is an A-composition series if and only if each of its factors is a nontrivial A-simple group.

Jordan–Hölder Theorem. Let $(G_i : 0 \leq i \leq n)$ and $(H_i : 0 \leq i \leq m)$ be A-composition series for G. Then $n = m$ and there exists a permutation σ of $\{i : 0 \leq i < n\}$ such that the representations of A on $G_{i\sigma+1}/G_{i\sigma}$ and H_{i+1}/H_i are equivalent for $0 \leq i < n$.

Proof. Let $H = H_{m-1}$ and $k = \min\{i : G_i \not\leq H\}$. I'll show:

(a) G/H is A-isomorphic to G_k/G_{k-1}, and
(b) $(X_i = G_{i\alpha} \cap H : 0 \leq i < n)$ is an A-composition series for H with X_{i+1}/X_i A-isomorphic to $G_{i\alpha+1}/G_{i\alpha}$ for $0 \leq i \leq n - 1$, where $i\alpha = i$ if $i < k - 1$ and $i\alpha = i + 1$ for $i \geq k - 1$.

Suppose (a) and (b) hold. By induction on n, $n - 1 = m - 1$, and there is a permutation β of $\{i : 0 \leq i < n - 1\}$ with $X_{i\beta+1}/X_{i\beta}$ A-isomorphic to H_{i+1}/H_i. Hence $n = m$ and the permutation σ of $\{i : 0 \leq i < n\}$ defined below does the trick in the Jordan–Hölder Theorem:

$$i\sigma = i\beta\alpha \quad \text{if } i < n - 1,$$

$$(n - 1)\sigma = k - 1,$$

So it remains to establish (a) and (b). First, as $G_i \leq H$ for $i < k$, $G_i = G_{i\alpha} = X_i$, so certainly X_i/X_{i-1}, is A-isomorphic to $G_{i\alpha}/G_{i\alpha-1}$.

If $G_k \cap H \not\leq G_{k-1}$, then, as $(G_k \cap H)G_{k-1}/G_{k-1}$ is an A-invariant normal subgroup of the A-simple group G_k/G_{k-1}, we have $G_k = (G_k \cap H)G_{k-1} \leq H$, contrary to the definition of k. So $G_k \cap H = G_{k-1}$. On the other hand, if $j \geq k$, then $G_j \not\leq H$, so a similar argument using 7.3 shows $G = HG_j$, and hence $G/H = HG_j/H$ is A-isomorphic to $G_j/(G_j \cap H)$. In particular $G_j/(G_j \cap H)$ is A-simple, so, if $G_j \cap H \leq G_{j-1}$, then $G_{j-1}/(G_j \cap H)$ is a proper A-invariant normal subgroup of the A-simple group $G_j/(G_j \cap H)$, and hence $G_{j-1} = G_j \cap H \leq H$. But then $j = k$ by definition of k. Moreover G/H is A-isomorphic to $G_k/(G_k \cap H) = G_k/G_{k-1}$, so (a) holds.

By the last paragraph, $G_j \cap H \not\leq G_{j-1}$ for $j > k$. So, as above, $G_j = (G_j \cap H)G_{j-1}$, and hence G_j/G_{j-1} is A-isomorphic to $(G_j \cap H)/(G_{j-1} \cap H)$, completing the proof of (b).

The Jordan–Hölder Theorem says that the factors of an A-composition series of G are (up to equivalence and order) independent of the series, and hence are an invariant of the representation π. These factors are the *composition factors of the representation π*. If $A = 1$, these factors are the *composition factors of G*.

(7.6) Let X be an A-invariant subnormal subgroup of the finite group G. Then

(1) The A-composition factors of X are a subfamily of the A-composition factors of G.

(2) If $X \trianglelefteq G$ then the A-composition factors of G are the union of the A-composition factors of X and G/X.

Proof. There is an A-invariant normal series containing X by 7.3, and as G is finite this series is contained in a maximal A-invariant series. Thus there is an A-composition series through X, so that the result is clear.

8 Characteristic subgroups and commutators

A subgroup H of a group G is *characteristic* in G if H is $\mathrm{Aut}(G)$-invariant. Write H char G to indicate that H is a characteristic subgroup of G.

(8.1) (1) If H char K and K char G, then H char G.

(2) If H char K and $K \trianglelefteq G$, then $H \trianglelefteq G$.

(3) If H char G and K char G, then HK char G and $(H \cap K)$ char G.

A group G is *characteristically simple* if G and 1 are the only characteristic subgroups of G. A *minimal normal subgroup* of G is a minimal member of the set of nonidentity normal subgroups of G, partially ordered by inclusion.

(8.2) If $1 \neq G$ is a characteristically simple finite group, then G is the direct product of isomorphic simple subgroups.

Proof. Let H be a minimal normal subgroup of G and M maximal subject to $M \trianglelefteq G$ and M the direct product of images of H under $\mathrm{Aut}(G)$. Now $X = \langle H\alpha : \alpha \in \mathrm{Aut}(G) \rangle$ is characteristic in G, so by hypothesis $X = G$. Hence, if $M \neq G$, there is $\alpha \in \mathrm{Aut}(G)$ with $H\alpha \nleq M$. As $H\alpha$ is a minimal normal subgroup of G and $H\alpha \cap M \trianglelefteq G$, we conclude $H\alpha \cap M = 1$. But then $M < M \times H\alpha \trianglelefteq G$, contradicting the maximality of M.

So $G = M$. Thus $G = H \times K$ for some $K \leq G$, so every normal subgroup of H is also normal in G. Thus H is simple by minimality of H, and the lemma is established.

(8.3) Minimal normal subgroups are characteristically simple.

Proof. This follows from 8.1.2.

(8.4) (1) The simple abelian groups are the groups of prime order.

(2) If G is characteristically simple, finite, and abelian, then $G \cong E_{p^n}$ for some prime p and some integer n.

For $x, y \in G$, write $[x, y]$ for the group element $x^{-1}y^{-1}xy$. $[x, y]$ is the *commutator* of x and y. For $X, Y \leq G$, define

$$[X, Y] = \langle [x, y] : x \in X, y \in Y \rangle.$$

For $z \in Z \leq G$ write $[x, y, z]$ for $[[x, y], z]$ and $[X, Y, Z]$ for $[[X, Y], Z]$.

(8.5) Let G be a group, $a, b, c \in G$, and $X, Y \leq G$. Then
   (1) $[a, b] = 1$ if and only if $ab = ba$.
   (2) $[X, Y] = 1$ if and only if $xy = yx$ for all $x \in X$ and $y \in Y$.
   (3) If $\alpha : G \to H$ is a group homomorphism then $[a, b]\alpha = [a\alpha, b\alpha]$ and $[X, Y]\alpha = [X\alpha, Y\alpha]$.
   (4) $[ab, c] = [a, c]^b[b, c]$ and $[a, bc] = [a, c][a, b]^c$.
   (5) $X \leq N_G(Y)$ if and only if $[X, Y] \leq Y$.
   (6) $[X, Y] = [Y, X] \trianglelefteq \langle X, Y \rangle$.

Proof. I prove (6) and leave the other parts as exercises. Notice $[a, b]^{-1} = [b, a]$, so $[X, Y] = [Y, X]$. Further, to prove $[X, Y] \trianglelefteq \langle X, Y \rangle$, it will suffice to show $[x, y]^z \in [X, Y]$ for each $x \in X$, $y \in Y$, and $z \in X \cup Y$. As $[x, y]^{-1} = [y, x]$, we may assume $z \in Y$. But, by (4), $[x, y]^z = [x, z]^{-1}[x, yz] \in [X, Y]$, so the proof is complete.

(8.6) Let G be a group, $x, y \in G$, and assume $z = [x, y]$ centralizes x and y. Then
   (1) $[x^n, y^m] = z^{nm}$ for all $n, m \in \mathbb{Z}$.
   (2) $(yx)^n = z^{n(n-1)/2}y^n x^n$ for all $0 \leq n \in \mathbb{Z}$.

Proof. Without loss $G = \langle x, y \rangle$, so $z \in Z(G)$. $z = [x, y]$ so $x^y = xz$. Then, for $n \in \mathbb{Z}$, $(x^n)^y = (x^y)^n = (xz)^n = x^n z^n$ as $z \in Z(G)$. Thus $[x^n, y] = z^n$. Similarly $[x, y^m] = z^m$, so $[x^n, y^m] = [x, y^m]^n = z^{mn}$, and (1) holds.

Part (2) is established by induction on n. Namely $(yx)^{n+1} = (yx)^n yx = z^{n(n-1)/2}y^n x^n yx$, while by (1) $x^n y = yx^n z^n$, so that the result holds.

(8.7) (Three-Subgroup Lemma) Let X, Y, Z be subgroups of a group G with $[X, Y, Z] = [Y, Z, X] = 1$. Then $[Z, X, Y] = 1$.

Proof. Let $x \in X$, $y \in Y$, and $z \in Z$. A straightforward calculation shows:

(*) $[x, y^{-1}, z]^y = x^{-1}y^{-1}xz^{-1}x^{-1}yxy^{-1}zy = a(x, y, z)^{-1}a(y, z, x),$

where $a(u, v, w) = uwu^{-1}vu$. Applying the permutations (x, y, z) and (x, z, y) to (*) and taking the product of (*) with these two images, we conclude:

(**) $$[x, y^{-1}z]^y[y, z^{-1}, x]^z[z, x^{-1}, y]^x = 1.$$

As $[X, Y, Z] = [Y, Z, X] = 1$, also $[x, y^{-1}, z] = [y, z^{-1}, x] = 1$, so by (**) we get $[z, x^{-1}, y] = 1$. Finally as $[Z, X]$ is generated by the commutators $[z, x^{-1}]$, $z \in Z, x \in X$, it follows from 8.5.1 that y centralizes $[Z, X]$. But then, by 8.5.2, $[Z, X, Y] = 1$.

The *commutator group* or *derived group* of a group G is the subgroup $G^{(1)} = [G, G]$. Extend the notation recursively and define $G^{(n)} = [G^{(n-1)}, G^{(n-1)}]$ for $n > 1$. Define $G^{(0)} = G$ and $G^\infty = \bigcap_{i=1}^\infty G^{(i)}$.

(8.8) Let G be a group and $H \leq G$. Then
   (1) $H^{(n)} \leq G^{(n)}$.
   (2) If $\alpha: G \to X$ is a surjective group homomorphism then $G^{(n)}\alpha = X^{(n)}$.
   (3) $G^{(n)}$ char G.
   (4) $G^{(1)} \leq H$ if and only if $H \trianglelefteq G$ and G/H is abelian.

A group G is *perfect* if $G = G^{(1)}$.

(8.9) Let X and L be subgroups of a group G with L perfect and $[X, L, L] = 1$. Then $[X, L] = 1$.

Proof. $[L, X, L] = [X, L, L]$ and by hypothesis both are 1. So by the Three-Subgroup Lemma, $[L, L, X] = 1$. But by hypothesis $L = [L, L]$, so $[L, X] = 1$.

9 Solvable and nilpotent groups

A group G is *solvable* if it possesses a normal series whose factors are abelian.

(9.1) A group G is solvable if and only if $G^{(n)} = 1$ for some positive integer n.

Proof. If $G^{(n)} = 1$, then $(G^{(n-i)}: 0 \leq i \leq n)$ is a normal series with abelian factors by 8.8.4. Conversely if $(G_i: 0 \leq i \leq n)$ is such a series then, by 8.8.4 and induction on i, $G^{(i)} \leq G_{n-i}$, so $G^{(n)} = 1$.

(9.2) A finite group is solvable if and only if all its composition factors are of prime order.

Proof. If all composition factors are of prime order then a composition series for G is a normal series all of whose factors are abelian. Conversely if $(G_i: 0 \leq i \leq n)$ is such a series then the composition factors of the abelian group G_{i+1}/G_i are also abelian, and hence, by 8.4.1, of prime order. Then, by 7.6, the composition factors of G are of prime order.

(9.3) (1) Subgroups and homomorphic images of solvable groups are solvable.

(2) If $H \trianglelefteq G$ with H and G/H solvable, then G is solvable.

(9.4) Solvable minimal normal subgroups of finite groups are elementary abelian p-groups.

Proof. Let G be finite and M a solvable minimal normal subgroup of G. By 9.1 and solvability of M, $M^{(1)} \neq M$. Next, by 8.3, M is characteristically simple. So, as $M^{(1)}$ char M, we conclude $M^{(1)} = 1$. Thus M is abelian by 8.8. Then 8.4.2 completes the proof.

Define $L_1(G) = G$, and, proceeding recursively, define $L_n(G) = [L_{n-1}(G), G]$ for $1 < n \in \mathbb{Z}$. G is said to be *nilpotent* if $L_n(G) = 1$ for some $1 \leq n \in \mathbb{Z}$. The *class* of a nilpotent group is $m - 1$, where $m = \min\{i: L_i(G) = 1\}$.

(9.5) (1) $L_n(G)$ char G for each $1 \leq n \in \mathbb{Z}$.

(2) $L_{n+1}(G) \leq L_n(G)$.

(3) $L_n(G)/L_{n+1}(G) \leq Z(G/L_{n+1}(G))$.

Proof. Part (1) follows from 8.5.3 by induction on n. Then (1) and 8.5.5 imply (2) while 8.5.1 and 8.5.3 imply (3).

Define $Z_0(G) = 1$ and proceeding recursively define $Z_n(G)$ to be the preimage in G of $Z(G/Z_{n-1}(G))$ for $1 \leq n \in \mathbb{Z}$. Evidently $Z_n(G)$ char G.

(9.6) G is nilpotent if and only if $G = Z_n(G)$ for some $0 \leq n \in \mathbb{Z}$. If G is nilpotent then the class of G is $k = \min\{n: G = Z_n(G)\}$.

Proof. I first show that if G is nilpotent of class m then $L_{m+1-i}(G) \leq Z_i(G)$ for $0 \leq i \leq m$. For $i = 0$ this follows directly from the definitions, while if $i > 0$ and $L_{m+2-i}(G) \leq Z_{i-1}(G)$ then $[L_{m+1-i}(G), G] = L_{m+2-i}(G) \leq Z_{i-1}(G)$, so, by 8.5, $L_{m+1-i}(G)Z_{i-1}(G)/Z_{i-1}(G) \leq Z(G/Z_{i-1}(G)) = Z_i(G)/Z_{i-1}(G)$, and hence $L_{m+1-i}(G) \leq Z_i(G)$. So the claim follows by induction on i, and we see $Z_m(G) = L_1(G) = G$, so $k \leq m$.

Next let's see that if $Z_n(G) = G$ for some $0 \leq n \in \mathbb{Z}$, then $L_{i+1}(G) \leq Z_{n-i}(G)$ for each $0 \leq i \leq n$. Again the case $i = 0$ is trivial, while if $i > 0$ and $L_i(G) \leq Z_{n-i+1}(G)$ then $L_{i+1}(G) = [L_i(G), G] \leq [Z_{n-i+1}(G), G] \leq Z_{n-i}(G)$ by 8.5.5, establishing the claim. In particular $L_{n+1}(G) \leq Z_0(G) = 1$, so G is nilpotent of class $m \leq n$, so that $m \leq k$.

(9.7) $1 \neq G$ is nilpotent of class m if and only if $G/Z(G)$ is nilpotent of class $m - 1$.

Proof. This is a direct consequence of 9.6.

(9.8) p-groups are nilpotent.

Proof. Let G be a minimal counter example. Then certainly $G \neq 1$, so, by 5.16, $Z(G) \neq 1$. Hence, by minimality of G, $G/Z(G)$ is nilpotent, so, by 9.7, G is nilpotent, contrary to the choice of G.

(9.9) Let G be nilpotent of class m. Then subgroups and homomorphic images of G are nilpotent of class at most m.

(9.10) If G is nilpotent and H is a proper subgroup of G, then H is proper in $N_G(H)$.

Proof. Assume $N_G(H) = H < G$. Then $Z(G) \leq N_G(H) = H$, so $H^* < G^* = G/Z(G)$. By 9.7 and induction on the nilpotence class of G, $H^* < N_{G^*}(H^*)$. But, as $Z(G) \leq H$, $N_{G^*}(H^*) = N_G(H)^*$, so $H < N_G(H)$, a contradiction.

(9.11) A finite group is nilpotent if and only if it is the direct product of its Sylow groups.

Proof. The direct product of nilpotent groups is nilpotent, so by 9.8 the direct product of p-groups is nilpotent. Conversely let G be nilpotent; we wish to show G is the direct product of its Sylow groups. Let $P \in \mathrm{Syl}_p(G)$. By Exercise 2.9 it suffices to show $P \trianglelefteq G$. If not, $M = N_G(P) < G$, so, by 9.10, $M < N_G(M)$. But, as $P \trianglelefteq M$, $\{P\} = \mathrm{Syl}_p(M)$, so P char M. Hence $N_G(M) \leq N_G(P) = M$, a contradiction.

10 Semidirect products

In this section A and G are groups and $\pi: A \to \mathrm{Aut}(G)$ is a representation of A as a group of group automorphisms of G.

Let $H \trianglelefteq G$. A *complement* to H in G is a subgroup K of G with $G = HK$ and $H \cap K = 1$. G is said to be an *extension* of a group X by a group Y if there exists $H \trianglelefteq G$ with $H \cong X$ and $G/H \cong Y$. The extension is said to *split* if H has a complement in G. The following construction can be used to describe split extensions.

Let S be the set product $A \times G$ and define a binary operation on S by

$$(a, g)(b, h) = (ab, g^{b\pi} h) \quad a, b \in A, g, h \in G$$

where $g^{b\pi}$ denotes the image of G under the automorphism $b\pi$ of G. We call S the *semidirect product* of G by A with respect to π. Denote S by $S(A, G, \pi)$.

(10.1) (1) $S = S(A, G, \pi)$ is a group.

(2) The maps $\sigma_A: A \to S$ and $\sigma_G: G \to S$ are injective group homomorphisms, where $\sigma_A: a \mapsto (a, 1)$ and $\sigma_G: g \mapsto (1, g)$.

(3) $G\sigma_G \trianglelefteq S$ and $A\sigma_A$ is a complement to $G\sigma_G$ in S.

(4) $(1, g)^{(a,1)} = (1, g^{a\pi})$ for $g \in G$, $a \in A$.

Observe that if π is the trivial homomorphism then the semidirect product is just the direct product of A and G.

(10.2) Let H be a group, $G \trianglelefteq H$, and B a complement to G in H. Let $\alpha: B \to \mathrm{Aut}(G)$ be the conjugation map (i.e., $b\alpha: g \to g^b$ for $b \in B$, $g \in G$; see Exercise 1.3). Define $\beta: S(B, G, \alpha) \to H$ by $(b, g)\beta = bg$. Then β is an isomorphism.

We see from 10.1 and 10.2 that the semidirect products of G by A are precisely the split extensions of G by A. Moreover the representation defining the semidirect product is a conjugation map.

Under the hypotheses of 10.2, I'll say that H is a semidirect product of G by B. Formally this means the map β of 10.2 is an isomorphism.

(10.3) Let $S_i = S(A_i, G_i, \pi_i)$, $i = 1, 2$, be semidirect products. Then there exists an isomorphism $\phi: S_1 \to S_2$ with $A_1\sigma_{A_1}\phi = A_2\sigma_{A_2}$ and $G_1\sigma_{G_1}\phi = G_2\sigma_{G_2}$ if and only if π_1 is quasiequivalent to π_2 in the category of groups and homomorphisms.

It is not difficult to see that semidirect products $S_1 = S(A, G, \pi_1)$ and $S_2 = S(G, A, \pi_2)$ can be isomorphic without π_1 being quasiequivalent to π_2. To investigate just when S_1 and S_2 are isomorphic we need to know more about how $\mathrm{Aut}(S_i)$ acts on its normal subgroups isomorphic to G, and how the stabilizer in $\mathrm{Aut}(S_i)$ of such a subgroup acts on its complements. This latter question is considered in chapter 6.

It's also easy to cook up nonsplit extensions, and it is of interest to generate conditions which insure that an extension splits. The following is perhaps the most important such condition:

(10.4) (Gaschütz' Theorem) Let p be a prime, V an abelian normal p-subgroup of a finite group G, and $P \in \mathrm{Syl}_p(G)$. Then G splits over V if and only if P splits over V.

Proof. Notice $V \le P$. Hence if H is a complement to V in G then by the Modular Property of Groups, 1.14, $P = P \cap G = P \cap VH = V(P \cap H)$ and $P \cap H$ is a complement to V in P.

Conversely suppose Q is a complement to V in P. Let $\bar{G} = G/V$ and observe $\bar{P} \cong \bar{Q} \cong Q$. Let X be a set of coset representatives for V in G. Then the map $x \mapsto \bar{x}$ is a bijection of X with $\bar{G}$ and I denote the inverse of this map by $a \mapsto x_a$. Then

(i) $\qquad x_a x_b = x_{ab} \gamma(a, b)$ for $a, b \in \bar{G}$, and for some $\gamma(a, b) \in V$.

Next using associativity in G and $\bar{G}$ we have $x_{abc}\gamma(a, bc)\gamma(b, c) = x_a x_{bc}\gamma(b, c) = x_a(x_b x_c) = (x_a x_b)x_c = x_{ab}\gamma(a, b)x_c = x_{ab}x_c\gamma(a, b)^{x_c} = x_{abc}\gamma(ab, c)\gamma(a, b)^{x_c}$, from which we conclude:

(ii) $\qquad \gamma(ab, c)\gamma(a, b)^{x_c} = \gamma(a, bc)\gamma(b, c) \quad$ for all $a, b, c \in \bar{G}$.

Now choose $X = QY$, where Y is a set of coset representatives for P in G. Then, for $g \in Q$ and $a \in \bar{G}$, $x_{\bar{g}a} = gx_a$, so:

(iii) $\qquad x_{\bar{g}} = g \quad$ and $\quad \gamma(\bar{g}, a) = 1 \quad$ for all $g \in Q \quad$ and $\quad a \in \bar{G}$.

Now (ii) and (iii) imply:

(iv) $\qquad \gamma(\bar{g}b, c) = \gamma(b, c) \quad$ for all $b, c \in \bar{G} \quad$ and $\quad g \in Q$.

Next for $c \in \bar{G}$ define $\beta(c) = \Pi_{\bar{y} \in \bar{Y}} \gamma(\bar{y}, c)$. By (iv), $\beta(c)$ is independent of the choice of the set $\bar{Y}$ of coset representatives. But if $b \in \bar{G}$ then $\bar{Y}b$ is another set of coset representatives for $\bar{Q}$ in $\bar{G}$, so:

(v) $\qquad\qquad \beta(c) = \prod_{\bar{y} \in \bar{Y}} \gamma(\bar{y}b, c) \quad$ for all $b, c \in \bar{G}$.

As V is abelian we conclude from (ii) that

$$\left(\prod_{\bar{y} \in \bar{Y}} \gamma(\bar{y}b, c)\right)\left(\prod_{\bar{y} \in \bar{Y}} \gamma(\bar{y}, b)\right)^{x_c} = \left(\prod_{\bar{y} \in \bar{Y}} \gamma(\bar{y}, bc)\right)\left(\prod_{\bar{y} \in \bar{Y}} \gamma(b, c)\right)$$

and then appealing to (v) we obtain

(vi) $\quad \beta(c)\beta(b)^{x_c} = \beta(bc)\gamma(b, c)^m \quad$ for all $b, c \in \bar{G}$, where $m = |G : P|$.

As $P \in \mathrm{Syl}_p(G)$, $(m, p) = 1$. Thus m is invertible mod $|V|$. Hence we can define $\alpha(c) = \beta(c)^{-m^{-1}}$, for $c \in \bar{G}$. Then taking the $-m^{-1}$ power of (vi) we obtain:

(vii) $\quad\quad \alpha(c)\alpha(b)^{x_c} = \alpha(bc)\gamma(b, c)^{-1} \quad$ for all $b, c \in \bar{G}$.

Finally define $y_a = x_a\alpha(a)$ for $\bar{a} \in \bar{G}$ and set $H = \{y_a : a \in \bar{G}\}$. H will be shown to be a complement to V in G. This will complete the proof.

It suffices to show $y_b y_c = y_{bc}$ for all $b, c \in \bar{G}$. But $y_b y_c = x_b\alpha(b)x_c\alpha(c) = x_b x_c \alpha(b)^{x_c}\alpha(c) = x_{bc}\gamma(b, c)\alpha(b)^{x_c}\alpha(c) = y_{bc}\alpha(bc)^{-1}\gamma(b, c)\alpha(b)^{x_c}\alpha(c)$. Then, as V is abelian, (vii) implies $y_b y_c = y_{bc}$, as desired.

The Schur–Zassenhaus Theorem in section 18 is another useful result on splitting.

11 Central products and wreath products

(11.1) Let $\{G_i : 1 \leq i \leq n\}$ be a set of subgroups of G. Then the following are equivalent:

(1) $G = \langle G_i : 1 \leq i \leq n \rangle$ and $[G_i, G_j] = 1$ for $i \neq j$.
(2) The map $\pi : (x_1, \ldots, x_n) \mapsto x_1 \ldots x_n$ is a surjective homomorphism of $G_1 \times \cdots \times G_n = D$ onto G with $D_i\pi = G_i$ and $D_i \cap \ker(\pi) = 1$, where D_i consists of those elements of D with 1 in all but the ith component.

If either of the equivalent conditions of 11.1 holds, then G is said to be a *central product* of the subgroups G_i, $1 \leq i \leq n$. Notice that the kernel of the homomorphism π of 11.1 is contained in the centre of $G_1 \times \cdots \times G_n$.

(11.2) Let $(G_i : 1 \leq i \leq n)$ be a family of groups such that $Z(G_1) \cong Z(G_i)$ and $\mathrm{Aut}(Z(G_i)) = \mathrm{Aut}_{\mathrm{Aut}(G_i)}(Z(G_i))$ for $1 \leq i \leq n$. Then, up to an isomorphism mapping G_i to G_i for each i, there exists a unique central product of the groups G_i in which $Z(G_1) = Z(G_i)$ for each i.

Proof. Adopt the notation of 11.1 and identify G_i with D_i. By hypothesis there are isomorphisms $\alpha_i : Z(D_1) \to Z(D_i)$, $1 \leq i \leq n$. Let E be the subgroup of D generated by $z(z^{-1}\alpha_i)$, $z \in Z(D_1)1 < i \leq n$. Observe E is a complement to $Z(D_i)$ in $Z = \langle Z(D_i) : 1 \leq i \leq n \rangle = Z(D)$ for each i. Thus D/E is a central product of the groups G_i with $Z(G_i) = Z(G_1)$ for each i, by 11.1.

Next assume G is a central product of the G_i with $Z(G_1) = Z(G_i)$ for each i, and let $\pi: D \to G$ be the surjective homomorphism supplied by 11.1. Let $\beta_i: Z(D_1) \to Z(D_i)$ be the isomorphism which is the composition of $\pi|_{Z(D_1)}$: $Z(D_1) \to Z(G_1)$ and $(\pi|_{Z(D_i)})^{-1}: Z(G_i) \to Z(D_i)$. Observe

$$\ker(\pi) = \langle z(z^{-1}\beta_i): z \in Z(D_1), 1 \le i \le n \rangle = A$$

is a complement to $Z(D_i)$ in Z for each i, and of course $G \cong D/A$. To complete the proof I exhibit $\gamma \in \mathrm{Aut}(D)$ with $D_i\gamma = D_i$ and $E\gamma = A$. Notice γ induces an isomorphism of D/E with D/A mapping G_i to G_i, demonstrating uniqueness.

Let $\delta_i = (\alpha_i)^{-1}\beta_i: Z(D_i) \to Z(D_i)$, so that $\delta_i \in \mathrm{Aut}(Z(D_i))$. By hypothesis there is $\gamma_i \in \mathrm{Aut}(D_i)$ with $\gamma_i|_{Z(D_i)} = \delta_i$. Define $\gamma: D \to D$ by

$$(x_1, \dots, x_n) \mapsto (x_1, x_2\gamma_2, \dots, x_n\gamma_n)$$

and observe $\gamma \in \mathrm{Aut}(D)$ with $(z(z^{-1}\alpha_i))\gamma = z(z^{-1}\beta_i)$, so that $E\gamma = A$. Thus the proof is complete.

Under the hypotheses of 11.2, we say G is the *central product of the groups G_i with identified centers*, and write $G = G_1 * G_2 * \cdots * G_n$.

Let L be a group and $\pi: G \to \mathrm{Sym}(X)$ a permutation representation of G on $X = \{1, \dots, n\}$. Form the direct product D of n copies of L. G acts as a group of automorphisms of D via the representation α defined by

$$g\alpha: (x_1, \dots, x_n) \mapsto (x_{1g^{-1}\pi}, \dots, x_{ng^{-1}\pi}).$$

The *wreath product* of L by G (with respect to π) is defined to be the semidirect product $S(G, D, \alpha)$. The wreath product is denoted by $L\mathrm{wr}\, G$ or $L\mathrm{wr}_\pi G$ or $L\mathrm{wr}_x G$.

(11.3) Let $W = L\mathrm{wr}_\pi G$ be the wreath product of L by G with respect to π. Then

(1) W is a semidirect product of D by G where $D = L_1 \times \cdots \times L_n$ is a direct product of n copies of L.

(2) G permutes $\Delta = \{L_i: 1 \le i \le n\}$ via conjugation and the permutation representation of G on Δ is equivalent to π. That is $(L_i)^g = L_{ig\pi}$ for each $g \in G$ and $1 \le i \le n$.

(3) The stabilizer G_i of i in G centralizes L_i.

Exercises for chapter 3

1. Let G and A be finite groups with $(|G|, |A|) = 1$, assume A is represented on G as a group of automorphisms, and $(G_i: 0 \le i \le n)$ is an A-invariant normal series for G such that A centralizes G_{i+1}/G_i for $0 \le i < n$. Prove

A centralizes G. Produce a counter example when $(|A|, |G|) \neq 1$. (Hint: Reduce to the case where A is a p-group and use 5.14.)

2. Let G be a finite group, p a prime, and X a p-subgroup of G. Prove

   (1) Either $X \in \mathrm{Syl}_p(G)$ or X is properly contained in a Sylow p-subgroup of $N_G(X)$.

   (2) If G is a p-group and X is a maximal subgroup of G, then $X \trianglelefteq G$ and $|G:X| = p$.

3. Prove lemmas 10.1 and 10.3. Exhibit a nonsplit extension.

4. Let p and q be primes with $p > q$. Prove every group of order pq is a split extension of $\mathbb{Z}_p$ by $\mathbb{Z}_q$. Up to isomorphism, how many groups of order pq exist? (Hint: Use 10.3 and Exercise 1.7. Prove $\mathrm{Aut}(\mathbb{Z}_p)$ is cyclic of order $p - 1$. You may use the fact that the multiplicative group of a finite field is cyclic.)

5. Let G be a central product of n copies G_i, $1 \leq i \leq n$, of a perfect group L and let α be an automorphism of G of order n permuting $\{G_i : 1 \leq i \leq n\}$ transitively. Prove $C_G(\alpha) = KZ$ where $K = C_G(\alpha)^{(1)} \cong L/U$ for some $U \leq Z(L)$ and $Z = C_{Z(G)}(\alpha)$. Further $N_{\mathrm{Aut}(G)}(G_1) \cap C(K) \leq C(G_1)$.

6. If A acts on a group G and centralizes a normal subgroup H of G then $[G, A] \leqslant C_G(H)$.

4

Linear representations

Chapter 4 develops the elementary theory of linear representations. Linear representations are discussed from the point of view of modules over the group ring. Irreducibility and indecomposability are defined, and we find that the Jordan–Hölder Theorem holds for finite dimensional linear representations. Maschke's Theorem is established in section 12. Maschke's Theorem says that, if G is a finite group and F a field whose characteristic does not divide the order of G, then the indecomposable representations of G over F are irreducible.

Section 13 explores the connection between finite dimensional linear representations and matrices. There is also a discussion of the special linear group, the general linear group, and the corresponding projective groups. In particular we find that the special linear group is generated by its transvections and is almost always perfect.

Section 14 contains a discussion of the dual representation which will be needed in section 17.

12 Modules over the group ring

Section 12 studies linear representations over a field F using the group ring of G over F. This requires an elementary knowledge of modules over rings. One reference for this material is chapter 3 of Lang [La].

Throughout section 12, V will be a vector space over F. The group of automorphisms of V in the category of vector spaces and F-linear transformations is the *general linear group* $\mathrm{GL}(V)$. Assume $\pi\colon G \to \mathrm{GL}(V)$ is a representation of G in this category. Such representations will be called *FG-representations* and V will be called the *representation module* for π. Representation modules for *FG-representations* will be termed *FG-modules*.

Let $R = F[G]$ be the vector space over F with basis G and define multiplication on R to be the linear extension of the multiplication of G. Hence a typical member of R is of the form $\sum_{g \in G} a_g g$, where $a_g \in F$ and at most a finite number of the coefficients a_g are nonzero. Multiplication becomes

$$\left(\sum_{g \in G} a_g g \right)\left(\sum_{h \in G} b_h h \right) = \sum_{g,h \in G} a_g b_h g h.$$

As is well known (and easy to check), this multiplication makes R into a ring with identity and, as the multiplication on R commutes with scalar multiplication from F, R is even an F-algebra. $R = F[G]$ is the *group ring* or *group algebra* of G over F.

Observe that V becomes a (right) R-module under the scalar multiplication:

$$v\left(\sum a_g g\right) = \sum a_g(v(g\pi)) \quad v \in V, \sum a_g g \in R$$

Conversely if U is an R-module then we have a representation $\alpha: G \to \mathrm{GL}(U)$ defined by $u(g\alpha) = ug$, where ug is the module product of $u \in U$ by $g \in R$.

Further if $\pi_i: G \to \mathrm{GL}(V_i)$, $i = 1, 2$, are FG-representations then $\beta: V_1 \to V_2$ is an equivalence of the representations precisely when β is an isomorphism of the corresponding R-modules. Indeed $\gamma: V_1 \to V_2$ is an FG-homomorphism if and only if γ is an R-module homomorphism of the corresponding R-modules. Here $\gamma: V_1 \to V_2$ is defined to be an *FG-homomorphism* if γ is an F-linear map commuting with the actions of G in the sense that $v(g\pi_1)\gamma = v\gamma(g\pi_2)$ for each $v \in V_1$, and $g \in G$. In the terminology of section 4, the FG-homomorphisms are the G-morphisms.

The upshot of these observations is that the study of FG-representations is equivalent to the study of modules for the group ring $F[G] = R$. I will take both points of view and appeal to various standard theorems on modules over rings. Lang [La] is a reference for such results.

Observe also that V is an abelian group under addition and π is a representation of G on V in the category of groups and homomorphisms. Indeed π induces a representation

$$\pi': F^\# \times G \to \mathrm{Aut}(V)$$

in that category defined by $(a, g)\pi': v \mapsto av(g\pi)$, for $a \in F^\#$, $g \in G$. Here $F^\#$ is the multiplicative group of F. Two FG-representations π and σ are equivalent if and only if π' and σ' are equivalent, so we can use the results of chapter 3 to study FG-representations. In the case where F is a field of prime order we can say even more.

(12.1) Let F be the field of integers modulo p for some prime p and assume V is of finite dimension. Then

(1) V is an elementary abelian p-group and π is a representation of G in the category of groups and homomorphisms.

(2) If U is an elementary abelian p-group written additively, then U is a vector space $_FU$ over F, where scalar multiplication is defined by $((p) + n)u = nu$, $n \in \mathbb{Z}, u \in U$.

(3) $\mathrm{GL}(_FU)$ is equal to the group $\mathrm{Aut}(U)$ of group automorphisms of U. Indeed if W is an elementary abelian p-group then the group homomorphisms from U into W are precisely the F-linear transformations from $_FU$ into $_FW$.

(4) The vector space $_F V$ defined using the construction in part (2) is precisely the vector space V.

As a consequence of 12.1, if F is a field of prime order, the FG-representations are the same as the representations of G on elementary abelian p-groups.

A vector subspace U of V is an *FG-submodule* of V if U is G-invariant. U is an FG-submodule if and only if U is an R-submodule of the R-module V. From 7.1 there are group representations of $F^\# \times G$ on U and V/U. These representations are also FG-representations and they correspond to the R-modules U and V/U.

V is *irreducible* or *simple* if 0 and V are the only FG-submodules. A *composition series* for V is a series

$$0 = V_0 \leq V_1 \leq \cdots \leq V_n = V$$

of FG-submodules such that each factor module V_{i+1}/V_i is a simple FG-module. This corresponds to the notion of composition series in section 7. The family $(V_{i+1}/V_i : 0 \leq i < n)$ is the family of *composition factors* of the series. If V is of finite dimension then it is easy to see that V possesses a composition series. Appealing to the Jördan–Hölder Theorem, established in section 7, and to remarks above, we get:

(12.2) (Jordan–Hölder Theorem for FG-modules) Let V be a finite dimensional FG-module. Then V possesses a composition series and the composition factors are independent (up to order and equivalence) of the choice of composition series.

The restrictions $\pi_i = \pi|_{V_i/V_{i-1}}$, $0 < i \leq n$, of π to the composition series $(V_i : 0 \leq i \leq n)$ of a finite dimensional FG-representation π are called the *irreducible constituents* of π. They are defined only up to order and equivalence but, subject to this constraint, they are well defined and unique by the Jordan–Hölder Theorem.

V is *decomposable* if there exist proper FG-submodules U and W of V with $V = U \oplus W$. Otherwise V is *indecomposable*. I'll write $\pi = \pi_1 + \pi_2$ if $V = V_1 \oplus V_2$ with V_1 and V_2 FG-submodules of V and $\pi|_{V_i}$ is equivalent to π_i. Observe that if $\alpha = \alpha_1 + \alpha_2$ is an FG-representation with α_i equivalent to π_i for $i = 1$ and 2, then π is equivalent to α.

As in section 4, an FG-module V is said to be the *extension* of a module X by a module Y if there exists a submodule U of V with $U \cong X$ and $V/U \cong Y$. A *complement* to U in V is an FG-submodule W with $V = U \oplus W$. The extension is said to *split* if U possesses a complement in V. As in chapter 3, we wish to investigate when extensions split.

An R-module V is *cyclic* if $V = xR = \{xr: r \in R\}$ for some $x \in V$. Equivalently $V = \langle xG \rangle$ is generated as a vector space by the images of x under G. The element x is said to be a *generator* for the cyclic module V. Notice that irreducible modules are cyclic.

(12.3) (1) If $V = xR$ is cyclic then the map $r \mapsto xr$ is surjective R-module homomorphism from R onto V with kernel $A(x) = \{r \in R: xr = 0\}$.

(2) Homomorphic images of cyclic modules are cyclic, so the cyclic R-modules are precisely the homomorphic images of R.

(3) V is irreducible if and only if $A(x)$ is a maximal right ideal of R.

Given R-modules U and V, $\mathrm{Hom}_R(U, V)$ denotes the set of all R-module homomorphisms of U into V. $\mathrm{Hom}_R(U, V)$ is an abelian group under the following definition of addition:

$$u(\alpha + \beta) = u\alpha + u\beta \quad u \in U, \alpha, \beta \in \mathrm{Hom}_R(U, V).$$

If R is commutative, $\mathrm{Hom}_R(U, V)$ is even an R-module when scalar multiplication is defined by

$$u(\alpha r) = (ur)\alpha \quad u \in U, r \in R, \alpha \in \mathrm{Hom}_R(U, V).$$

Finally $\mathrm{Hom}_R(V, V) = \mathrm{End}_R(V)$ is a ring, where multiplication is defined to be composition. That is

$$u(\alpha \cdot \beta) = (u\alpha)\beta \quad u \in V, \alpha, \beta \in \mathrm{End}_R(V).$$

In the language of section 4, $\mathrm{Hom}_R(U, V) = \mathrm{Mor}_G(U, V)$.

(12.4) (Schur's Lemma) Let U and V be R-modules and $\alpha \in \mathrm{Hom}_R(U, V)$. Then
  (1) If U is simple either $\alpha = 0$ or α is an injection.
  (2) If V is simple either $\alpha = 0$ or α is a surjection.
  (3) If U and V are simple then either $\alpha = 0$ or α is an isomorphism.
  (4) If V is simple then $\mathrm{End}_R(V)$ is a division ring.

The module V is a *semisimple* R-module if V is the direct sum of simple submodules. The *socle* of V is the submodule $\mathrm{Soc}(V)$ generated by all the simple submodules of V.

(12.5) Assume Ω is a set of simple submodules of V and $\Delta \subseteq \Omega$ such that $V = \langle \Omega \rangle$ and $\langle \Delta \rangle = \bigoplus_{A \in \Delta} A$. Then there exists $\Gamma \subseteq \Omega$ with $\Delta \subseteq \Gamma$ such that $V = \bigoplus_{b \in \Gamma} B$.

Proof. Let S be the set of $\Gamma \subseteq \Omega$ with $\Delta \subseteq \Gamma$ and $\langle \Gamma \rangle = \bigoplus_{B \in \Gamma} B$. Partially order S by inclusion. Check that if C is a chain in S then $\bigcup_{\Gamma \in C} \Gamma$ is an upper bound for C in S. Hence by Zorn's Lemma there is a maximal member Γ of S. Finally prove $V = \langle \Gamma \rangle$.

(12.6) The following are equivalent:
  (1) V is semisimple.
  (2) $V = \mathrm{Soc}(V)$.
  (3) V splits over every submodule of V.

Proof. The equivalence of (1) and (2) follows from 12.5.

Assume (3) holds but $V \neq \mathrm{Soc}(V)$. By (3) there is a complement U to $\mathrm{Soc}(V)$ in V. Let $x \in U^{\#}$, I a maximal right ideal of R containing $A(x) = \{r \in R: xr = 0\}$, and W the image of I in xR under the homomorphism of 12.3.1. By (3) there is a complement Z to W in V. By the Modular Property of Groups, 1.14, $Z \cap xR = M$ is a complement to W in xR. Then $M \cong xR/W$, so M is simple by 12.3.3. Hence $M \leq \mathrm{Soc}(V)$, so $0 \neq M \leq \mathrm{Soc}(V) \cap U = 0$, a contradiction. Thus (3) implies (2).

Finally assume V is semisimple and U is a submodule of V with no complement in V. Now $\mathrm{Soc}(U) = \bigoplus_{A \in \Delta} A$ for some set of simple submodules, so by 12.4 there is a set Γ of simple submodules of V with $\Delta \subseteq \Gamma$ and $V = \bigoplus_{B \in \Gamma} B$. Then $W = \langle \Gamma - \Delta \rangle$ is a complement to $\mathrm{Soc}(U)$ in V. Hence $U \neq \mathrm{Soc}(U)$. By the Modular Property of Groups, 1.14, $U = \mathrm{Soc}(U) \oplus (U \cap W)$. Thus $U \cap W$ has no simple submodules.

Choose $x \in (U \cap W)^{\#}$ so that x has nonzero projection on a minimal number n of members of Γ, and let $A \in \Gamma$ such that $x\alpha \neq 0$, where $\alpha: xR \to A$ is the projection of xR onto A. For $0 \neq y \in xR$, the set $\mathrm{supp}(y)$ of members of Γ upon which y projects nontrivially is a subset of $\mathrm{supp}(x)$. Thus by minimality of n, $\mathrm{supp}(x) = \mathrm{supp}(y)$. Therefore $\alpha: xR \to A$ is an injection, and hence, by 12.4.2, α is an isomorphism. But then $xR \cong A$ is simple, whereas it has already been observed that $U \cap W$ has no simple submodules.

(12.7) Submodules and homomorphic images of semisimple modules are semisimple.

(12.8) Assume G is finite, let U be an FG-submodule of V, and if $\mathrm{char}(F) = p > 0$ assume there is an FP-complement W to U in V for some $P \in \mathrm{Syl}_p(G)$. Then V splits over U.

Proof. Let W be a vector subspace of V with $V = U \oplus W$, and if $\mathrm{char}(F) = p > 0$ choose W to be P-invariant for some $P \in \mathrm{Syl}_p(G)$. If $\mathrm{char}\,(F) = 0$ let

$P = 1$. Let X be a set of coset representatives for P in G and let $\pi: V \to U$ be the projection of V on U with respect to the decomposition $V = U \oplus W$. Let $n = |G : P|$ and define $\theta: V \to V$ by $\theta = (\sum_{x \in X} x^{-1} \pi x)/n$, where the sum takes place in $\text{End}_F(V)$. As $(p, n) = 1$, $1/n$ exists in F. Also x, x^{-1}, and π are in $\text{End}_F(V)$, so θ is a well-defined member of $\text{End}_F(V)$. As W is P-invariant, $h\pi = \pi h$ for all $h \in P$, so if $x \mapsto h_x$ is a map from X into P then $\theta = (\sum_{x \in X} x^{-1}(h_x)^{-1} \pi h_x x)/n$. That is θ is independent of the choice of coset representatives X of P in G.

Claim $\theta \in \text{End}_R(V)$. As the multiplication in R is a linear extension of that in G and $\theta \in \text{End}_F(V)$, it suffices to show $g\theta = \theta g$ for all $g \in G$. But $g\theta = (\sum_{x \in X} (xg^{-1})^{-1} \pi x g^{-1})g/n = \theta g$ as Xg^{-1} is a set of coset representatives for P in G and θ is independent of the choice of X.

It remains to observe that, as π is the identity on U, as $W = \ker(\pi)$, and as U is G-invariant, we also have θ the identity on U and $U = V\theta$. Hence $\theta^2 = \theta$. Therefore $V = V\theta \oplus \ker(\theta)$. As $\theta \in \text{End}_R(G)$, $\ker(\theta)$ is an FG-submodule. Hence, as $U = V\theta$, $\ker(\theta)$ is a complement to U in V. That is V splits over U.

(12.9) (Maschke's Theorem) Assume G is a finite group and $\text{char}(F)$ does not divide the order of G. Then every FG-module is semisimple and every FG-extension splits.

Proof. This is a direct consequence of 12.6 and 12.8.

Using 12.9 and notation and terminology introduced earlier in this section we have:

(12.10) Assume G is a finite group and $\text{char}(F)$ does not divide the order of G. Let $\pi: G \to GL(V)$ be a finite dimensional FG-representation. Then

(1) $\pi = \sum_{i=1}^{r} \pi_i$ is the sum of its irreducible constituents $(\pi_i: 1 \le i \le r)$.

(2) If $\alpha = \sum_{i=1}^{s} \alpha_i$ is another finite dimensional FG-representation with irreducible constituents $(\alpha_i: 1 \le i \le s)$ then π is equivalent to α if and only if $r = s$ and there is a permutation σ of $\{1, 2, \ldots, r\}$ with π_i equivalent to $\alpha_{i\sigma}$ for each i.

So, in this special case, the study of FG-representations is essentially reduced to the study of irreducible FG-representations.

Let V be a semisimple R-module and S a simple R-module. The *homogeneous component of V determined by S* is $\langle U: U \le V, U \cong S \rangle$. V is *homogeneous* if it is generated by isomorphic simple submodules.

(12.11) Let V be a semisimple R-module. Then

(1) If V is homogeneous then every pair of simple submodules of V is isomorphic.

(2) V is the direct sum of its homogeneous components.

Proof. As V is semisimple, $V = \bigoplus_{A \in \Omega} A$ for some set Ω of simple submodules of V. Let T be a simple submodule of V and supp(T) the set of submodules in Ω upon which T projects nontrivially. If $A \in$ supp(T) then the projection map $\alpha: T \to A$ is an isomorphism by Schur's Lemma. But if V is homogeneous then, by 12.5, we may choose $A \cong S$ for some simple R-module S and all $A \in \Omega$. Hence (1) holds.

Similarly if S is a simple R-module, H the homogeneous component of V determined by S, and K the submodule of V generated by the remaining homogeneous components, then, as V is semisimple, $V = H + K$. Further if $H \cap K \neq 0$, we may choose $T \leq H \cap K$ by 12.7. But now, by (1), $S \cong T \cong Q$ for some simple R-module Q determining a homogeneous component distinct from that of S, a contradiction.

(12.12) Let $H \trianglelefteq G$ and U a simple FH-submodule of V. Then

(1) Ug is a simple FH-submodule of V for each $g \in G$.

(2) If $g \in C_G(H)$ then U is FH-isomorphic to Ug.

(3) If X and Y are isomorphic FH-submodules of V then Xg and Yg are FH-isomorphic submodules for each $g \in G$.

(12.13) (Clifford's Theorem) Let V be a finite dimensional irreducible FG-module and $H \trianglelefteq G$. Then

(1) V is a semisimple FH-module.

(2) G acts transitively on the FH-homogeneous components of V.

(3) Let U be an FH-homogeneous component of V. Then $N_G(U)$ is irreducible on U and $HC_G(H) \leq N_G(U)$.

Proof. By 12.12.1, G acts on the socle of V, regarded as an FH-module. Thus (1) holds by 12.6 and the irreducible action of G. By 12.12.3, G permutes the homogeneous FH-components of V. Then, by 12.11.2 and the irreducible action of G, G is transitive on those homogeneous components. By 12.12.2, $HC_G(H)$ acts on each homogeneous component. Then 12.11.2 and the irreducible action of G completes the proof of (3).

Observe that π can be extended to a representation of R on V (that is to an F-algebra homomorphism of R into End$_F(V)$) via $\pi: \sum a_g g \mapsto \sum a_g(g\pi)$. Indeed for $r \in R$ and $v \in V$, $v(r\pi) = vr$ is just the image of v under the module product of v by r in the R-module V. Further

$$\ker(\pi) = \{r \in R : vr = 0 \quad \text{for all } v \in V\}.$$

V is said to be a *faithful* R-module if π is an injection on R. As G generates R as an F-algebra, $R\pi$ is the subalgebra of $\text{End}_F(V)$ generated by $G\pi$. We call $R\pi$ the *enveloping algebra* of the representation π.

(12.14) $\text{End}_R(V) = C_{\text{End}_F(V)}(R\pi) = C_{\text{End}_F(V)}(G\pi)$.

(12.15) If G is finite and π is irreducible then $Z(G\pi)$ is cyclic.

Proof. Let $E = \text{End}_F(V)$. As π is irreducible, $D = \text{End}_R(V)$ is a division ring by Schur's Lemma. $Z = Z(G\pi) \leq C_E(G\pi) = D$ by 12.14. Also $D \leq C_E(G\pi) \leq C_E(Z)$, so $Z \leq Z(D)$. Thus the sub-division-ring K of D generated by Z is a field. Now Z is a finite subgroup of the multiplicative group of the field K, and hence K is cyclic.

I conclude this section by recording two results whose proofs can be found in section 3 of chapter 17 of Lang [La].

(12.16) Let $\pi : G \to GL(V)$ be an irreducible finite dimensional FG-representation. Then $R\pi$ is isomorphic as an F-algebra to the ring of all m by m matrices over the division ring $\text{End}_{FG}(V) = D$, where $m = \dim_D(V)$. Further F is in the centre of D.

(12.17) (Burnside) Assume F is algebraically closed and $\pi : G \to GL(V)$ is an irreducible finite dimensional FG-representation. Then $\text{End}_F(V) = R\pi$ and $F = \text{End}_{FG}(V)$.

13 The general linear group and special linear group

In this section F is a field, n is a positive integer, and V is an n-dimensional vector space over F. Recall the group of vector space automorphisms of V is the general linear group $GL(V)$. As the isomorphism type of V depends only on n and F, the same is true for $GL(V)$, so we can also write $GL_n(F)$ for $GL(V)$.

(13.1) Let $F^{n \times n}$ denote the F-algebra of all n by n matrices over F, let $X = (x_1 \ldots, x_n)$ be an ordered basis for V, and for $g \in \text{End}_F(V)$ let $M_X(g) = (g_{ij})$ be the matrix defined by $x_i g = \sum_j g_{ij} x_j$, $g_{ij} \in F$. Then

(1) The map $M_X : g \mapsto M_X(g)$ is an F-algebra isomorphism of $\text{End}_F(V)$ with $F^{n \times n}$. Hence the map restricts to a group isomorphism of $GL(V)$ with the group of all nonsingular n by n matrices over F.

(2) Let $Y = (y_1, \ldots, y_n)$ be a second ordered basis of V, let h be the unique element of GL(V) with $x_i h = y_i$, $1 \leq i \leq n$, and $B = M_Y(h)$. Then $M_X = h^* M_Y = M_Y B^*$ is the composition of h^* with M_Y and of M_Y with B^*, where h^* and B^* are the conjugation automorphisms induced by h and B on $\mathrm{End}_F(V)$ and $F^{n \times n}$, respectively.

Because of 13.1, we can think of subgroups of GL(V) as groups of matrices if we choose. I take this point of view when it is profitable. Similarly an *FG*-representation π on V can be thought of as a homomorphism from G into the group of all n by n nonsingular matrices over G, by composing π with the isomorphism M_X. π is equivalent to $\pi' \colon G \to \mathrm{GL}(V)$ if and only if $\pi' = \pi h^*$ for some $h \in \mathrm{GL}(V)$, and by 13.1.2 this happens precisely when $\pi' M_X = \pi M_X B^*$ for some nonsingular matrix B. This gives a notion of equivalence for 'matrix representations'. Namely two homomorphisms α and α' of G into the group of all n by n nonsingular matrices over F are equivalent if there exists a nonsingular matrix B with $\alpha' = \alpha B^*$.

Let's see next what the notions of reducibility and decomposability correspond to from the point of view of matrices.

(13.2) Let $\pi \colon G \to \mathrm{GL}(V)$ be an *FG*-representation, U an *FG*-submodule of V, $\bar{V} = V/U$, and $X = (x_i \colon 1 \leq i \leq n)$ a basis for V with $Y = (x_i \colon 1 \leq i \leq m)$ a basis for U. Then, for $g \in G$,

$$M_X(g\pi) = \begin{bmatrix} M_Y(g\pi|_U) & 0 \\ A(g) & M_{\bar{X}}(g\pi(g\pi|_{\bar{V}})) \end{bmatrix}$$

for some $n - m$ by m matrix $A(g)$.

Of course there is a suitable converse to 13.2.

(13.3) Let $\pi \colon G \to \mathrm{GL}(V)$ be an *FG*-representation, U and W *FG*-submodules of V with $V = U \oplus W$, and $X = (x_i \colon 1 \leq i \leq n)$ a basis for V such that $Y = (x_i \colon 1 \leq i \leq m)$ and $Z = (x_i \colon m < i \leq n)$ are basis for U and W, respectively. Then for $g \in G$,

$$M_X(g\pi) = \begin{bmatrix} M_Y(g\pi|_U) & 0 \\ 0 & M_Z(g\pi|_W) \end{bmatrix}.$$

Again there is a suitable converse to 13.3.

Recall the notion of geometry defined in section 3. We associate a geometry PG(V) to V, called the *projective geometry* of V. The objects of PG(V) are the proper nonzero subspaces of V, with incidence defined by inclusion. If U

is a subspace of V, the *projective dimension* of U is $\text{Pdim}(U) = \dim_F(U) - 1$. The type function for $\text{PG}(V)$ is the projective dimension function

$$\text{Pdim}: \text{PG}(V) \to I = \{0, 1, \ldots, n-1\}.$$

$\text{PG}(V)$ is said to be of dimension $n-1$. The subspaces of projective dimension 0, 1, and $n-2$ are referred to as *points*, *lines*, and *hyperplanes*, respectively.

For $g \in \text{GL}(V)$ define $gP: \text{PG}(V) \to \text{PG}(V)$ by $gP: U \mapsto Ug$, for $U \in \text{PG}(V)$. Evidently $P: \text{GL}(V) \to \text{Aut}(\text{PG}(V))$ is a representation of $\text{GL}(V)$ in the category of geometries. (See the discussion in section 3.) Denote the image of $\text{GL}(V)$ under P by $\text{PGL}(V)$. $\text{PGL}(V)$ is the *projective general linear group*. The notation $\text{PGL}_n(F)$ is also used for $\text{PGL}(V)$.

A *scalar transformation* of V is a member g of $\text{End}_F(V)$ such that $vg = av$ for all v in V and some a in F independent of v. A *scalar matrix* is a matrix of the form aI, $a \in F$, where I is the identity matrix.

(13.4) (1) $Z(\text{End}_F(V))$ is the set of scalar transformations. The image of $Z(\text{End}_F(V))$ under M_X is the set of scalar matrices.

(2) $Z(\text{GL}(V))$ is the set of nonzero scalar transformations.

(3) $Z(\text{GL}(V)) = \ker(P)$.

By 13.4, the projective general linear group $\text{PGL}(V)$ is isomorphic to the group of all n by n nonsingular matrices modulo the subgroup of scalar matrices. Often it will be convenient to regard these groups as the same.

Given any FG-representation $\pi: G \to \text{GL}(V)$, π can be composed with P to obtain a homomorphism $\pi P: G \to \text{PGL}(V)$. Observe that πP is a representation of G on the projective geometry $\text{PG}(V)$.

For $y \in \text{End}_F(V)$ define the *determinant* of y to be $\det(x) = \det(M_X(y))$. That is the determinant of y is the determinant of its associated matrix. Similarly define the *trace* of y to be $\text{Tr}(y) = \text{Tr}(M_X(y))$. So the trace of y is the trace of its associated matrix. If A is a matrix and B is a nonsingular matrix then $\det(A^B) = \det(A)$ and $\text{Tr}(A^B) = \text{Tr}(A)$, so $\det(y)$ and $\text{Tr}(y)$ are independent of the choice of basis X by 13.1.2.

Define the *special linear group* $\text{SL}(V)$ to be the set of elements of $\text{GL}(V)$ of determinant 1. The determinant map is a homomorphism of $\text{GL}(V)$ onto the multiplicative group of F with $\text{SL}(V)$ the kernel of this homomorphism, so $\text{SL}(V)$ is a normal subgroup of $\text{GL}(V)$ and $\text{GL}(V)/\text{SL}(V) \cong F^{\#}$. Also write $\text{SL}_n(F)$ for $\text{SL}(V)$. The image of $\text{SL}(V)$ under P is denoted by $\text{PSL}(V)$ or $\text{PSL}_n(F)$. The group $\text{PSL}(V)$ is the *projective special linear group*. Sometimes $\text{PSL}_n(F)$ is denoted by $\text{L}_n(F)$.

Prove the next lemma for GL(V) and then use 5.20 to show the result holds for SL(V). See section 15 for the definition of 2-transitivity.

(13.5) SL(V) is 2-transitive on the points of PGL(V).

For v in V and α in $\mathrm{End}_F(V)$, $[v, \alpha] = v\alpha - v$ is the *commutator* of v with α. This corresponds with the notion of commutator in section 8. Indeed we can form the semidirect product of V by GL(V) with respect to the natural representation, and in this group the two notions agree. Similarly, for $G \leq$ GL(V),

$$[V, G] = \langle [v, g]: v \in V, g \in G \rangle$$

and, for $g \in G$, $[V, g] = [V, \langle g \rangle]$.

A *transvection is an element* t of GL(V) such that $[V, t]$ is a point of PG(V), $C_V(t)$ is a hyperplane of PG(V), and $[V, t] \leq C_V(t)$. $[V, t]$ and $C_V(t)$ are called the *center* and *axis* of t, respectively. Let $x_n \in V - C_V(t)$. Then $[x_n, t] = x_1$ generates $[V, t]$ and we choose $x_i \in C_V(t)$, $1 < i < n$, so that $X = (x_i : 1 \leq i \leq n)$ is a basis of V. Then

$$M_X(t) = \begin{pmatrix} 1 & 0 & 0 \\ 0 & I & 0 \\ 1 & 0 & 1 \end{pmatrix},$$

so evidently t is of determinant 1 and 13.1.2 implies GL(V) is transitive on its transvections. Write diag($a_1, \ldots, a_n$) for the diagonal matrix whose (i, i)-th entry is a_i. If $n > 2$ let

$$A = \{\mathrm{diag}(1, a, 1, \ldots, 1): a \in F^\#\}$$

and if $n = 2$ let

$$A = \{\mathrm{diag}(a, a): a \in F^\#\}.$$

Then $A \leq C_{\mathrm{GL}(V)}(t)$ and either det: $A \to F^\#$ is a surjection or $n = 2$ and some element of F is not a square in F. In the first case GL(V) = ASL(V), so, as GL(V) is transitive on its transvections, so is SL(V) by 5.20. Further if $n > 2$ and s is the transvection with $C_V(s) = C_V(t)$ and $[x_n, s] = x_2$, then st is also a transvection. So, as SL(V) is transitive on its transvections, $t = s^{-1}(st)$ is a commutator of SL(V). On the other hand, if $n = 2$, then for $b \in F^\#$ let $t(b)$ be the transvection with $x_2 t(b) = x_2 + bx_1$ and $g = \mathrm{diag}(a, a^{-1})$. Then $t(b)^g = t(a^2 b)$. Further if $|F| > 3$ then a can be chosen with $a^2 \neq 1$. Thus, setting $b = (a^2 - 1)^{-1}$, we have $[t(b), g] = t$, and again t is a commutator of SL(V).

We have shown:

(13.6) (1) Transvections are of determinant 1.

(2) The transvections form a conjugacy class of $GL(V)$.

(3) Either the transvections form a conjugacy class of $SL(V)$ or $n = 2$ and F contains nonsquares.

(4) If $|F| > 3$ or $n > 2$, then each transvection is in the commutator group of $SL(V)$.

(13.7) $SL(V)$ is generated by its transvections.

Proof. Let Ω be the set of n-tuples $\omega = (x_1, \ldots, x_{n-1}, \langle x_n \rangle)$ such that $(x_1, \ldots, x_n)$ is a basis for V. Let T be the subgroup of $G = SL(V)$ generated by the transvections of G. I'll show T is transitive on Ω. Then, by 5.20, $G = TG_\omega$. But $G_\omega = 1$, so the lemma holds.

It remains to show T is transitive on Ω. Pick $\alpha = (y_1, \ldots, y_{n-1}, \langle y_n \rangle) \in \Omega$ such that $\alpha \notin \omega T$, $y_i = x_i$ for $i \leq m$, and, subject to these constraints, with m maximal. Let $U = \langle x_i : i \leq m \rangle$, $x = x_{m+1}$, $y = y_{m+1}$, and $W = \langle U, x, y \rangle$. Then $\dim(W/U) = k = 1$ or 2.

Suppose $k = 2$ and let H be a hyperplane of V containing U and $x - y$ but not x. Let t be the transvection with axis H and $[y, t] = x - y$. Then $yt = x$ and $x_i t = x_i$ for $i \leq m$, so $\alpha t \in \omega T$ by maximality of m. Then $\alpha \in \omega T$, contrary to the choice of α.

So $k = 1$. Suppose $m = n - 1$. As $k = 1$, $ax - y \in U$ for some $a \in F^\#$. As $m = n - 1$ and $\alpha \neq \omega$, $ax - y \neq 0$. So there is a transvection t with axis U and $[y, t] = ax - y$. Now $\alpha t = \omega$, contradicting $\alpha \notin \omega T$. So $m < n - 1$, and hence there is $z \in V - W$. An argument in the last paragraph shows there are transvections s and t with $U \leq C_V(t) \cap C_V(s)$, $ys = z$, and $zt = x$. But now $x_i st = x_i$ for $i \leq m$ and $yst = x$, contradicting the choice of α.

(13.8) If $n \geq 2$ then $SL_n(F)$ is perfect unless $n = 2$ and $|F| = 2$ or 3.

Proof. Let $G = SL_n(F)$. By 13.7 it suffices to show transvections are contained in $G^{(1)}$, and this follows from 13.6.4.

14 The dual representation

In section 14, V continues to be an n-dimensional vector space over F and $\pi : G \to GL(V)$ is an FG-representation of a group G.

Let $(V_i : -\infty < i < \infty)$ be a sequence of FG-modules and

$$\cdots \to V_{-1} \xrightarrow{\alpha_{-1}} V_0 \xrightarrow{\alpha_0} V_1 \to \cdots$$

a sequence of FG-homomorphisms. The latter sequence is said to be *exact* if $\ker(\alpha_{i+1}) = V_i\alpha_i$ for each i. A *short exact sequence* is an exact sequence of the form $0 \rightarrow U \xrightarrow{\alpha} V \xrightarrow{\beta} W \rightarrow 0$. The maps $0 \rightarrow U$ and $W \rightarrow 0$ are forced to be trivial. Observe that the hypothesis that the sequence is exact is equivalent to requiring that α be an injection, β a surjection, and $U\alpha = \ker(\beta)$. Hence $W \cong V/U\alpha$ and the sequence is essentially $0 \rightarrow U\alpha \rightarrow V \rightarrow V/U\alpha \rightarrow 0$ with $U\alpha \rightarrow V$ inclusion and $V \rightarrow V/U\alpha$ the natural map. The sequence is said to *split* if V splits over $U\alpha$. As is well known, the sequence splits if and only if there is $y \in \text{Hom}_{FG}(W, V)$ with $\gamma\beta = 1$, and this condition is equivalent in turn to the existence of $\delta \in \text{Hom}_{FG}(V, U)$ with $\alpha\delta = 1$.

Let $V^* = \text{Hom}_F(V, F)$ and recall from section 13 that V^* is a vector space over F. We call V^* the *dual space* of V. It is well known that $n = \dim_F(V^*)$. If U is an F-space and $\alpha \in \text{Hom}_F(U, V)$ define $\alpha^* \in \text{Hom}_F(V^*, U^*)$ by $x\alpha^* = \alpha x, x \in V^*$.

(14.1) Let U, V, and W be finite dimensional F-spaces, $\alpha \in \text{Hom}_F(U, V)$, and $\beta \in \text{Hom}_F(V, W)$. Then

(1) The map $\gamma \mapsto \gamma^*$ is an F-space isomorphism of $\text{Hom}_F(U, V)$ with $\text{Hom}_F(V^*, U^*)$.

(2) $(\alpha\beta)^* = \beta^*\alpha^*$.

(3) If $U \xrightarrow{\alpha} V \xrightarrow{\beta} W$ is exact then so is $W^* \xrightarrow{\beta^*} V^* \xrightarrow{\alpha^*} U^*$.

If $\pi: G \rightarrow \text{GL}(V)$ is an FG-representation then, from 14.1, $\pi^*: G \rightarrow \text{GL}(V^*)$ is also an FG-representation, where $\pi^*: g \mapsto (g^{-1}\pi)^*$. The representation π^* is called the *dual* of π and the representation module V^* of π^* is called the *dual* of the representation module V of π.

Given a basis $X = (x_i: 1 \leq i \leq n)$ for V, the *dual basis* $\hat{X} = (\hat{x}_i: 1 \leq i \leq n)$ of X is defined by $\hat{x}_i: x_j \mapsto \delta_{ij}$. Notice $\sum_i a_i\hat{x}_i$ is the unique member of V^* mapping x_i to a_i for each i.

(14.2) Let $\pi: G \rightarrow \text{GL}(V)$ be an FG-representation and X a basis for V. Then $M_{\hat{X}}(g\pi^*) = (M_X(g\pi)^{-1})^{\text{T}}$, where B^{T} denotes the transpose of a matrix B.

By 14.2, if π is viewed as a matrix representation, then π^* is just the composition of π with the transpose-inverse map on $\text{GL}_n(F)$. As the transpose-inverse map is of order 2, we conclude

(14.3) $(\pi^*)^*$ is equivalent to π for each finite dimensional FG-representation π.

There is a more concrete way to see this.

48   *Linear representations*

(14.4) Let U and V be finite dimensional F-spaces. Then
  (1) For each $v \in V$ there exists a unique element $v\theta \in (V^*)^*$ with $xv\theta = vx$ for all $x \in V^*$.
  (2) The map $\theta: v \mapsto v\theta$ is an F-isomorphism of V with $(V^*)^*$.
  (3) For each $\alpha \in \mathrm{Hom}_F(U, V)$, $\alpha\theta = \theta(\alpha^*)^*$.
  (4) θ defines an equivalence of π and $(\pi^*)^*$.

Proof. To prove (1) let $X = (x_i: 1 \le i \le n)$ be a basis for V, $\hat{X} = (\hat{x}_i: 1 \le i \le n)$ its dual basis, and $\tilde{X} = (\tilde{x}_i : 1 \le i \le n)$ the dual basis of $\hat{X}$ in $(V^*)^*$. Let $v = \sum_i a_i x_i \in V$ and $\hat{v} = \sum_i b_i \tilde{x}_i \in (V^*)^*$. Then $x\hat{v} = vx$ for all $x \in V^*$ if and only if $\hat{x}_i \hat{v} = v\hat{x}_i$ for all i. Further $\hat{x}_i \hat{v} = b_i$ and $v\hat{x}_i = a_i$, so $v\theta = \sum_i a_i \tilde{x}_i$ is uniquely determined.
  As $\theta: \sum a_i x_i \mapsto \sum a_i \tilde{x}_i$, (2) holds. Part (4) follows directly from (2) and (3). To prove (3) we must show $(u\theta)(\alpha^*)^* = (u\alpha)\theta$ for each $u \in U$. But, for $x \in V^*$, $x(u\theta)(\alpha^*)^* = x\alpha^*(u\theta) = \alpha x(u\theta) = u(\alpha x) = (u\alpha)x = x((u\alpha)\theta)$, completing the proof.

Notice 14.4 gives a constructive proof of 14.3. It will also be useful in the proof of the next lemma.

(14.5) Let G be a group and U, V, and W finite dimensional FG-modules. Then
  (1) $\mathrm{Hom}_{FG}(U^*, V^*) = \{\alpha^*: \alpha \in \mathrm{Hom}_{FG}(V, U)\}$.
  (2) $0 \to U \xrightarrow{\alpha} V \xrightarrow{\beta} W \to 0$ is an exact sequence of FG-modules if and only if $0 \to W^* \xrightarrow{\beta^*} V^* \xrightarrow{\alpha^*} U^* \to 0$ is. The first sequence splits if and only if the second splits.
  (3) V is irreducible, indecomposable, semisimple, and homogeneous if and only if V^* has the respective property.

Proof. Part (1) follows from 14.1.1 and 14.1.2. The first part of (2) follows from 14.1.3 and 14.4. The second part follows from the remark about splitting at the beginning of this section and 14.1.2. Part (3) follows from (2), since the properties in (3) can be described in terms of exact sequences and the splitting of such sequences.

(14.6) (1) Let $\alpha: V \to U$ be an FG-homomorphism of finite dimensional modules. Then $[G, U] \le V\alpha$ if and only if $\ker(\alpha^*) \le C_{U^*}(G)$, while $[G, V^*] \le U^*\alpha^*$ if and only if $\ker(\alpha) \le C_V(G)$.
  (2) If U is a finite dimensional FG-module then $U = [U, G]$ if and only if $C_{U^*}(G) = 0$, while $U^* = [U^*, G]$ if and only if $C_U(G) = 0$.

Proof. We have the exact sequence

$$V \xrightarrow{\alpha} U \to U/V\alpha \to 0$$

so, by 14.1.3,

$$0 \to (U/V\alpha)^* \to U^* \xrightarrow{\alpha^*} V^*$$

is also exact. Let π be the representation of G on $U/V\alpha$. Then $[G, U] \le V\alpha$ if and only if $G\pi = 1$. This is equivalent to $G\pi^* = 1$ which in turn holds if and only if G centralizes $(U/V\alpha)^*$. As $(U/V\alpha)^*$ is FG-isomorphic to $\ker(\alpha^*)$ by the exactness of the second series above, the first part of (1) holds, while the second follows from the first and 14.4.3.

Let $U \ne [U, G] = V$ and $\alpha: V \to U$ the inclusion. By (1), $\ker(\alpha^*) \le C_{U^*}(G)$. As α is not a surjection, $\ker(\alpha^*) \ne 0$ by 14.5.2. Thus $0 \ne C_{U^*}(G)$. Hence by 14.4.3, if $U^* \ne [U^*, G]$ then $0 \ne C_U(G)$.

Similarly, if $0 \ne C_U(G)$, let $\beta: U \to U/C_U(G)$ be the natural map. By (1), $[G, U^*] \le (U/C_U(G))^*\beta^*$. As β is not an injection, β^* is not a surjection by 14.5.2. So $U^* \ne [G, U^*]$. Applying 14.4.3 we see that if $0 \ne C_{U^*}(G)$ then $U \ne [G, U]$, completing the proof of (2).

The *character* of an FG-representation π is the map $\chi: G \to F$ defined by $\chi(g) = \mathrm{Tr}(g\pi)$. Remember $\mathrm{Tr}(g\pi)$ is the trace of the matrix $M_X(g\pi)$ and is independent of the choice of the basis X for the representation module of π.

(14.7) Let π be an FG-representation, π^* the dual of π, and χ and χ^* the characters of π and π^*, respectively. Then $\chi^*(g) = \chi(g^{-1})$ for each $g \in G$.

Proof. By 14.2, $M_{\hat{X}}(g\pi^*) = M_X(g^{-1}\pi)^{\mathrm{T}}$, so the lemma follows as $\mathrm{Tr}(A) = \mathrm{Tr}(A^{\mathrm{T}})$ for each n by n matrix A.

Since characters have now been introduced I should probably record two more properties of characters which are immediate from 13.1 and the fact that conjugate matrices have the same trace.

(14.8) (1) Equivalent FG-representations have the same character.

(2) If χ is the character of an FG-representation then $\chi(g^h) = \chi(g)$ for each $g, h \in G$.

Remarks. The stuff in sections 12 and 13 is pretty basic but section 14 is more specialized. Section 14 is included here to prepare the way for the 1-cohomology in section 17. That section is also specialized. Both can be safely skipped or postponed by the casual reader. If so, lemma 17.10 must be assumed in proving the Schur-Zassenhaus Theorem in section 18. But that's no problem.

The reader who is not familiar with the theory of modules over rings might want to bone up on modules before beginning section 12.

Exercises for chapter 4

1. Let G be a finite subgroup of $GL(V)$, where V is a finite dimensional vector space over a field F with (char (F), $|G|$) = 1. Prove
   (1) $V = [G, V] \oplus C_V(G)$.
   (2) If G is abelian then $V = \langle C_V(D): D \in \Delta \rangle$, where Δ is the set of subgroups D of G with G/D cyclic.
   (3) If $G \cong E_{p^n}$, $n > 0$, and $V = [V, G]$, then $V = \bigoplus_{H \in \Gamma} C_V(H)$, where Γ is the set of subgroups of G of index p.

2. Let V be a finite dimensional vector space over a field F, $g \in \text{End}_F(V)$, and U a g-invariant subspace of V. Prove
   (1) g centralizes V/U if and only if $[V, g] \le U$.
   (2) The map $v \mapsto [v, g]$ is a surjective linear transformation of V onto $[V, g]$ with kernel $C_V(g)$.
   (3) $\dim_F(V) = \dim_F(C_V(g)) + \dim_F([V, g])$.

3. Let G be a finite group, F a field of prime characteristic p, and π an irreducible FG-representation. Prove $O_p(G\pi) = 1$.

4. Let F be a field, r and q be primes, X a group of order r acting irreducibly on a noncyclic elementary abelian q-group Q, and $V = [V, Q]$ a faithful irreducible FXQ-module. Then $\dim_F(V) = rk$ where $k = \dim_F(C_V(X)) = \dim_F(C_V(H))$ for some hyperplane H of Q.

5. Let $\alpha: G \to \text{Sym}(I)$ be a permutation representation of a finite group G on a finite set I, and let F be a field and V an F-space with basis $X = (x_i: i \in I)$. The *FG-representation π induced by* α is the representation on V with $g\pi: x_i \mapsto x_{ig\alpha}$ for each $g \in G$ and $i \in I$. V is called the *permutation module* of α. Let χ be the character of π. Prove
   (1) $\chi(g)$ is the number of fixed points of $g\alpha$ on I for each $g \in G$.
   (2) $(\sum_{g \in G} \chi(g))/|G|$ is the number of fixed points of G on I.
   (3) If G is transitive on I then $(\sum_{g \in G} \chi(g)^2)/|G|$ is the permutation rank of G on I. (See section 15 for the definition of permutation rank.)

6. Assume the hypothesis of the previous exercise with G transitive on I. Let $z = \sum_{i \in I} x_i$, $Z = \langle z \rangle$, and

$$U = \left\{ \sum_{i \in I} \alpha_i x_i: \sum_{i \in I} = 0, a_i \in F \right\}.$$

U is the *core* of the permutation module V. Prove
   (1) $Z = C_V(G)$ and $U = [V, G]$.

(2) If W is an FG-module, $i \in I$, $H = G_i$ is the stabilizer in G of i, $w \in C_W(H)$, and $W = \langle wG \rangle$, then there is a surjective homomorphism of V onto W.

(3) Assume $p = \text{char}(F)$ is a prime divisor of $|I|$. Then V does not split over U, V does not split over Z, and if $O^P(G) = G$ then $H^1(G, U/Z) \neq 0$. (See section 17 for a discussion of the 1-cohomology group H^1; in particular use 17.11.)

7. Let F be a field, U a 2-dimensional vector space over F with basis $\{x, y\}$, $G = GL(U)$, and $V = F[x, y]$ the polynomial ring in x and y over F. Prove
   (1) π is an FG-representation on V where π is defined by $f(x, y)g\pi = f(xg, yg)$ for $f \in V$ and $g \in G$.
   (2) G acts on the $(n+1)$-dimensional subspace V_n of homogeneous polynomials of degree n. Let π_n be the restriction of π to V_n.
   (3) If $\text{char}(F) = p > 0$, prove π_n is not irreducible for $p \leq n \not\equiv -1 \bmod p$, but π_n is irreducible for $0 \leq n < p$.
   (4) $\ker(\pi_n)$ is the group of scalar transformations aI of U with $a \in F$ and $a^n = 1$.
   (Hint: In (3) let T be the group of transvections in G with center $\langle x \rangle$ and for $i \leq n$ let M_i be the subspace of $M = V_n$ generated by $y^j x^{n-j}, 0 \leq j \leq i$. Prove $[y^i x^{n-1} + M_{i-2}, T] = M_{i-1}/M_{i-2}$ for all $1 \leq i \leq n$. Conclude M_0 is contained in any nonzero FG-submodule of M and then, as M_0 is conjugate to $\langle y^n \rangle$ under G, conclude M_i is contained in any such submodule for all i.)

8. Let V be a vector space over a field F and $0 = V_0 \leq V_1 \cdots \leq V_n = V$ a sequence of subspaces. Let G be a subgroup of $GL(V)$ centralizing V_{i+1}/V_i for each i, $0 \leq i < n$. Prove
   (1) G is nilpotent of class at most $n - 1$.
   (2) If $0 \neq U$ is a G-invariant subspace of V then $C_U(G) \neq 0$ and $[U, G] < U$.

9. Let V be an n-dimensional vector space over a field F with $n \geq 2$, $G = GL(V)$, $X = (x_i: 1 \leq i \leq n)$ a basis for V, $V_i = \langle x_j : 1 \leq j \leq i \rangle$, $T = \{V_i: 1 \leq i < n\}$, and $Y = \{\langle x_i \rangle: 1 \leq i \leq n\}$. Prove
   (1) T is a flag of $PG(V)$ of type $I = \{0, \ldots, n-1\}$. $G_T = B$ is the group of lower triangular matrices and B is the semidirect product of the subgroups U and H where U consists of the matrices in B with 1 on the main diagonal and H is the group of diagonal matrices.
   (2) $H = G_Y$ is the direct product of n copies of $F^\#$. U is nilpotent and $Z(U)$ is the root group of a transvection. (The *root group* of a transvection t consists of those $g \in GL(V)$ with $C_V(t) \leq C_V(G)$ and $[V, g] \leq [V, t]$.)
   (3) $N_G(Y)$ is the semidirect product of H by S_n. If $|F| > 2$ then $N_G(Y) = N_G(H)$.

(4) If F is finite of characteristic p then $U \in \mathrm{Syl}_p(G)$.

(5) $B = N_G(U)$. Indeed V_i is the unique object of type i fixed by U.

(6) The residue Γ_S of a flag S of corank 1 is isomorphic to the projective line over F, and $(G_S)^{\Gamma_S} \cong \mathrm{PGL}_2(F)$. (See section 3 for the definition of residue.)

10. Let F be a field and $\Gamma = F \cup \{\infty\}$ the *projective line* over F. Let $G = \mathrm{GL}_2(F)$ be the group of invertible 2 by 2 matrices over F, and for

$$A = (a_{i,j}) = \begin{pmatrix} a_{1,1} & a_{1,2} \\ a_{2,1} & a_{2,2} \end{pmatrix} \in \mathrm{GL}_2(F)$$

define $\phi(A) \colon \Gamma \to \Gamma$ by

$$\phi(A) \colon z \mapsto \frac{a_{1,1}z + a_{2,1}}{a_{1,2}z + a_{2,2}},$$

where by convention $a/\infty = 0$ for $a \in F^{\#}$ and $\infty \phi(A) = a_{1,1}/a_{1,2}$. Pick a basis $B = \{x_1, x_2\}$ for a 2-dimensional vector space V over F, and identify $\mathrm{GL}(V)$ with G via the isomorphism $M_B \colon \mathrm{GL}(V) \to G$. Let Ω be the points of the projective geometry of V. Prove

(1) For $A \in G$, $\phi(A)$ is a permutation of Γ.

(2) $G^* = \{\phi(A) \colon A \in G\}$ is a subgroup of $\mathrm{Sym}(\Gamma)$, and $\phi \colon G \to G^*$ is a surjective group homomorphism with kernel $Z(G)$, so ϕ induces an isomorphism $\bar{\phi} \colon \bar{G} \to G^*$, where $\bar{G} = \mathrm{PGL}(V)$.

(3) Define $\alpha \colon \Omega \to \Gamma$ by $\alpha \colon Fx_1 \mapsto \infty$ and $\alpha \colon F(\lambda x_1 + x_2) \mapsto \lambda$ for $\lambda \in F$. Then α is a bijection such that $(\omega \bar{g})\alpha = (\omega \alpha)\bar{\phi}(\bar{g})$ for each $\omega \in \Omega$ and $\bar{g} \in \bar{G}$. Hence α is an equivalence of the permutation representations of $\bar{G}$ on Ω and G^* on Γ.

5

Permutation groups

This chapter derives a number of properties of the alternating and symmetric groups A_n and S_n of finite degree n. For example the conjugacy of elements in A_n and S_n is determined, and it is shown that A_n is simple if $n \geq 5$. Section 15 also contains a brief discussion of multiply transitive permutation groups. Section 16 studies rank 3 permutation groups.

15 The symmetric and alternating groups

Let X be a set and S the symmetric group on X. A *permutation group* on X is a subgroup of S. Let G be a permutation group on X. In this section X is assumed to be of finite order n. Thus S is of order $n!$, so S and G are finite.

Suppose $g \in S$ and let $H = \langle g \rangle$. Then g is of finite order m and $H = \{g^i : 0 \leq i < m\}$. Further H has a finite number of orbits $(x_i H : 1 \leq i \leq k)$, and the orbit $x_i H$ is of finite order l_i. Let $H_i = H_{x_i}$ be the stabilizer in H of x_i. By 5.11, $l_i = |H : H_i|$, so, as $H = \langle g \rangle$ is cyclic, $H_i = \langle g^{l_i} \rangle$ and $\{g^j : 0 \leq j < l_i\}$ is a set of coset representatives for H_i in H. Hence, by 5.8, $x_i H = \{x_i g^j : 0 \leq j < l_i\}$. Therefore g acts on $x_i H$ as the following cycle:

$$g|_{x_i H} = \left(x_i, x_i g, x_i g^2, \ldots, x_i g^{l_i - 1} \right).$$

This notation indicates that $g : x_i g^j \mapsto x_i g^{j+1}$ for $0 \leq j < l_i - 1$ and $g : x_i g^{l_i - 1} \mapsto x_i$. The last fact holds as g^{l_i} fixes x_i. Further, as the orbits of H partition X, we can describe the action of g on X with the following notation:

$$\left(x_1, x_1 g, \ldots, x_1 g^{l_1 - 1} \right) \left(x_2, x_2 g, \ldots, x_2 g^{l_2 - 1} \right) \cdots \left(x_k, x_k g, \ldots, x_k g^{l_k - 1} \right).$$

This is the *cycle notation* for the permutation g. It describes g, and the description is unique up to a choice of representative x_i for the ith orbit and the ordering of the orbits. For example, if X is the set of integers $\{1, 2, \ldots, n\}$, the representative x_i could be chosen to be the minimal member of the ith orbit, and the orbits ordered so that $x_1 < x_2 < \cdots < x_k$. If so, g can be written uniquely in cycle notation, and conversely each partition of X and each ordering of the partition and the members of the partition, subject to these constraints, defines some member of S in the cycle notation.

By convention the terms (x_i) corresponding to orbits of H of length 1 are omitted. Thus for example if $n = 5$ we would write $g = (1, 2)(3, 4)(5)$ as

$g = (1, 2)(3, 4)$. Notice g is still uniquely described in this modified cycle notation.

Subject to this convention, $g_i = (x_i, x_i g, \ldots, x_i g^{l_i - 1})$ is a member of S. The elements $g_1, \ldots, g_k$ are called the *cycles* of g. Also g is said to be a *cycle* if H has at most one orbit of length greater than 1. Notice the two uses of the term 'cycle' are compatible.

Given a subset A of X let $\mathrm{Mov}(A)$ be the set of points of X moved by A. Here x in X is *moved* by A if $ax \neq x$ for some $a \in A$. Notice $\mathrm{Mov}(A) = \mathrm{Mov}(\langle A \rangle)$ and X is the disjoint union of $\mathrm{Mov}(A)$ and $\mathrm{Fix}(A)$. Cycles c and d in S are said to be *disjoint* if $\mathrm{Mov}(c) \cap \mathrm{Mov}(d)$ is empty.

(15.1) Let $A, B \subseteq S$ with $\mathrm{Mov}(A) \cap \mathrm{Mov}(B)$ empty. Then $ab = ba$ for all $a \in A$ and $b \in B$.

(15.2) Let $g_1, \ldots, g_r$ be the nontrivial cycles of $g \in S^{\#}$. Then
   (1) $g_i g_j = g_j g_i$ for $i \neq j$.
   (2) $g = g_1 \ldots g_r$ is the product in S of its nontrivial cycles.
   (3)   If $g = c_1 \ldots c_s$ with $\{c_1, \ldots, c_s\}$ a set of nontrivial disjoint cycles then $\{c_1, \ldots, c_s\} = \{g_1, \ldots, g_r\}$.
   (4)   The order of g is the least common multiple of the lengths of its cycles.

By 15.2, each member of $S^{\#}$ can be written uniquely as the product of non-trivial disjoint cycles, and these cycles commute, so the order of the product is immaterial.

For $g \in S$ define Cyc_g to be the function from $\mathbb{Z}^+$ into $\mathbb{Z}$ such that $\mathrm{Cyc}_g(i)$ is the number of cycles of g of length i. Permutations g and h are said to have the same *cycle structure* if $\mathrm{Cyc}_g = \mathrm{Cyc}_h$.

(15.3) Let $g, h \in S$ with

$$g = (a_1, \ldots, a_\alpha)(b_1, \ldots, b_\beta) \ldots$$

Then
   (1) $g^h = (a_1 h, \ldots, a_\alpha h)(b_1 h, \ldots, b_\beta h) \ldots.$
   (2) $s \in S$ is conjugate to g in S if and only if s and g have the same cycle structure.

A *transposition* is an element of S moving exactly two points of X. That is a transposition is a cycle of length 2.

(15.4) S is generated by its transpositions.

Proof. By 15.2 it suffices to show each cycle is a product of transpositions. But $(1, 2, \ldots, m) = (1, 2)(1, 3) \ldots (1, m)$.

A permutation is said to be an *even permutation* if it can be written as the product of an even number of transpositions, and to be an *odd permutation* if it can be written as the product of an odd number of transpositions. Denote by Alt(X) the set of all even permutations of X.

(15.5) (1) Alt(X) is a normal subgroup of Sym(X) of index 2.

(2) A permutation is even if and only if it has an even number of cycles of even length. A permutation is odd if and only if it has an odd number of cycles of even length.

Proof. Without loss $X = \{1, 2, \ldots, n\}$. Consider the polynomial ring $R = \mathbb{Z}[x_1, \ldots, x_n]$ in n variables x_i over the ring $\mathbb{Z}$ of integers. For $s \in S$ define $s\alpha: R \to R$ by $f(x_1, \ldots, x_n)s\alpha = f(x_{1s}, \ldots, x_{ns})$. Check that $\alpha: S \to \text{Aut}(R)$ is a representation of S in the category of rings and ring homomorphisms.

Consider the polynomial $P(x_1, \ldots, x_n) = P \in R$ defined by

$$P = P(x_1, \ldots, x_n) = \prod_{1 \le i < j \le n} (x_j - x_i).$$

For $s \in S$, $s\alpha$ permutes the factors $x_j - x_i$ of P up to a change of sign, so $Ps\alpha = P$ or $-P$. Moreover if t is the transposition $(1, 2)$, then $Pt\alpha = -P$. Thus $\Delta = \{P, -P\}$ is the orbit of P under S. Let $A = S_P$ be the stabilizer of P in S. A is also the kernel of α so $A \trianglelefteq S$ and, by 2.11, $|S:A| = |\Delta| = 2$. By 15.3, the transpositions form a conjugacy class of S, so, as $t \in S - A$ and $A \trianglelefteq S$, each transposition is in $S - A$. But, as $|S:A| = 2$, the product of m elements of $S - A$ is in A if and only if m is even. So $A = \text{Alt}(X)$ and $S - A$ is the set of odd permutations. This proves (1), and, since we saw during the proof of 15.4 that a cycle of length m is the product of $m - 1$ transpositions, (2) also holds.

The group Alt(X) is the *alternating group* on X. Evidently the isomorphism type of Sym(X) and Alt(X) depends only on the cardinality of X, so we may write S_n and A_n for Sym(X) and Alt(X), respectively, when $|X| = n$. The groups S_n and A_n are the *symmetric* and *alternating groups* of degree n.

If m is a positive integer, denote by X^m the set product of m copies of X. If G is a permutation group on X, then G is also a permutation group on X^m via $g: (x_1, \ldots, x_m) \mapsto (x_1 g, \ldots, x_m g)$ for $g \in G$ and $x_i \in X$. Assume G is transitive on X. Then the orbits of G on X^2 are called the *orbitals* of G. One orbital is the *diagonal orbital* $\{(x, x): x \in X\}$. The *permutation rank* of a transitive permutation group G is defined to be the number of orbitals of G.

(15.6) Let G be a transitive permutation group on X, $x \in G$, and $(x_i: 1 \le i \le r)$ representatives for the action of $H = G_x$ on X. Then $\{(x, x_i): 1 \le i \le r\}$ is a set of representatives for the orbitals of G and $(x, y) \in (x, x_i)G$ if and only if $y \in x_i H$. In particular r is the permutation rank of G.

The *regular permutation representation* of a group H is the representation of H on itself by right multiplication. A permutation representation $\pi: H \to \mathrm{Sym}(Y)$ is *semiregular* if and only if the identity element is the only element of H fixing points of Y. Equivalently $H_y = 1$ for all y in Y.

(15.8) A permutation representation π of finite degree is semiregular if and only if the transitive constituents of π are regular.

A *regular normal subgroup* of G is a normal subgroup of G which is regular on X.

(15.9) Let G be transitive on X, $x \in X$, and $H \le G$. Then H is regular on X if and only if G_x is a complement to H in G.

(15.10) Let H be the split extension of a normal subgroup K by a subgroup A of H and let π be the representation of H on the cosets of A. Then $K \cong K\pi$ and $K\pi$ is a regular normal subgroup of $H\pi$.

(15.11) Let H be a regular normal subgroup of G and $x \in X$. Then the map $\alpha: H \to X$ defined by $\alpha: h \mapsto xh$ is an equivalence of the representation of G_x on H via conjugation with the representation of G_x on X.

Given a positive integer m, G is said to be *m-transitive* on X if G acts transitively on the subset of X^m consisting of the m-tuples all of whose entries are distinct. Notice G is 2-transitive if and only if it is transitive of permutation rank 2. Also m-transitivity implies k-transitivity for $k \le m$.

(15.12) (1) Let $m \ge 2$ and $x \in X$. Then G is m-transitive on X if and only if G is transitive and G_x is $(m-1)$-transitive on $X - \{x\}$.
  (2) $\mathrm{Sym}(X)$ is n-transitive on X.
  (3) $\mathrm{Alt}(X)$ is $(n-2)$-transitive on X.
  (4) If G is $(n-2)$-transitive on X then $G = \mathrm{Sym}(X)$ or $\mathrm{Alt}(X)$.

Proof. The first three statements are straightforward. Prove the fourth by induction on n using (1) and the following observation: If G is transitive on X and $G_x = \mathrm{Sym}(X)_x$ or $\mathrm{Alt}(X)_x$ then $G = \mathrm{Sym}(X)$ or $\mathrm{Alt}(X)$, respectively. The observation follows from 5.20 plus the fact that, if $G_x \le \mathrm{Alt}(X)$ and $n \ge 4$, then $G = \langle G_y : y \in X \rangle \leq \text{Alt}(X)$, since $\text{Alt}(X) \trianglelefteq \text{Sym}(X)$ and G is 2-transitive on X.

(15.13) Let G be m-transitive on X and H a regular normal subgroup of G. Then

(1) If $m = 2$ then n is a power of some prime p and H is an elementary abelian p-group.

(2) If $m = 3$ then either n is a power of 2 or $n = 3$ and $G = \text{Sym}(X)$.

(3) If $m \geq 4$ then $m = 4 = n$ and $G = \text{Sym}(X)$.

Proof. We may take $m \geq 2$. Let $x \in X$ and $K = G_x$. By 15.12, K is $(m-1)$-transitive on $X - \{x\}$, and then, by 15.11, K acts $(m-1)$-transitively on $H^\#$ via conjugation. In particular K is transitive on $H^\#$.

Let p be a prime divisor of n. As H is regular on X, $n = |H|$. So by Cauchy's Theorem there is $h \in H$ of order p. Thus, as K is transitive on $H^\#$, every element of $H^\#$ is of order p. We conclude from Cauchy's Theorem that H is a p-group. So $n = |H|$ is a power of p. By 9.8, H is solvable and as K is transitive on $H^\#$, H is a minimal normal subgroup of G. So, by 9.4, H is elementary abelian.

This completes the proof of (1), so we may take $m \geq 3$. Let $y = xh$. By 15.2, K_y is transitive on $X - \{x, y\}$ and so, by 15.11, $K_y = C_K(h)$ is transitive on $H - \{1, h\}$ via conjugation. But $C_K(h) = C_K(\langle h \rangle)$, so either $\langle h \rangle = \{1, h\}$ or $n = 3$. In the first case $p = 2$ and in the second $G = \text{Sym}(X)$ by 15.12.4.

This completes the proof of (2), so we may take $m \geq 4$. Let $g \in H - \langle h \rangle$. By 15.11 and 15.12, $C_K(g) \cap C_K(h) = J$ is transitive on $H - \{1, g, h\}$ via conjugation. But J centralizes gh, so $n = 4$. Hence $G = \text{Sym}(X)$ by 15.12.4.

Recall the definition of a primitive representation from section 5.

(15.14) 2-transitive representations are primitive.

(15.15) If G is primitive on X and $1 \neq H \trianglelefteq G$ then H is transitive on X and $G = G_x H$ for each $x \in X$.

Proof. Let $x \in X$. By 5.19, $M = G_x$ is a maximal subgroup of G, while as $H \trianglelefteq G$, MH is a subgroup of G containing M by 1.7. Thus $MH = M$ or G. In the latter case H is transitive on X by 5.20. In the former $H \leq M$, so $H \leq \ker(\pi)$ by 5.7, where π is the representation of G on X. But, as $G \leq S$, π is the identity map on G and in particular is faithful. This is impossible as $H \neq 1$.

(15.16) The alternating group A_n is simple if $n \geq 5$.

Proof. Let $n \geq 5$, $G = \text{Alt}(X)$, and $1 \neq H \trianglelefteq G$. We must show $H = G$. By 15.12, G is $(n - 2)$-transitive on X, so, by 15.14 and 15.15, H is transitive on X and $G = KH$, where $K = G_x$ and $x \in X$. If $1 = H \cap K$ then, by 15.9, H is regular on X. But this contradicts 15.13 and the fact that G is $(n - 2)$-transitive with $n \geq 5$.

So $1 \neq H \cap K$. But $K = \text{Alt}(Y)$, where $Y = X - \{x\}$, so, by induction on n, either K is simple or $n = 5$. In the former case, as $1 \neq H \cap K \trianglelefteq K$, $K = H \cap K \leq H$, so $G = KH = H$. Thus we may take $n = 5$. Here at least $H \cap K$ is transitive on Y by 15.12, 15.14, and 15.15. So $4 = |Y| = |H \cap K : (H \cap K)_y|$ for $y \in Y$ by 5.11. Similarly $5 = |X| = |H : H \cap K|$, so 20 divides the order of H. But $|S| = 5! = 120$ while $|S : G| = 2$, so $|G| = 60$. Thus, as $|H|$ divides $|G|$, $|H| = 20$ or 60. In the latter case $H = G$, so we may assume the former.

By Exercise 2.6, H has a unique Sylow 5-group P. Hence P char $H \trianglelefteq G$, so, by 8.1, $P \trianglelefteq G$. This is impossible as we have just shown that 4 divides the order of any nontrivial normal subgroup of G.

(15.17) (Jordan) Let G be a primitive permutation group on a finite set X and suppose Y is a nonempty subset of X such that $|X - Y| > 1$ and G_Y is transitive on $X - Y$. Then

    (1) G is 2-transitive on X, and

    (2) if G_Y is primitive on $X - Y$ then G_x is primitive on $X - \{x\}$ for $x \in X$.

Proof. Let $\Gamma = X - Y$ and induct on $|Y|$. If $|Y| = 1$ the result is trivial. So assume $|Y| > 1$ and let x and y be distinct points of Y. By Exercise 5.5, there is $g \in G$ with $x \in Yg$ and $y \notin Yg$. Let $H = \langle G_Y, G_{Yg} \rangle$ and $\Omega = \Gamma \cup \Gamma g$. Then $H \leq G_x$ and H acts on Ω. Suppose $|\Gamma| > |Y|$. Then $\Gamma \cap \Gamma g$ is nonempty so H is transitive on Ω. As $H \leq G_x$, $\Gamma \cup \{y\}$ is contained in an orbit of G_x. Since this holds for each $y \in Y - \{x\}$, G is 2-transitive on X by 15.12.1. Further if G_Y is primitive on Γ and Q is a G_x-invariant partition of $X' = X - \{x\}$, then for $Z \in Q$ either $|Z \cap \Gamma| \leq 1$ or $\Gamma \subseteq Z$. As $|\Gamma| \geq |Y|$ it follows that $|Z| = 1$ or $|X'|$. Hence G_x is primitive on X'.

So assume $|\Gamma| \leq |Y|$ and let α, γ be distinct points of Γ. By Exercise 5.5 there is $h \in G$ with $\gamma \in \Gamma h$ but $\alpha \notin \Gamma h$. Let $K = \langle G_Y, G_{Yh} \rangle$, $\Gamma' = \Gamma \cup \Gamma h$, and $Y' = X - \Gamma'$. Then $K \leq G_{Y'}$, and as $\gamma \in \Gamma \cap \Gamma h$, K is transitive on Γ' and $|\Gamma| > |\Gamma' - \Gamma|$. As $|\Gamma| \leq |Y|$ and $\gamma \in \Gamma \cap \Gamma h$, we have $Y' \neq \emptyset$. If G_Y is primitive on Γ then, as $|\Gamma| > |\Gamma' - \Gamma|$, K is primitive on Γ' by an argument in the last paragraph. So, replacing Y by Y' and applying induction on $|Y|$, the result holds.

Jordan's Theorem is a useful tool for investigating finite alternating groups and symmetric groups. See for example Exercises 5.6, 5.7, and 16.2.

16 Rank 3 permutation groups

In this section G is a transitive permutation group on a finite set X of order n.

Recall the definition of an orbital in the preceding section. Given an orbital Δ of G, the *paired orbital* Δ^p of Δ is

$$\Delta^p = \{(y, x): (x, y) \in \Delta\}.$$

Evidently Δ^p is an orbital of G with $(\Delta^p)^p = \Delta$. The orbital Δ is said to be *self paired* if $\Delta = \Delta^p$.

(16.1) (1) A nondiagonal orbital Δ of G is self paired if and only if (x, y) is a cycle in some $g \in G$, for $(x, y) \in \Delta$.

(2) G possesses a nondiagonal self paired orbital if and only if G is of even order.

(3) If G is of even order and (permutation) rank 3 then both nondiagonal orbitals of G are self paired.

Recall the definition of a graph from section 3.

(16.2) Let Δ be a self paired orbital of G. Then Δ is a symmetric relation on X, so $\mathcal{G} = (X, \Delta)$ is a graph and G is a group of automorphisms of $\mathcal{G}$ transitive on the edges of $\mathcal{G}$.

In the remainder of this section assume G is of even order and of permutation rank 3. Hence G has two nondiagonal orbitals Δ and Γ. By 16.1.3 both Δ and Γ are self paired. For $x \in X$, G_x has two orbits $\Delta(x)$ and $\Gamma(x)$ on $X - \{x\}$, where $\Delta(x) = \{y \in X: (x, y) \in \Delta\}$ and $\Gamma(x) = \{y \in X: (x, y) \in \Gamma\}$; this holds by 15.6. By 16.2, $\mathcal{G} = (X, \Delta)$ is a graph and G is a group of automorphisms of $\mathcal{G}$ transitive on the edges of $\mathcal{G}$. Notice $\Delta(x)$ is the set of vertices adjacent to x in this graph. Define $x^\perp = \{x\} \cup \Delta(x), k = |\Delta(x)|, l = |\Gamma(x)|, \lambda = |\Delta(x) \cap \Delta(y)|$ for $y \in \Delta(x)$, and $\mu = |\Delta(x) \cap \Delta(z)|$ for $Z \in \Gamma(x)$. As G_x is transitive on $\Delta(x)$ and $\Gamma(x)$, λ and μ are well defined. As G is transitive on X these definitions are independent of the choice of x.

(16.3) (1) $n = 1 + k + l$.
(2) $\mu l = k(k - \lambda - 1)$.

Proof. Part (1) is trivial. To prove (2) count $|\Omega|$ in two different ways, where

$$\Omega = \{(a, b): b, x \in \Delta(a), b \in \Gamma(x)\}$$

for fixed $x \in X$.

(16.4) If $k \leq l$ then the following are equivalent:

(1) G is imprimitive.
(2) $\lambda = k - 1$.
(3) $\mu = 0$.
(4) $x^{\perp} = y^{\perp}$ for $y \in \Delta(x)$, $\{(x^{\perp})g : g \in G\} = \Omega$ is a system of imprimitivity for G, and G is 2-transitive on Ω.

Proof. If S is a system of imprimitivity for G and $x \in \theta \in S$, then, by 5.18, G_x acts on θ. So, as G_x is transitive on $\Delta(x)$ and $\Gamma(x)$, either $\theta = x^{\perp}$ or $\theta = \{x\} \cup \Gamma(x)$. By 5.18, $|\theta|$ divides n, so as $k \leq l$, $\theta = x^{\perp}$. Now (4) holds. If $x^{\perp} = y^{\perp}$ then (2) and (3) hold. Also (2) is equivalent to (3). Finally if (2) holds then $x^{\perp} = y^{\perp}$, so Ω is a system of imprimitivity for G and hence (1) holds.

(16.5) If $\mu \neq 0$ or k then G is primitive.

Proof. By 16.4 we may take $k > l$. Let $\bar{\mu} = l - (k - \lambda - 1)$. If $\bar{\mu} = 0$ then, by 16.3.2, $k = \mu$, contrary to hypothesis. Hence, by 16.4 and symmetry between Δ and Γ, G is primitive.

(16.6) If G is primitive then $\mathscr{G}$ is connected.

Proof. By 5.19, G_x is maximal in G, so $G = \langle G_x, G_y \rangle$ for $x \neq y$. Now Exercise 5.2 completes the proof.

(16.7) Assume G is primitive. Then either

(1) $k = l$ and $\mu = \lambda + l = k/2$, or
(2) $d = (\lambda - \mu)^2 + 4(k - \mu)$ is a square and setting
$D = 2k + (\lambda - \mu)(k + l)$ we have:
 (a) $d^{1/2}$ divides D but $2d^{1/2}$ does not if n is even, while
 (b) $2d^{1/2}$ divides D if n is odd.

Proof. Let A be the incidence matrix for $\mathscr{G}$. That is $A = (a_{xy})$ is the n by n matrix whose rows and columns are indexed by X, and with $a_{xy} = 1$ if (x, y) is an edge of $\mathscr{G}$ while $a_{xy} = 0$ otherwise. Let B be the incidence matrix for (X, Γ), J the n by n matrix all of whose entries are 1, and I the n by n identity matrix. Observe:

(i) A is symmetric.
(ii) $I + A + B = J$.

(iii) $AJ = kJ$, so $(A - kI)J = 0$.

(iv) $A^2 = kI + \lambda A + \mu B$.

The first two statements are immediate from the definitions. As $|\Delta(x)| = k$, each row of A has k entries equal to 1. Thus (iii) holds. By (i) the (x, y)-th entry of A^2 is the inner product of the xth and yth rows of A. But this inner product just counts $|\Delta(x) \cap \Delta(y)|$, so (iv) holds.

Next (ii), (iii), and (iv) imply:

(v) $(A - kI)(A^2 - (\lambda - \mu)A - (k - \mu)I) = 0$,

so the minimal polynomial of A divides

$$p(x) = (x - k)(x^2 - (\lambda - \mu)x - (k - \mu)).$$

The roots of $p(x)$ are k, s, and t, where $s = ((\lambda - \mu) + d^{1/2})/2$ and $t = ((\lambda - \mu) - d^{1/2})/2$. Let m_e be the multiplicity of the eigenvalue e.

Claim $m_k = 1$. Indeed for $c = (c_1, \ldots, c_n)$, $cA = kc$ if and only if $c_i = c_1$ for all i. To prove this we can take $|c_1| \geq |c_i|$ for each i. As $cA = kc$, $kc_1 = \Sigma_i c_i a_{i1}$, so, as exactly k of the a_{i1} are 1 and the rest are 0, $c_1 = c_i$ for each $i \in \Delta(1)$. But now as $\mathcal{G}$ is connected it follows that $c_1 = c_i$ for all $i \in X$, completing the proof of the claim.

As $m_k = 1$ it follows that:

(vi) $m_s + m_t = n - 1 = k + l$.

Also, as A is of trace 0:

(vii) $k + m_s s + m_t t = 0$.

Now (vi) and (vii) imply:

(viii) $m_s = ((k + l)t + k)/(t - s)$.

Of course $t - s = -d^{1/2}$, $t = ((\lambda - \mu) - d^{1/2})/2$, and m_s is an integer. Thus

(ix) $(Dd^{-1/2} - (k + l))/2$ is an integer, where

$$D = 2k + (\lambda - \mu)(k + l).$$

In particular either d is a square or $D = 0$. If $D = 0$ then $\mu = \lambda + 1$ and $l = k$, so 16.7.1 holds by 16.3.2. If d is a square then 16.7.2 holds by (ix).

Remarks. Wielandt [Wi 2] is a good place to find more information about permutation groups. The material in section 16 comes from Higman [Hi].

Section 16 is somewhat technical and can be safely omitted by the novice. On the other hand the results in section 15 are reasonably basic.

Exercises for chapter 5

1. (1) Prove A_5 has no faithful permutation representation of degree less than 5.

   (2) Prove that, up to equivalence, A_5 has unique transitive representations of degree 5 and 6. Prove both are 2-transitive.

   (3) Prove $\mathrm{Aut}(A_5) = S_5$.

   (4) Prove there are exactly two conjugacy classes of subgroups of S_6 isomorphic to A_5. Prove the same for A_6.

2. Let G be a transitive permutation group on X, Δ a self paired orbital of G, $(x, y) \in \Delta$, $\mathcal{G} = (X, \Delta)$ the graph on X determined by Δ, and $H = \langle G_x, G_y \rangle$. Prove

   (1) $xH \cup yH$ is the connected component of $\mathcal{G}$ containing x, and

   (2) $\mathcal{G}$ is connected precisely when one of the following holds:

      (i) $G = H$, or

      (ii) $|G : H| = 2$ and $\mathcal{G}$ is bipartite with partition $\{xH, yH\}$ (i.e. $\{xH, yH\}$ is a partition of X and $\Delta \subseteq (xH \times yH) \cup (yH \times xH)$).

3. Let S and A be the symmetric and alternating groups of degree n, respectively, and let $a \in A$. Prove $a^A \neq a^S$ precisely when (*) holds

   (*) $\mathrm{Cyc}_a(2m) = 0$ and $\mathrm{Cyc}_a(2m - 1) \leq 1$ for each positive integer m.

   in which case $a^S = a^A \cup b^A$ for some $b \in A$ with $\langle a \rangle = \langle b \rangle$ and $|a^A| = |b^A|$.

4. Let A be the alternating group on a set X of finite order $n > 3$ and let V be the set of subsets of X of order 2. Prove

   (1) A is a rank 3 permutation group on V.

   (2) A_v is a maximal subgroup of A for $v \in V$.

   (3) If G is a primitive rank 3 group on a set Ω of order 10 then $G \cong A_5$ or S_5 and the representation of G on Ω is equivalent to its representation on V.

5. Let G be a primitive permutation group on a finite set X and let x and y be distinct points of X. Let Y be a nonempty proper subset of X and define $S(x) = \{Yg : g \in G, x \in Yg\}$ and $T(x) = \bigcap_{Z \in S(x)} Z$. Prove

   (1) $T(x) = \{x\}$, and

   (2) there exists $g \in G$ with $x \in Yg$ but $y \notin Yg$.

6. Let G be a primitive permutation group on a set X of finite order n. Prove

   (1) If $Y \subseteq X$ such that G_Y is primitive on $X - Y$ and $|Y| = m$ with $1 \leq m \leq n - 2$, then G is $(m + 1)$-transitive on X.

   (2) If G contains a transposition or a cycle of length 3 then G contains the alternating group on X.

   (3) If $a, b \in G$ with $|\mathrm{Mov}(a) \cap \mathrm{Mov}(b)| = 1$ then $[a, b]$ is a cycle of length 3.

   (4) Let $Y \subseteq X$ be of minimal order subject to $G_Y = 1$. Prove either G contains the alternating group on X or $|Y| \leq n/2$.

    (5) If G does not contain the alternating group on X, prove $|\text{Sym}(X) : G| \geq [(n+1)/2]!$

7. Let X be a set of finite order $n \leq 5$, A the alternating group on X, and G a proper subgroup of A. Prove one of the following holds:

    (1) $|A : G| > n$.

    (2) $|A : G| = n$ and $G = A_x$ for some $x \in X$.

    (3) $|A : G| = n = 6$ and $G \cong A_5$.

8. Prove that either

    (1) A_n has a unique conjugacy class of subgroups isomorphic to A_{n-1}, or

    (2) $n = 6$ and A_n has exactly two such classes.

6

Extensions of groups and modules

Chapter 6 considers various questions about extensions of groups and modules, most particularly the conjugacy of complements to some fixed normal subgroup in a split group extension. Suppose G is represented on an abelian group or F-space V and form the semidirect product GV. Section 17 shows there is a bijection between the set of conjugacy classes of complements to V in GV and the 1-cohomology group $H^1(G, V)$. If V is an F-space so is $H^1(G, V)$. Moreover if $C_V(G) = 0$ there is a largest member of the class of FG-modules U such that $C_U(G) = 0$ and U is the extension of V by a module centralized by G. Indeed it turns out that if U is the largest member of this class then $U/V \cong H^1(G, V)$. Further the dual of the statement is also true: that is if $V = [V, G]$ then there is a largest FG-module U^* such that $U^* = [U^*, G]$ and U^* is the extension of an FG-module Z by V with G centralizing Z. In this case $Z \cong H^1(G, V^*)$.

These results together with Maschke's Theorem are then used to prove the Schur–Zassenhaus Theorem, which gives reasonably complete information about extensions of a finite group B by a finite group A when the orders of A and B are relatively prime. The Schur–Zassenhaus Theorem is then used to prove Phillip Hall's extended Sylow Theorem for solvable groups. Hall's Theorem supplies a good illustration of how restrictions on the composition factors of a finite group can be used to derive strong information about the group.

I have chosen to discuss 1-cohomology from a group theoretical point of view. Homological algebra is kept to a minimum. Still the arguments in section 17 have a different flavor than most in this book.

17 1-cohomology

In this section p is a prime, F is a field of characteristic p, V is an abelian group written additively, G is a finite group, and $\pi: G \to \text{Aut}(V)$ is a representation of G on V.

Form the semidirect product $S(G, V, \pi)$ of V by G with respect to π and identify G and V with subgroups of $S(G, V, \pi)$ via the injections of 10.1. Then $S(G, V, \pi) = GV$ with $V \trianglelefteq GV$ and G is a complement to V to GV.

A *cocycle* from G into V is a function $\gamma: G \to V$ satisfying the *cocycle condition*

$$(gh)\gamma = (g\gamma)^h + h\gamma \quad g, h \in G.$$

Notice the cocycle condition forces each cocycle to map the identity of G to the identity of V.

Let $\Gamma(G, V)$ denote the set of cocycles from G to V and make $\Gamma(G, V)$ into a group (again written additively) via:

$$g(\gamma + \delta) = g\gamma + g\delta \quad \gamma, \delta \in \Gamma(G, V), g \in G.$$

Let $A = \mathrm{Aut}(GV)$ be the group of automorphisms of GV, and let $U(G, V) = C_A(GV/V) \cap C_A(V)$.

If V is a vector space over F and π is an FG-representation then $\Gamma(G, V)$ is also a vector space over F via:

$$g(a\gamma) = a(g\gamma) \quad \gamma \in \Gamma(G, V), g \in G, a \in F.$$

(17.1) For $\gamma \in \Gamma(G, V)$ define $S(\gamma) = \{gg\gamma : g \in G\}$. Then the map $S: \gamma \mapsto S(\gamma)$ is a bijection of $\Gamma(G, V)$ with the set of complements to V in GV.

Proof. The cocycle condition says that $S(\gamma)$ is a subgroup of GV. Evidently $S(\gamma)$ is a complement to V in GV and S is injective. Conversely if C is a complement then, for $g \in G$, $gV \cap C$ contains a unique element gv and $\gamma: g \mapsto v$ is a cocycle with $S(\gamma) = C$.

(17.2) For $\gamma \in \Gamma(G, V)$ define $\gamma\phi: GV \to GV$ by

$$\gamma\phi: gv \mapsto g((g\gamma) + v).$$

Then the map $\phi: \gamma \mapsto \gamma\phi$ is a group isomorphism of $\Gamma(G, V)$ with $U(G, V)$. For $u \in U(G, V)$ and $g \in G$, $u\phi^{-1}: g \mapsto g^{-1}g^u$.

Proof. The cocycle condition implies $\gamma\phi$ is a homomorphism. $(-\gamma)\phi$ is an inverse for $\gamma\phi$, so $\gamma\phi \in \mathrm{Aut}(GV)$. By definition $\gamma\phi$ centralizes V and GV/V, so $\gamma\phi \in U(G, V)$. For $u \in U$ define $u\psi: g \mapsto g^{-1}g^u$. An easy check shows $u\psi$ is a cocycle and $\psi = \phi^{-1}$, so ϕ is an isomorphism.

(17.3) $U(G, V)$ acts regularly on the set of complements to V in GV. Indeed $G^u = S(u\phi^{-1})$ for each $u \in U(G, V)$.

Proof. As $U = U(G, V)$ acts on V it permutes the complements to V in GV. By definition of the maps ϕ and S, $G^u = S(u\phi^{-1})$ for $u \in U$. Hence, as S and ϕ are bijections, the action of U is regular.

Lemmas 17.1 and 17.3 give descriptions of the complements to V in GV in terms of $\Gamma(G, V)$ and $U(G, V)$, while 17.2 and 17.3 give the correspondence between the two descriptions. The next few lemmas describe the representations of G on $\Gamma(G, V)$ and $U(G, V)$, and show these representations are equivalent.

(17.4) For $g \in G$ and $\gamma \in \Gamma(G, V)$ define $\gamma^g: G \to V$ by $h(\gamma^g) = [(h^{g^{-1}})\gamma]^g$, for $h \in G$. Then $\tilde{\pi}$ is a representation of G on $\Gamma(G, V)$, where $g\tilde{\pi}: \gamma \mapsto \gamma^g$. If π is an FG-representation so is $\tilde{\pi}$.

(17.5) For $v \in V$ define $v\alpha: G \to V$ by $g(v\alpha) = [g, v] = v - v^g$. Then the map $\alpha: v \mapsto v\alpha$ is a G-homomorphism of V into $\Gamma(G, V)$ with kernel $C_V(G)$. If π is an FG-representation then α is an FG-homomorphism.

Proof. Lemma 8.5.4 says $v\alpha$ is a cocycle. The rest is straightforward.

If V is an FG-module then so is $\Gamma(G, V)$ and ϕ induces an F-space structure on $U(G, V)$ which makes ϕ into an F-space isomorphism. That is, for $u \in U(G, V)$ and $a \in F$, $au = (a(u\phi^{-1}))\phi$. Equivalently if $g^u = gv$ then $g^{au} = g(av)$. This is the F-space structure on $U(G, V)$ implicit in the remainder of the section.

(17.6) Let $c: GV \to \text{Aut}(GV)$ and $d: \text{Aut}(GV) \cap N(V) \to \text{Aut}(U(G, V))$ be the conjugation maps. Then

is a commutative diagram, the maps c, α, and ϕ are G-homomorphisms, and if V is an FG-module the maps are FG-homomorphisms. Here ϕ, α, and $\tilde{\pi}$ are defined in 17.2, 17.5, and 17.4, respectively, and $\pi, \tilde{\pi}$, and cd are the representations of G on V, $\Gamma(G, V)$, and $U(G, V)$, respectively.

Proof. Let $v, w \in V, u \in U(G, V)$, and $g, h \in G$. Then

$$vc: gw \mapsto g^v w = g(v - v^g + w) = (gw)(v\alpha)\phi,$$

so $c = \alpha\phi$ and the diagram commutes.

By 17.5, α is a G-homomorphism, and even an FG-homomorphism if V is an FG-module. By 17.2, ϕ is a homomorphism.

Next $u^{gcd}: hv \mapsto (hv)^{g^{-1}ug} = h^{g^{-1}ug}v$. Hence $(hv)(\gamma^g\phi) = h(h\gamma^g + v) = h(((h^{g^{-1}})\gamma)^g + v) = (h^{g^{-1}}(h^{g^{-1}})\gamma)^g v = ((h^{g^{-1}})(\gamma\phi))^g v = (hv)((\gamma\phi)^{gcd})$, so $\gamma^g\phi = (\gamma\phi)^{gcd}$, and therefore ϕ is a G-homomorphism.

Suppose V is an FG-module. For $a \in F$, $(au)^{gcd} = ((a(u\phi^{-1}))\phi)^{gcd} = ((a(u\phi^{-1}))^g)\phi = (a((u^{gcd})\phi^{-1}))\phi = a(u^{gcd})$, so cd is an FG-representation.

Finally, as $c = \alpha\phi$ and α and ϕ are G-homomorphisms (or FG-homomorphisms), so is c. Therefore the proof is complete.

By 17.6, $\Gamma(G, V)/V\alpha \cong U(G, V)/Vc$. The *first cohomology group* of the representation π is $H^1(G, V) \cong \Gamma(G, V)/V\alpha \cong U(G, V)/Vc$. This is an additive group and, if V is an FG-module, $H^1(G, V)$ is even a vector space over F.

The next lemma says that $H^1(G, V)$ is in one to one correspondence with the set of conjugacy classes of complements to V in GV.

(17.7) $H^1(G, V)$ acts regularly on the set of conjugacy classes of complements to V in GV via

$$(Vc)u: (G^w)^V \mapsto (G^{wu})^V \quad u, w \in U(G, V).$$

In particular the number of conjugacy classes of complements to V in GV is $|H^1(G, V)|$.

Proof. This is a consequence of 17.3 and the fact that $U(G, V)$ acts on V.

(17.8) $[Gcd, U(G, V)] \leq Vc$.

Proof. For $g \in G$, $u \in U(G, V)$, $[gcd, u] = [gc, u] \in U(G, V) \cap GVc$, as $U(G, V)$ and GVc are normal in Aut(GV). But $U(G, V) \cap GVc = Vc$.

(17.9) Assume either $C_V(G) = 0$ or $G = O^p(G)$. Then $C_{U(G,V)}(Gcd) = 0$ and if $C_V(G) = 0$ then $c: V \to U(G, V)$ is an injection.

Proof. If $u \in C_{U(G,V)}(Gcd)$ then $1 = [u, Gcd] = [u, Gc]$, so $[u, G] \leq \ker(c) = Z(GV) = C_{Z(G)}(V) \times C_V(G)$. I'll show $[u, G] \leq G$, so that, by 17.3, $u = 1$, and hence $C_{U(G,V)}(Gcd) = 0$.

If $C_V(G) = 0$ then $[u, G] \leq C_{Z(G)}(V) \leq G$. If $G = O^p(G)$ then $G = O^p(GZ(GV))$ char $GZ(GV)$, so $[u, G] \leq G$. So the claim is established.

As $C_V(G)$ is the kernel of c, c is an injection if $C_V(G) = 0$.

(17.10) If G is a finite p'-group and V an FG-module, then $H^1(G, V) = 0$. Hence G is transitive on the complements to V in GV in this case.

Proof. By Maschke's Theorem, the FG-module $U = U(G, V)$ splits over Vc. Let W be an FG-complement to Vc in U. By 17.8, $[Gcd, W] \leq W \cap Vc = 0$. So, by 17.9, $W = 0$. Thus $H^1(G, V) \cong W = 0$, and 17.7 completes the proof.

Lemma 17.10 will be used to prove the Schur–Zassenhaus Theorem in the next section.

(17.11) Let V be an FG-module, $\beta: V \to W$ an injective FG-homomorphism, and assume $[G, W] \leq V\beta$ and $C_W(G) = 0$. Then there exists an injective FG-homomorphism $\gamma: W \to U(G, V)$ making the following diagram commute:

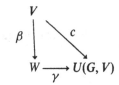

In particular $\dim_F(W/V\beta) \leq \dim_F(H^1(G, V))$.

Proof. Let π' be the representation of G on W and consider the semidirect products $H = S(F^\# \times G, W, \pi')$ and $S = S(F^\# \times G, V, \pi)$. As β is an injective FG-homomorphism it induces by 10.3 an injective group homomorphism $\beta: S \to H$ which is the identity on $F^\# \times G$. As $[G, W] \leq V\beta$, $(GV)\beta \trianglelefteq H$, so the conjugation map e of H on $(GV)\beta$ composed with $(\beta^{-1})^*: \mathrm{Aut}((GV)\beta) \to \mathrm{Aut}(GV) = A$ maps H into A. As $C_W(G) = 0$, the restriction γ of $e(\beta^{-1})^*$ to W is an injection of W into $U = U(G, V)$. As γ is the composition of $(F^\# \times G)$-homomorphisms, γ is an $(F^\# \times G)$-homomorphism, and hence is a G-homomorphism preserving the multiplication by F. In particular if the multiplication $u \mapsto u^a$, $a \in F^\#$, $u \in U$ of $F^\#$ on U is that of the F-space structure on U defined earlier, then γ is an FG-homomorphism. For this we need to show $g^{u^a} = g^{au}$ for $g \in G$. But if $g^u = gv$ then $g^{au} = gav$, while $g^{u^a} = g^{a^{-1}ua} = g^{ua} = (gv)^a = gav$ as $[a, g] = 1$ and $v^a = av$.

It remains to show $\beta\gamma = c$. Keeping in mind that $\beta: S \to H$ is an injective homomorphism trivial on $F^\# \times G$, we see that for $v, w \in V$ and $g \in G$ we have $(gw)^{v\beta\gamma} = (((gw)\beta)^{v\beta})\beta^{-1} = (g^{v\beta}w\beta)\beta^{-1} = (g[g, v\beta]w\beta)\beta^{-1} = g[g, v]w = g^v w = (gw)^v = (gw)^{vc}$. So $\beta\gamma = c$ as desired.

Lemma 17.11 says that if $C_V(G) = 0$ then $U(G, V)$ is the largest extension W of V such that $[W, G] \leq V$ and $C_W(G) = 0$.

Recall the definition of the dual V^* of a finite dimensional FG-module V given in section 14, and define $U^*(G, V)$ to be $(U(G, V^*))^*$.

(17.12) Let V be a finite dimensional FG-module, $\beta\colon W \to V$ a surjective FG-homomorphism, and assume $\ker(\beta) \leq C_W(G)$ and $W = [W, G]$. Then there exists a surjective FG-homomorphism $\gamma\colon U^*(G, V) \to W$ making the following diagram commute:

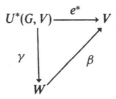

Here e^* is the dual of the conjugation map $e\colon V^* \to U(G, V^*)$ and e^* is a surjective FG-homomorphism with $H^1(G, V^*) \cong \ker(e^*) \leq C_{U^*(G,V)}(G)$. Thus $\dim_F(\ker(\beta)) \leq \dim_F(H^1(G, V^*))$.

Proof. As $W = [W, G]$, $C_{W^*}(G) = 0$ by 14.6. Similarly, as $\ker(\beta) \leq C_W(G)$, $[G, W^*] \leq V^*\beta^*$ by 14.6. Let $e\colon V^* \to U(G, V^*)$ be the conjugation map. By 17.11 there is an injective FG-homomorphism $\delta\colon W^* \to U(G, V^*)$ such that the diagram commutes:

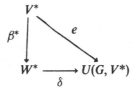

Then applying 14.1.2 and 14.4.3 we conclude the following diagram commutes,

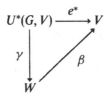

where $\gamma = \delta^*$. As δ is injective, γ is surjective, so $e^* = \gamma\beta$ is surjective since β is surjective by hypothesis. $H^1(G, V^*) = U(G, V^*)/V^*e$, so $H^1(G, V^*) \cong H^1(G, V^*)^* \cong \ker(e^*)$ by 14.5. As $[U(G, V^*), G] \leq V^*e$, G centralizes $\ker(e^*)$ by 14.6.

Lemma 17.12 says that if $V = [V, G]$ then $U^*(G, V)$ is the largest extension W of a module Z by V with $Z \leq C_W(G)$ and $W = [W, G]$.

18 Coprime action

(18.1) (Schur–Zassenhaus Theorem) Let G be a finite group, $H \trianglelefteq G$, and assume

(i) $(|H|, |G/H|) = 1$, and
(ii) either H or G/H is solvable.

Then
(1) G splits over H, and
(2) G is transitive on the complements to H in G.

Proof. Let G be a minimal counterexample. Assume first that H is solvable and let M be a minimal normal subgroup of G contained in H. By 9.4, M is an elementary abelian p-group for some prime p. Let $\bar{G} = G/M$. By minimality of G, there is a complement $\bar{X}$ to $\bar{H}$ in $\bar{G}$ and $\bar{G}$ is transitive on the complements to $\bar{H}$. Also if Y is a complement to H in G then $\bar{Y}$ is a complement to $\bar{H}$ in $\bar{G}$, so it suffices to show X splits over M and X is transitive on its complements to M. Hence, by minimality of G, $G = X$ and $H = M$. Now Gaschütz' Theorem, 10.4, says that G splits over H. Let Y be a complement to H in G.

By 12.1 the conjugation map $c: Y \to \mathrm{Aut}(M)$ is a FY-representation, where F is the field of order p. Hence by 17.10, G is transitive on the complements to $H = M$ in G.

Assume next that G/H is solvable. Let $G^* = G/H$ and K^* a minimal normal subgroup of G^*. By 9.4, K^* is an elementary abelian p-group for some prime p. Let $P \in \mathrm{Syl}_p(K)$ and observe that P is a complement to H in K. By a Frattini Argument, 6.2, $G = KN_G(P)$, so as $K = HP$, also $G = HN_G(P)$.

If Y is a complement to H in G then $K = K \cap G = K \cap HY = H(K \cap Y)$ by the Modular Property of Groups, 1.14. Then $R = K \cap Y$ is a complement to H in K, so $R \in \mathrm{Syl}_p(K)$. Hence, by Sylow's Theorem, there is $k \in K$ with $R^k = P$. As $K \trianglelefteq G$, $R = K \cap Y \trianglelefteq Y$, so $Y^k \leq N_G(P)$. Thus, if $N_G(P)$ splits over $N_H(P)$ and is transitive on its complements to $N_H(P)$, then the theorem holds. So by minimality of G, $P \trianglelefteq G$.

Finally let $\tilde{G} = G/P$. If Y is a complement to H in G, then Y contains a Sylow p-group of G and hence $P \leq Y$. Also $\tilde{Y}$ is a complement to $\tilde{H}$ in $\tilde{G}$. Moreover, by minimality of G, there is a complement $\tilde{C}$ to $\tilde{H}$ in $\tilde{G}$ and $\tilde{G}$ is transitive on its complements to $\tilde{H}$. Now C is a complement to H in G and there is $g \in G$ with $\tilde{C}^g = \tilde{Y}$, so $C^g = Y$, completing the proof.

Remarks. The Odd Order Theorem of Feit and Thompson [FT] says that groups of odd order are solvable. Notice that if $(|A|, |G|) = 1$ then either $|A|$ or $|G|$ is odd, so A or G is solvable. Thus the Odd Order Theorem says hypothesis (ii) of the Schur–Zassenhaus Theorem can be removed.

Let π be a set of primes. The π-*part* of a positive integer n is $n_\pi = \prod_{p\in\pi} p^{e_p}$, where $\prod_p p^{e_p} = n$ is the prime factorization of n. Given a finite group G, $\pi(G)$ denotes the set of prime factors of $|G|$. G is a π-*group* if $\pi(G)$ is a subset of π. A *Hall* π-*subgroup* of G is a subgroup of order $|G|_\pi$. π' denotes the set of primes not in π.

The following lemma gives a useful characterization of Hall π-subgroups.

(18.2) H is a Hall π-subgroup of the finite group G if and only if H is a π-subgroup of G and $|G:H|_\pi = 1$.

(18.3) Let G be a finite group and H a π-subgroup of G. Then

(1) If $\alpha: G \to G\alpha$ is a homomorphism then $H\alpha$ is a π-subgroup of $G\alpha$. If H is a Hall π-subgroup of G and α is surjective then $H\alpha$ is a Hall π-subgroup of $G\alpha$.

(2) If H is a Hall π-subgroup of G and $H \le K \le G$ then H is a Hall π-subgroup of K.

(18.4) If p and q are distinct primes and H and K are Hall p'- and q'-subgroups of a finite group G, respectively, then

(1) $G = HK$, and

(2) $H \cap K$ is a Hall $\{p, q\}'$-subgroup of G, a Hall p'-subgroup of K, and a Hall q'-subgroup of H.

Proof. This follows from 1.7.3.

(18.5) (Phillip Hall's Theorem) Let G be a finite solvable group and π a set of primes. Then

(1) G possesses a Hall π-subgroup.

(2) G acts transitively on its Hall π-subgroups via conjugation.

(3) Any π-subgroup of G is contained in some Hall π-subgroup of G.

Proof. Let G be a minimal counterexample and M a minimal normal subgroup of G. By 9.4, M is a p-group for some prime p. Let $G^* = G/M$. By minimality of G, G^* satisfies the theorem. In particular G^* possesses a Hall π-subgroup H^*. Also if X is a π-subgroup of G then so is X^*, so X^* is contained in some conjugate H^{*g} of H^*, and hence $X \le H^g$.

Suppose $p \in \pi$. Then H is a Hall π-subgroup of G. Further X is contained in the Hall π-subgroup H^g of G. Indeed if X is a Hall π-subgroup of G, then $|X| = |H^g|$ so $X = H^g$ is a conjugate of H.

Thus we may assume $p \notin \pi$. By the Schur–Zassenhaus Theorem, 18.1, there is a complement K to M in H, and H is transitive on such complements.

But as $|H^*| = |G|_{\pi'}$ these complements are precisely the Hall π-subgroups of G contained in H. Therefore G possesses Hall π-subgroups and H is transitive on the Hall π-subgroups of G contained in H. So, as $X^g \leq H$, X is conjugate to K if X is a Hall π-subgroup of G. This shows G is transitive on its Hall π-subgroups. Finally if $H \neq G$ then, by minimality of G, X is contained in a Hall π-subgroup of H^g, which is also a Hall π-subgroup of G.

So assume $H = G$. Then $G = KM$ so $XM = XM \cap G = XM \cap KM = (XM \cap K)M$ by the Modular Property of Groups, 1.14. Now X and $XM \cap K$ are Hall π-subgroups of XM, so, by (2), there is $y \in XM$ with $X^y = XM \cap K \leq K$. Thus the proof is complete.

A converse of Phillip Hall's Theorem also holds, as we will see soon. In particular the hypothesis of solvability is necessary to insure the validity of the theorem. Hall's Theorem is a good example of how restrictions on the composition factors of a finite group can lead to significant restrictions of the global structure of the group.

Now to a proof of a converse to Phillip Hall's Theorem. The proof will appeal to Burnside's $p^a q^b$-Theorem that finite groups G with $|\pi(G)| = 2$ are solvable. The proof of Burnside's Theorem will be postponed until the chapter on character theory.

(18.6) Let G be a finite group possessing a Hall p'-subgroup for each $p \in \pi(G)$. Then G is solvable.

Proof. Let G be a minimal counterexample. By 9.8, G is not a p-group. By Burnside's $p^a q^a$ Theorem, 35.13, $|\pi(G)| \neq 2$. Thus $|\pi(G)| > 2$.

Let $p \in \pi(G)$ and H a Hall p'-subgroup of G. By 18.4, $H \cap K$ is a Hall q'-subgroup of H for each prime q distinct from p and each Hall q'-subgroup of K of G. Therefore H satisfies the hypothesis of the lemma, and hence H is solvable by minimality of G. Let M be a minimal normal subgroup of H. By 9.4, M is an r-group for some prime r. As $|\pi(G)| > 2$ there is $q \in \pi(G) - \{p, r\}$. Let K be a Hall q'-subgroup of G. As $q \neq r$, K contains a Sylow r-subgroup of G. Hence, by Sylow's Theorem, M is contained in some conjugate of K, which we may take to be K. As $q \neq p$, $G = HK$ by 18.4. So, as $M \trianglelefteq H$, $X = \langle M^G \rangle = \langle M^{HK} \rangle = \langle M^K \rangle \leq K$. Hence, as subgroups of solvable groups are solvable, X is solvable. Of course $X \trianglelefteq G$ and we conclude from 18.3 that G/X satisfies the hypothesis of the lemma. So, by minimality of G, G/X is solvable. Therefore G is solvable by 9.3.

(18.7) (Coprime Action) Let A and H be finite groups with $(|A|, |H|) = 1$. Assume A is represented as a group of automorphisms of H and either A or H is solvable. Let p be a prime. Then

(1)   There exists an A-invariant Sylow p-subgroup of H.

(2) $C_H(A)$ is transitive on the A-invariant Sylow p-subgroups of G.

(3)   Every A-invariant p-subgroup of H is contained in an A-invariant Sylow p-subgroup of H.

(4)   Let K be an A-invariant normal subgroup of H and $H^* = H/K$. Then $C_{H^*}(A) = N_{H^*}(A) = C_H(A)^*$.

Proof. Form the semidirect product G of H by A with respect to the representation of A on H, and identify A and H with subgroups of G via the injections of 10.1. Then $H \trianglelefteq G$, A is a complement to H in G, and the action of A on H is via conjugation.

$N_G(A) = N_G(A) \cap AH = AN_H(A)$ by the Modular Property of Groups, 1.14. Also $[A, N_H(A)] \leq A \cap H = 1$, so $N_H(A) = C_H(A)$ and $N_G(A) = A \times C_H(A)$.

Let $P \in \mathrm{Syl}_p(H)$. By a Frattini Argument, 6.3, $G = HN_G(P)$. Then by Schur–Zassenhaus there is a complement B to $N_H(P)$ in $N_G(P)$ and $B^g = A$ for some $g \in G$. Hence $P^g = Q$ is an A-invariant Sylow p-subgroup of H. $N_G(Q)$ is transitive on $A^G \cap N_G(Q)$ by Schur–Zassenhaus, so, by 5.21, $N_G(A)$ is transitive on the A-invariant Sylow p-subgroups of H. As $N_G(A) = AC_H(A)$ and $A \leq N_G(Q)$, $C_H(A)$ is also transitive on the A-invariant Sylow p-subgroups of H by 5.20.

Let R be a maximal A-invariant p-subgroup of H. If $R \notin \mathrm{Syl}_p(H)$ then, by Exercise 3.2, $R \notin \mathrm{Syl}_p(N_H(R))$. But $N_H(R)$ is A-invariant so by (1) there is an A-invariant Sylow p-subgroup of $N_H(R)$, contradicting the maximality of R. This establishes (3).

Assume the hypothesis of (4) and let $X^* = C_{H^*}(A)$. We have already shown $X^* = N_{H^*}(A)$. Now $AK \trianglelefteq AX$ and, by Schur–Zassenhaus, $A^{AX} = A^{AK}$, so by a Frattini Argument, 6.3, $AX = KN_{AX}(A)$. Hence $X = KN_X(A)$ by the Modular Property of Groups, 1.14. So as $C_H(A)^* \leq C_{H^*}(A)$, we conclude $C_{H^*}(A) = X^* = C_X(A)^* = C_H(A)^*$, completing the proof of (4).

Again the hypothesis that A or G is solvable can be removed from the statement of 18.7, modulo the Odd Order Theorem.

Remarks. The material in section 18 is basic, while that in section 17 is more specialized. Thus the reader may wish to skip or postpone section 17. If so, 17.10 must be assumed in proving the Schur–Zassenhaus Theorem.

74 *Extensions of groups and modules*

Exercises for chapter 6

1. Let A be a solvable group acting on $G = XY$ with $Y \trianglelefteq G$, X and YA-invariant, and $(|A|, |G|) = 1$. Then $C_G(A) = C_X(A)C_Y(A)$.

2. Prove 18.7 with p replaced by a set of primes π, under the assumption that H is solvable.

3. Let G be the alternating group on a set I of finite order $n > 2$, let $F = \mathrm{GF}(2)$, let V be the permutation module of the representation on I, and define Z and the core U of V as in Exercise 4.6. Prove
   (1) $0, Z, U,$ and V are the only FG-submodules of V. In particular $\bar{U} = (U + Z)/Z$ is an irreducible FG-module.
   (2) If n is odd prove $H^1(G, \bar{U}) = 0$.
   (3) If n is even prove V is an indecomposable FG-module, $H^1(G, \bar{U}) \cong F$, and $\bar{V} = U(G, \bar{U})$.
   (Hint: In (2) and (3) let H be the stabilizer of a point x of I and proceed by induction on n. If n is odd prove H centralizes $w \in U(G, \bar{U}) - \bar{U}$, and appeal to Exercise 4.6. If n is even prove H centralizes a complement W to $\bar{U}$ in $\bar{U}(G, \bar{U})$ and W is a hyperplane of $C_{U(G,\bar{U})}(H_y)$ for $x \neq y \in I$. Use this to conclude $\dim(W) \leq 1$.)

4. (Alperin–Gorenstein) Let F be a field of characteristic p, G a finite group, V an FG-module, and Δ a G-invariant collection of p'-subgroups such that:
   (1) $V = [V, X]$ for each $X \in \Delta$, and
   (2) the graph on Δ obtained by joining X to Y if $[X, Y] = 1$ is connected.
   Prove $H^1(G, V) = 0$.

5. Let A be an abelian r-group acting on an r'-group G. Then $G = \langle C_G(B): B \leq A, A/B$ cyclic$\rangle$.

7

Spaces with forms

Chapter 7 considers pairs (V, f) where V is a finite dimensional vector space over a field F and f is a nontrivial sesquilinear, bilinear, or quadratic form on V. We'll be primarily interested in the situation where $\text{Aut}(V, f)$ is large; in that event f satisfies one of several symmetry conditions (cf. Exercises 7.9, 7.10, 9.1, and 9.9). Under suitable restrictions on F, such pairs are determined up to isomorphism in sections 19, 20, and 21. For example if F is finite the isomorphism types are listed explicitly in section 21.

It turns out that such spaces satisfy the Witt property: that is, if X and Y are subobjects of (V, f) and $\alpha : X \to Y$ is an isomorphism, then α extends to an automorphism of (V, f). As a result the representation of $\text{Aut}(V, f)$ on (V, f) is particularly useful in investigating $\text{Aut}(V, f)$.

The groups $\text{Aut}(V, f)$, certain normal subgroups of these groups, and their images under the projective map of section 13 are called the *classical groups*. Section 22 derives various properties of the classical groups. For example for suitable fields they are essentially generated by their transvections or reflections, and are essentially perfect. It will develop much later in section 41 that if G is a perfect finite classical group then the projective group PG is simple. Conversely the Classification Theorem for finite simple groups says that, by some measure, most of the finite simple groups are classical groups.

This chapter is one of the longest and most complicated in the book. Moreover the material covered here is in some sense specialized and tangential to much of the other material in this book. Still, as I've indicated, the classical groups and their representations on the associated spaces (V, Q) are very important, so the effort seems warranted.

19 Bilinear, sesquilinear, and quadratic forms

In this section V is an n-dimensional vector space over a field F and θ is an automorphism of F. A *sesquilinear form* on V with respect to θ is a map $f : V \times V \to F$ such that, for all $x, y, z \in V$ and all $a \in F$:

$$f(x + y, z) = f(x, z) + f(y, z) \qquad f(ax, y) = af(x, y)$$
$$f(x, y + z) = f(x, y) + f(x, z) \qquad f(x, ay) = a^{\theta} f(x, y).$$

The form f is said to be *bilinear* if $\theta = 1$. Usually I'll write (x, y) for $f(x, y)$.

I'll always assume θ is of order at most 2. There is little loss of generality in this assumption since we are interested in forms with big symmetry groups. Exercises 7.10 and 9.9 make this comment more precise.

f is *symmetric* if f is bilinear and $f(x, y) = f(y, x)$ for all x, y in V. f is *skew symmetric* if f is bilinear and $f(x, y) = -f(y, x)$ for all x, y in V. Finally f is *hermitian symmetric* if θ is an involution and $f(x, y) = f(y, x)^\theta$ for all x, y in V. I'll always assume that f has one of these three symmetry conditions. One consequence of this assumption is that

(*)   For all x, y in V, $f(x, y) = 0$ if and only if $f(y, x) = 0$.

On the other hand if (*) holds then Exercise 7.10 shows that f (essentially) satisfies one of the three symmetry conditions. Further if our form has a big group of automorphisms then Exercises 7.9 and 9.1 show we may as well take f to satisfy one of the conditions.

If $f(x, y) = 0$ I'll write $x \perp y$ and say that x and y are *orthogonal*. For $X \subseteq V$ define

$$X^\perp = \{v \in V : x \perp v \quad \text{for all } x \in X\}$$

and observe that $X^\perp$ is a subspace of V and that $X^\perp = \langle X \rangle^\perp$. Indeed

(19.1) For $x \in V$, $x^\perp = \ker(\alpha)$, where $\alpha \in \mathrm{Hom}_F(V, F)$ is defined by $y\alpha = (y, x)$. Hence $\dim(x^\perp) \geq n - 1$ with equality precisely when $x \notin V^\perp$.

$V^\perp$ is called the *radical* of V. Write $\mathrm{Rad}(V)$ for $V^\perp$. We say f is *nondegenerate* if $\mathrm{Rad}(V) = 0$.

The form f will be said to be *orthogonal* if f is nondegenerate and symmetric, and if in addition, when $\mathrm{char}(F) = 2$, $f(x, x) = 0$ for all x in V. The form f is said to be *symplectic* if f is nondegenerate and skew symmetric, and in addition when $\mathrm{char}(F) = 2$, $f(x, x) = 0$ for all x in V. Finally f is said to be *unitary* if f is nondegenerate and hermitian symmetric.

A few words to motivate these definitions. I've already indicated why the symmetry assumptions are appropriate. For any space V, $V = \mathrm{Rad}(V) \oplus U$ for some subspace U such that the restriction of f to U is nondegenerate. Thus there is little loss in assuming f to be nondegenerate. Besides, from Exercise 9.1, if (V, f) admits an irreducible group of automorphisms then (essentially) f is forced to be nondegenerate. Observe that if $\mathrm{char}(F) = 2$ then symmetry and skew symmetry are the same. Also, if $\mathrm{char}(F) \neq 2$ and f is skew symmetric, notice $f(x, x) = 0$ for all $x \in V$. This motivates the requirement that $f(x, x) = 0$ for all $x \in V$ when $\mathrm{char}(F) = 2$ and f is orthogonal or symplectic. Indeed Exercise 7.9.3 shows that this assumption leads to little loss of generality and is always satisfied if (V, f) admits an irreducible group of automorphisms.

(19.2) Let f be nondegenerate and $U \leq V$. Then $\dim(U^\perp) = \text{codim}(U)$.

Proof. The proof is by induction on $m = \dim(U)$. The lemma is trivial if $m = 0$, so take $m > 0$. Then as f is nondegenerate there exists $x \in V - U^\perp$. By 19.1, $W = U \cap x^\perp$ is a hyperplane of U. By induction on m, $\dim(W^\perp) = n - m + 1$. Let $u \in U - W$; then $U^\perp = W^\perp \cap u^\perp$. As $x \in W^\perp - u^\perp$, $U^\perp$ is a hyperplane of $W^\perp$ by 19.1. So $\dim(U^\perp) = \dim(W^\perp) - 1 = n - m$, completing the proof.

A vector $x \in V$ is *isotropic* if $f(x, x) = 0$. I've already observed that, if f is skew symmetric and $\text{char}(F)$ is not 2, then every vector is isotropic. Recall that this is part of the defining hypothesis of a symplectic or orthogonal form when $\text{char}(F) = 2$,

A subspace U of V is *totally isotropic* if the restriction of f to U is trivial, or equivalently if $U \leq U^\perp$. U is *nondegenerate* if the restriction of f to U is nondegenerate, or equivalently $U \cap U^\perp = \text{Rad}(U) = 0$.

(19.3) Let f be nondegenerate and U a subspace of V. Then
  (1) U is nondegenerate if and only if $V = U \oplus U^\perp$.
  (2) $(U^\perp)^\perp = U$.
  (3) If U is totally isotropic then each complement to U in $U^\perp$ is nondegenerate.
  (4) If U is totally isotropic then $\dim(U) \leq n/2$.

Proof. Parts (1), (2), and (4) are easy consequences of 19.2, while (2) implies (3).

Assume f is symmetric. A *quadratic form* on V associated to f is a map $Q: V \to F$ such that, for all $x, y \in V$ and $a \in F$, $Q(ax) = a^2 Q(x)$ and

$$Q(x + y) = Q(x) + Q(y) + f(x, y).$$

Observe that if $\text{char}(F) \neq 2$ this definition forces $Q(x) = f(x, x)/2$, so the quadratic form is uniquely determined by f, and hence adds no new information. On the other hand, if $\text{char}(F) = 2$ there are many quadratic forms associated to f. Observe also that f is uniquely determined by Q, since

$$f(x, y) = Q(x + y) - Q(x) - Q(y).$$

A *symplectic space* (V, f) is a pair consisting of a vector space V and a symplectic form f on V. A *unitary space* is a pair (V, f) with f a unitary form. An *orthogonal space* is a pair (V, Q) where Q is a quadratic form on V with associated bilinear form f.

In the remainder of this section assume (V, f) is a symplectic or unitary space, or Q is a quadratic form on V with associated orthogonal form f and (V, Q) is an orthogonal space. The *type* of V is symplectic, unitary, or orthogonal, respectively.

A vector v in V is *singular* if v is isotropic and also $Q(v) = 0$ when V is an orthogonal space. A subspace U of V is *totally singular* if U is totally isotropic and also each vector of U is singular. The *Witt index* of V is the maximum dimension of a totally singular subspace of V. Notice 19.3 says the Witt index of V is at most $n/2$.

An *isometry* of spaces (V, f) and (U, g) is a nonsingular linear transformation $\alpha: V \to U$ such that $g(x\alpha, y\alpha) = f(x, y)$ for all $x, y \in V$. A *similarity* is a nonsingular linear transformation $\alpha: V \to U$ such that $g(x\alpha, y\alpha) = \lambda(\alpha)f(x, y)$ for all $x, y \in V$ and some $\lambda(\alpha) \in F^{\#}$ independent of x and y. If (V, Q) and (U, P) are orthogonal I'll also require $P(x\alpha) = Q(x)$ or $P(x\alpha) = \lambda(\alpha)Q(x)$, in the respective case. Forms f and g (or P and Q) on V are said to be *equivalent* if (V, f) and (V, g) (or (V, Q) and (U, P)) are isometric. f and g (or P and Q) are *similar* if the corresponding spaces are similar. $O(V, f)$ (or $O(V, Q)$) denotes the group of isometries of the space, while $\Delta(V, f)$ (or $\Delta(V, Q)$) denotes the group of similarities. Evidently $O(V, f) \trianglelefteq \Delta(V, f)$.

Let $X = (x_i : 1 \le i \le n)$ be a basis of V. Define $J = J(X, f)$ to be the n by n matrix $J = (J_{ij})$ with $J_{ij} = f(x_i, x_j)$. Observe that J uniquely determines the form f.

Suppose $Y = (y_i : 1 \le i \le n)$ is a second basis for V, let $y_i = \sum_j a_{ij} x_j$, $a_{ij} \in F$, and $A = (a_{ij})$. Set $A^\theta = (a_{ij}^\theta)$ and let A^T be the transpose of A. Observe $A^{\theta T} = A^{T\theta}$ and $J(Y, f) = AJA^{T\theta}$. Further

(19.4) A form g on V is similar to f if and only if there exists a basis $Y = (y_i : 1 \le i \le n)$ of V with $J(Y, g) = \lambda J(X, f)$ for some $\lambda \in F^{\#}$. Equivalence holds precisely when λ can be chosen to be 1. If Q and P are quadratic forms associated to f and g, respectively, then Q is similar to P precisely when Y can be chosen so that $J(Y, g) = \lambda J(X, f)$ and $P(y_i) = \lambda Q(x_i)$ for each i, with $\lambda = 1$ in case of equivalence.

Proof. Let $\alpha: (V, f) \to (V, g)$ be a similarity and let $y_i = x_i \alpha$ and $Y = (y_i : 1 \le i \le n)$. Then $J(Y, g) = \lambda(\alpha)J(X, f)$. Conversely if $J(Y, g) = \lambda J(X, f)$ let $\alpha \in GL(V)$ with $x_i \alpha = y_i$. Then α is a similarity with $\lambda(\alpha) = \lambda$. Of course the same arguments extend to quadratic forms.

(19.5) A form g on V is similar to f if and only if $J(X, g) = \lambda A J(X, f) A^{T\theta}$ for some nonsingular matrix A and some $\lambda \in F^{\#}$, with $\lambda = 1$ in case of equivalence.

Proof. This follows from 19.4 and the remark immediately preceding it.

One can see from the preceding discussion that equivalence of forms corresponds to equivalence of the associated defining matrices of the forms.

Given $\alpha \in GL(V)$ define $M_X(\alpha)$ to be the n by n matrix (α_{ij}) defined by $x_i \alpha = \sum_j \alpha_{ij} x_j$.

(19.6) Let $\alpha \in GL(V)$. Then $\alpha \in \Delta(V, f)$ if and only if $(x_i \alpha, x_j \alpha) = \lambda(\alpha)$ (x_i, x_j) for all i and j, and some $\lambda(\alpha) \in F^{\#}$, with $\alpha \in O(V, f)$ precisely when $\lambda(\alpha) = 1$. If (V, Q) is orthogonal then $\alpha \in \Delta(V, Q)$ if and only if $\alpha \in \Delta(V, f)$ and $Q(x_i \alpha) = \lambda(\alpha) Q(x_i)$ for each i, with $\lambda(\alpha) = 1$ for equivalence.

(19.7) Let $\alpha \in GL(V)$, $A = M_X(\alpha)$, and $J = J(X, f)$. Then $\alpha \in O(V, f)$ if and only if $J = AJA^{T\theta}$. $\alpha \in \Delta(V, f)$ if and only if $\lambda J = AJA^{T\theta}$ for some $\lambda \in F^{\#}$.

(19.8) If V is not a symplectic space then V contains a nonsingular vector.

Proof. Assume otherwise. Let $x \in V^{\#}$ and $y \in V - x^{\perp}$. Then $1 = ((x, y)^{-1}x, y)$, so without loss $(x, y) = 1$. Now $(ax + by, ax + by) = ab^{\theta} + ba^{\theta}$ as x and y are singular and $(x, y) = 1$. If char$(F) \neq 2$ take $a = b = 1$ to get $ax + by$ nonsingular. If char$(F) = 2$ and V is unitary take $a = 1$ and $b \neq b^{\theta}$. Finally if V is orthogonal and char$(F) = 2$ then as x and y are singular, $Q(x) = Q(y) = 0$, so $Q(x + y) = (x, y) = 1$, and hence $x + y$ is nonsingular.

(19.9) Assume V is not symplectic, and if V is orthogonal assume char$(F) \neq 2$. Then there exists a basis $X = (x_i: 1 \leq i \leq n)$ of V such that the members of X are nonsingular and distinct members are orthogonal.

Proof. By 19.8 there is a nonisotropic vector $x_1 \in V$. By 19.3, $V = \langle x_1 \rangle \oplus (x_1)^{\perp}$ with $(x_1)^{\perp}$ nondegenerate. By induction on n there is a corresponding basis $(x_i: 1 < i \leq n)$ of $(x_1)^{\perp}$.

A basis like the one of 19.9 will be termed an *orthogonal basis*. An *orthonormal basis* for V is a basis X such that $J(X, f) = I$.

(19.10) If V is unitary then (x, x) is in the fixed field Fix(θ) of θ for each $x \in V$.

(19.11) Assume V is not symplectic, and if V is orthogonal assume char$(F) \neq 2$. Assume further that the fixed field Fix(θ) of θ satisfies Fix$(\theta) = \{aa^\theta : a \in F\}$. Then

(1) V possesses an orthonormal basis.

(2) All forms on V of each of the prescribed types are equivalent.

Proof. Notice (1) and 19.4 imply (2). To prove (1), choose an orthogonal basis X as in 19.9. Then by hypothesis and Lemma 19.10, $(x_i, x_i) = a_i^{-(1+\theta)}$ for some $a_i \in F^\#$. Now replacing x_i by $a_i x_i$, we obtain our orthonormal basis.

V is a *hyperbolic plane* if $n = 2$ and V possesses a basis $X = (x_1, x_2)$ such that x_1 and x_2 are singular and $(x_1, x_2) = 1$. Such a basis will be termed a *hyperbolic pair*.

(19.12) Let $x \in V^\#$ be singular and $y \in V - x^\perp$. Then $\langle x, y \rangle = U$ is a hyperbolic plane and x is contained in a hyperbolic pair of U.

Proof. Let $b = (x, y)^{-\theta}$. Then $(x, by) = 1$, so without loss $(x, y) = 1$. Observe U is nondegenerate, so if y is singular we are done. Thus we may assume each member of $U - \langle x \rangle$ is nonsingular, so in particular V is not symplectic. Thus, unless V is orthogonal and char$(F) = 2$, $0 \neq (ax + y, ax + y) = a + a^\theta + (y, y)$. However if char$(F) \neq 2$ we may take $a = -(y, y)/2$, and use 19.10 to obtain a contradiction.

Thus char$(F) = 2$. Suppose V is unitary. Let $d \in F - $ Fix(θ). Then $e = d + d^\theta \neq 0$. Let $c = (y, y)/e$ and $a = cd$. By 19.10, $c \in$ Fix(θ), so $a + a^\theta = ce = (y, y)$, and hence $ax + y$ is singular.

This leaves the case V orthogonal. Then choosing $a = Q(y)$, $ax + y$ is singular, completing the proof.

Here's an immediate corollary to 19.12 and 19.4:

(19.13) Let dim$(V) = 2$. If $V^\#$ possesses a singular vector, then V is a hyperbolic plane. In particular, up to equivalence, there is a unique nondegenerate form on V of each type possessing a nontrivial singular vector.

(19.14) Let U be a totally singular subspace of V, $R = (r_i : 1 \leq i \leq m)$ a basis for U, and W a complement to U in $U^\perp$. Then there exists $S = (s_i : 1 \leq i \leq m) \subseteq V$ such that r_i, s_i is a hyperbolic pair for the hyperbolic plane $U_i = \langle r_i, s_i \rangle$ and $W^\perp$ is the orthogonal direct sum of the planes $(U_i : 1 \leq i \leq m)$.

Proof. By 19.3 we may take $W = 0$. Thus $U = U^\perp$. Let $U_0 = \langle r_i : 1 < i \leq m \rangle$. By 19.2, U is a hyperplane of $(U_0)^\perp$. Then there exists a complement $U_1 = \langle r_1, s_1 \rangle$ to U_0 in $U_0^\perp$ and by 19.3, U_1 is nondegenerate. By 19.12 we may assume r_1, s_1 is a hyperbolic pair for U_1. By 19.3, $V = U_1 \oplus U_1^\perp$. Finally by induction on m we may choose $(s_i : 1 < i \leq m)$ in $(U_1)^\perp$ to satisfy the lemma.

Define V to be *hyperbolic* if V is the orthogonal direct sum of hyperbolic planes. A *hyperbolic basis* for a hyperbolic space V is a basis $X = (x_i : 1 \leq i \leq m)$ such that V is the orthogonal sum of the hyperbolic planes $\langle x_{2i-1}, x_{2i} \rangle$ with hyperbolic pair x_{2i-1}, x_i. We say V is *definite* if V possesses no nontrivial singular vectors.

As a consequence of 19.14 and 19.3 we have:

(19.15) Let U be a maximal hyperbolic subspace of V. Then $V = U \oplus U^\perp$ and $U^\perp$ is definite. Moreover every totally singular subspace of V of dimension m is contained in a hyperbolic subspace of dimension $2m$ and Witt index m.

(19.16) All symplectic spaces are hyperbolic. In particular all symplectic spaces are of even dimension and, up to equivalence, each space of even dimension admits a unique symplectic form.

Proof. This is immediate from 19.15, 19.4, and the fact that all vectors in a symplectic space are singular.

If $\text{char}(F) = 2$ and (V, Q) is orthogonal then (V, f) is symplectic, where f is the bilinear form determined by Q. Hence by 19.16:

(19.17) If V is orthogonal and $\text{char}(F) = 2$, then V is of even dimension.

20 Witt's Lemma

This section is devoted to a proof of Witt's Lemma. I feel Witt's Lemma is probably the most important result in the theory of spaces with forms. Here it is:

Witt's Lemma. Let V be an orthogonal, symplectic, or unitary space. Let U and W be subspaces of V and suppose $\alpha : U \to W$ is an isometry. Then α extends to an isometry of V.

Before proving Witt's Lemma let me interject an aside. Define an object X in a category $\mathscr{C}$ to possess the *Witt property* if, whenever Y and Z are subobjects of X and $\alpha: Y \to Z$ is an isomorphism, then α extends to an automorphism of X. Witt's Lemma says that orthogonal spaces, symplectic spaces, and unitary spaces have the Witt property in the category of spaces with forms and isometries. All objects in the category of sets and functions have the Witt property. But in most categories few objects have the Witt property; those that do are very well behaved indeed. If X is an object with the Witt property and G is its group of automorphisms, then the representation of G on X is usually an excellent tool for studying G.

Now to the proof of Witt's Lemma. Continue the hypothesis and notation of the previous section. The proof involves a number of steps. Assume the lemma is false and let V be a counterexample with n minimal.

(20.1) Let $H \leq U$ and suppose $\alpha|_H$ extends to an isometry β of V. Then $\gamma = \alpha\beta^{-1}: U \to W\beta^{-1}$ is an isometry with $\gamma|_H = 1$, and α extends to an isometry of V if and only if γ does.

(20.2) Assume $0 \neq H \leq U$ with H nondegenerate. Then
   (1) If $H^{\perp} \cong (H\alpha)^{\perp}$ then α extends to an isometry of V.
   (2) If $H\alpha = H$ then α extends to an isometry of V.

Proof. Notice (1) implies (2). As H is nondegenerate, so is $H\alpha$, and $V = H \oplus H^{\perp} = H\alpha \oplus (H\alpha)^{\perp}$. Let $\beta: H^{\perp} \to (H\alpha)^{\perp}$ be an isometry. By minimality of n, $(\alpha|_{U \cap H^{\perp}})\beta^{-1}$ extends to an isometry γ of $H^{\perp}$. Then $\gamma\beta: H^{\perp} \to (H\alpha)^{\perp}$ is an isometry extending $\alpha|_{U \cap H^{\perp}}$ and $\alpha|_H + \gamma\beta$ is an isometry of V extending α.

(20.3) If H is a totally singular subspace of Rad(U) and K a complement to H in Rad(U), then there exist subspaces U' and W' of V with $K = $ Rad(U') and $U \leq U'$ such that α extends to an isometry $\alpha: U' \to W'$. If $U = H^{\perp}$ then $U' = V$.

Proof. Let $(r_i: 1 \leq i \leq m)$ be a basis for H, X a complement to H in $H^{\perp}$ containing K, and X' a complement to $H\alpha$ in $(H\alpha)^{\perp}$ containing $(X \cap U)\alpha$. By 19.14 there is $(s_i: 1 \leq i \leq m)$ and $(s_i': 1 \leq i \leq m)$ such that $X^{\perp}$ and $(X')^{\perp}$ are the orthogonal sum of hyperbolic planes $\langle r_i, s_i \rangle$ and $\langle r_i\alpha, s_i' \rangle$, respectively. Extend α to

$$U' = \langle U, s_i: 1 \leq i \leq m \rangle$$

by defining $s_i\alpha = s_i'$.

(20.4) V is not symplectic.

Proof. By 20.3 we may assume U is nondegenerate. As $U \cong W$, $\dim(U) = \dim(W)$, so $\dim(U^{\perp}) = \dim(W^{\perp})$. Hence, by 19.16, $U^{\perp} \cong W^{\perp}$. Then 20.2 contradicts the choice of V as a counterexample.

(20.5) If there exists a totally singular subspace $0 \neq H = H\alpha$ of $\mathrm{Rad}(U)$ then α extends to V.

Proof. Let $L = H^{\perp}$ and $\bar{L} = L/H$. Then f (or Q) induces a form $\bar{f}$ of type f (or $\bar{Q}$) on $\bar{L}$ defined by $\bar{f}(\bar{x}, \bar{y}) = f(x, y)$ and the induced map $\bar{\alpha} : \bar{U} \to \bar{W}$ is an isometry, so, by minimality of n, $\bar{\alpha}$ extends to an isometry $\bar{\beta}$ of $\bar{L}$. Let X be a basis of L with $X \cap H$ and $X \cap U$ bases for H and U, respectively, and let $\beta \in \mathrm{GL}(L)$ be a map with $\beta|_U = \alpha$ and $\overline{x\beta} = \bar{x}\bar{\beta}$ for $x \in X - U$. By construction

$$(x, y) = (\bar{x}, \bar{y}) = (\bar{x}\bar{\beta}, \bar{y}\bar{\beta}) = (x\beta, y\beta)$$

for $x, y \in X$, so β is an isometry of L. Now, by 20.3, β, and hence also α, extends to an isometry of V.

(20.6) Assume H is a hyperplane of U with $\alpha|_H = 1$. Assume also that $H = 0$ if V is unitary or $\mathrm{char}(F) \neq 2$. Then α extends to an isometry of V.

Proof. Let $u \in U - H$ and set $K = U + W$. Assume α does not extend.

Suppose $U = W$. Then $\mathrm{Rad}(U) \neq 0$ by 20.2, and as α acts on $\mathrm{Rad}(U)$, $\mathrm{Rad}(U)$ is not totally singular by 20.5. Thus $\mathrm{char}(F) = 2$ and V is orthogonal. As α does not extend, $\alpha|_U \neq 1$, so $u \neq u\alpha$. Now $u\alpha = au + h$, for some $a \in F^{\#}$ and $h \in H$. As $\alpha|_H = 1$, α acts on $X = \langle u, h \rangle$. By 20.2, $\mathrm{Rad}(X) \neq 0$. Hence as each member of $V^{\#}$ is isotropic (because V is orthogonal and $\mathrm{char}(F) = 2$), X is totally isotropic. Hence as $Q(u) = Q(u\alpha)$, $z = u + u\alpha$ is singular. Therefore either X is totally singular or $\langle z \rangle$ is the unique singular point in X, and hence is α-invariant. By 20.2, H contains no nondegenerate subspaces, so $f|_H = 0$ by 19.12. Thus $h \in \mathrm{Rad}(U)$, so, by 20.5, h is nonsingular. So $\langle z \rangle = \langle z\alpha \rangle$ is singular and $z \notin H$. Hence we may assume $z = u$. Again by 20.5 there is $h' \in H - z^{\perp}$. Now α acts on $X' = \langle h', z \rangle$ and, as X' is nondegenerate, 20.2 supplies a contradiction.

So $U \neq W$. Let $c = 1$ if $(u, u\alpha) = 0$ and $c = (u, u\alpha)^{\theta}/(u, u\alpha)$ otherwise. Observe we can extend α to an isometry α' of K with $(u\alpha)\alpha' = cu$, by 19.6. So, by the first argument in the previous paragraph, $\mathrm{char}(F) = 2$ and V is orthogonal. By definition of α', α' fixes $z = u + u\alpha$. Now $H' = \langle H, z \rangle$ is a hyperplane of K with $\alpha'|_{H'} = 1$, so, by the previous paragraph, α', and hence also α, extends to an isometry of V.

(20.7) V is orthogonal and char$(F) = 2$.

Proof. Assume not. By 20.3, we may take U to be nondegenerate. By 19.8 and 20.4 there is a nonsingular point L in U. By 20.6 applied to L in the role of U, $\alpha|_L$ extends to an isometry of V. Then by 20.1 we may take $\alpha|_L = 1$. But now 20.2 supplies a contradiction.

We are now in a position to complete the proof of Witt's Lemma. Choose U of minimal dimension so that an isometry $\alpha: U \to W$ does not extend to V. Let H be a hyperplane of U. By minimality of U, $\alpha|_H$ extends to an isometry of V, so by 20.1 we may take $\alpha|_H = 1$. Now 20.6 supplies a contradiction and completes the proof.

I close this section with some corollaries to Witt's Lemma.

(20.8) (1) The isometry group of V is transitive on the maximal totally singular subspaces of V, and on the maximal hyperbolic subspaces of V.

(2) V is the orthogonal direct sum of a hyperbolic space H and a definite space. Moreover H is a maximal hyperbolic space and this decomposition is unique up to an isometry of V.

(3) The dimension of a maximal hyperbolic subspace of V is twice the Witt index of V.

Proof. These remarks are a consequence of Witt's Lemma and 19.15.

(20.9) (1) If K is a quadratic Galois extension of F and $N_F^K: K \to F$ is the norm of K over F, then (K, N_F^K) is a 2-dimensional definite orthogonal space over F.

(2) Every 2-dimensional definite orthogonal space over F is similar to a space (K, N_F^K) for some quadratic Galois extension K of F.

Proof. If K is a quadratic Galois extension of F then Gal$(K/F) = \langle \sigma \rangle$ is of order 2 and $N_F^K(a) = aa^\sigma$ for $a \in K$. It is straightforward to prove (K, N_F^K) is a definite orthogonal space.

Next a proof of (2). Let (V, Q) be a definite orthogonal space and $\{x, y\}$ a basis for V. If char$(F) \neq 2$ then by 19.9 we can choose $(x, y) = 0$, while if char$(F) = 2$ choose $(x, y) = 1$. Replacing Q by a scalar multiple if necessary, we can assume $Q(x) = 1$. Let $Q(y) = b$ and $P(t) = t^2 + t(x, y) + b$, so that P is a quadratic polynomial over F. As V is definite, P is irreducible. Let K be the splitting field for P over F and $c \in K$ a root of P. Then the map $x \mapsto 1, y \mapsto c$ induces an isometry of (V, Q) with (K, N_F^K).

(20.10) Assume F is algebraically closed. Then
    (1) If char$(F) \neq 2$ then, up to equivalence, V admits a unique nondegenerate quadratic form. Moreover V has an orthonormal basis with respect to that form.
    (2) If char$(F) = 2$ then V admits a nondegenerate quadratic form if and only if n is even. The form is determined up to equivalence and V is a hyperbolic space with respect to this form.

Proof. Part (1) follows from 19.11. To prove part (2) it suffices by 19.17 and 20.8 to take V an orthogonal space of dimension 2 and prove V is not definite. But as F is algebraically closed it possesses no quadratic extensions, so V is not definite by 20.9.

(20.11) If V is an orthogonal space of dimension at least 2 then V has a nondegenerate 2-dimensional subspace.

Proof. If char$(F) \neq 2$ this is a consequence of 19.9. If char$(F) = 2$ then by 19.16 the underlying symplectic space is hyperbolic and hence possesses a hyperbolic plane, which is a nondegenerate subspace of the orthogonal space V.

21 Spaces over finite fields

In this section the hypothesis and notation of section 19 continue; in particular V is an orthogonal, symplectic, or unitary space over F. In addition assume F is a finite field of characteristic p.

(21.1) Assume $n = 2$. Then up to equivalence there is a unique nondegenerate definite quadratic form Q on V. Further there is a basis $X = \{x, y\}$ of V such that:

(1) If p is odd then $(x, y) = 0$, $Q(x) = 1$, and $-Q(y)$ is a generator of $F^{\#}$.
(2) If $p = 2$ then $(x, y) = 1$, $Q(x) = 1$, $Q(y) = b$, and $P(t) = t^2 + t + b$ is an irreducible polynomial over F.

Proof. By 20.9 and its proof, Q is at least similar to such a form. It is then an easy exercise to prove forms similar to Q are even equivalent to Q. As F is finite, it has a unique quadratic extension, so Q is unique by 20.9.2.

Denote by $D = D_+$ and $Q = D_-$ the (isometry type of the) hyperbolic plane and the 2-dimensional definite orthogonal space over F, respectively. Write $D^m Q^k$ for the orthogonal direct sum of m copies of D with k copies of Q.

(21.2) Let F be a finite field. Then
    (1) D^m is a hyperbolic space of Witt index m.
    (2) $D^{m-1} Q$ is of Witt index $m - 1$.
    (3) D^{2m} is isometric to Q^{2m}.
    (4) Every $2m$-dimensional orthogonal space over F is isometric to exactly one of D^m or $D^{m-1} Q$.

Proof. By construction D^m is hyperbolic and, by 19.3.4, D^m is of Witt index m. By construction Q is of Witt index 0. Let $V \cong D^{m-1} Q$, $D^{m-1} \cong U \leq V$, and $Q \cong U^\perp = W$. Let X be a maximal totally singular subspace of U. As $\dim(X) = m - 1$, $X^\perp = X \oplus W$. For $w \in W^{\#}$, $Q(w) \neq 0$, so, for $x \in X$ $Q(x + w) = Q(w) \neq 0$. Thus X is a maximal totally singular subspace of $X^\perp$, so X is also a maximal totally singular subspace of V. Hence, by 20.8, $D^{m-1} Q$ is not isometric to D^m, as they have different Witt indices.

Let V be a $2m$-dimensional orthogonal space over F. By 20.11 V has a nondegenerate plane U. By 21.1 and induction on m, $U \cong D$ or Q and $U^\perp \cong D^{m-1}$ or $D^{m-2} Q$. So $V \cong D^m$, $D^{m-1} Q$, or $D^{m-2} Q^2$. Thus to complete the proof of 21.2 it remains to show $Q^2 \cong D^2$. This will follow from 20.8 if we can show Q^2 has a 2-dimensional totally singular subspace. So take $U \cong U^\perp \cong Q$. Let $\{x, y\}$ and $\{u, v\}$ be bases for U and $U^\perp$, respectively. If $\mathrm{char}(F) = 2$ then by 21.1 we may choose $(x, y) = (u, v) = 1$, $Q(x) = Q(u) = 1$, and $Q(y) = Q(v)$. Then $\langle x + u, y + v \rangle$ is a totally singular plane of V. So take $\mathrm{char}(F)$ to be odd. Then by 21.1 we may take x, y, u, and v to be orthogonal with $Q(u) = -Q(x)$ and $Q(v) = -Q(y)$. Then again $\langle x + u, y + v \rangle$ is a totally singular plane.

If F is finite and $n = 2m$ is even, then 21.2 says that, up to equivalence, there are exactly two quadratic forms on V, and that the corresponding orthogonal spaces have Witt index m and $m - 1$, respectively. Define the *sign* of these spaces to be $+1$ and -1, respectively, and write $\mathrm{sgn}(Q)$ or $\mathrm{sgn}(V)$ for the sign of the space. Thus the isometry type of an even dimensional orthogonal space over a finite field is determined by its sign. If V is an orthogonal space of odd dimension over F, then, by 19.17, $\mathrm{char}(F)$ is odd. Let's look at such spaces next.

(21.3) Let V be an orthogonal space of odd dimension over a finite field F. Then V possesses a hyperplane which is hyperbolic.

Proof. The proof is by induction on n. The remark is trivial if $n = 1$, so take $n \geq 3$. By 19.9, V possesses a nondegenerate subspace U of codimension 2. By induction on n, U possesses a hyperbolic hyperplane K. If $n > 3$ then, by induction on n, $K^\perp$ possesses a hyperbolic plane W. Then $K \oplus W$ is a hyperbolic hyperplane of V.

So $n = 3$. Choose a basis $X = (x_i : 1 \leq i \leq 3)$ for V as in 19.9. We may assume V is definite. Thus $\langle x_1, x_2 \rangle$ is definite, and hence possesses a vector y such that $Q(y)$ has the same quadratic character as $-Q(x_3)$. Thus there is $a \in F$ with $a^2 Q(y) = -Q(x_3)$. Now $ay + x_3$ is singular and 19.12 completes the proof.

If V is an odd dimensional orthogonal space over a finite field then, by 21.3, V possesses a hyperbolic hyperplane H, and, by 20.8, H is determined up to conjugacy under the isometry group of V. Let x be a generator of $H^\perp$ and define the *sign* of V (or Q) to be $+1$ if $Q(x)$ is a quadratic residue in F, and -1 if $Q(x)$ is not a quadratic residue. Then evidently when n is odd there are orthogonal spaces of sign $\varepsilon = +1$ and -1, and, by the uniqueness of H (up to conjugacy), these spaces are not isometric. On the other hand if c is a generator for the multiplicative group $F^\#$ of F, then cQ is similar to Q under the scalar transformation cI. Moreover (H, Q) is similar to (H, cQ), so (H, cQ) is also hyperbolic. Hence, as $(cQ)(x) = cQ(x)$ has different quadratic character from $Q(x)$, $\mathrm{sgn}(cQ) \neq \mathrm{sgn}(Q)$. Thus we have shown:

(21.4) Let F be a field of odd order, n an odd integer, and c a generator of the multiplicative group $F^\#$ of F. Then

(1) Up to equivalence there are exactly two nondegenerate quadratic forms Q and cQ on an n-dimensional vector space V over F.

(2) $\mathrm{sgn}(Q) = +1$ and $\mathrm{sgn}(cQ) = -1$.

(3) Q and cQ are similar via the scalar transformation cI, so $O(V, Q) = O(V, cQ)$.

(21.5) Let F be a finite field of square order. Then up to equivalence V admits a unique unitary form f. Further (V, f) possesses an orthonormal basis.

Proof. As F is of square order it possesses a unique automorphism θ of order 2. Moreover $\mathrm{Fix}(\theta) = \{aa^\theta : a \in F\}$, so 19.11 completes the proof.

The final lemma of this section summarizes some of the previous lemmas in this chapter, and provides a complete description of forms over finite fields.

(21.6) Let V be an n-dimensional space over a finite field F of order q and characteristic p. Then

(1) V admits a symplectic form f if and only if n is even, in which case f is unique up to equivalence and (V, f) is hyperbolic.

(2) V admits a unitary form f if and only if q is a square, in which case f is unique up to equivalence and (V, f) has a orthonormal basis.

(3) If n is even then V admits exactly two equivalence classes of nondegenerate quadratic forms. Two forms are equivalent precisely when they have the same sign. If P is such a form then (V, P) is isometric to $D^{n/2}$ or $D^{(n/2)-1}Q$ of sign $+1$ and -1, respectively.

(4) If n is odd then V admits a nondegenerate quadratic form precisely when p is odd, in which case there are two equivalence classes of forms. All forms are similar.

22 The classical groups

In section 22, continue to assume the hypothesis and notation of section 19. Section 22 considers the isometry groups $O(V, f)$ and $O(V, Q)$, certain normal subgroups of these groups, and the images of such groups under the projective map P of section 13. Notice that one can also regard the general linear group as the isometry group $O(V, f)$, where f is the trivial form $f(u, v) = 0$ for all $u, v \in V$. The groups G and PG, as G ranges over certain normal subgroups of $O(V, f)$, are called the *classical groups* (where f is trivial, orthogonal, symplectic, or unitary). We'll be particularly concerned with classical groups over finite fields.

Observe that if two spaces are isometric then their isometry groups are isomorphic. This is a special case of an observation made in section 2. As a matter of fact the isometry groups are isomorphic if the spaces are only similar, which is relevant because of 21.6.4. The upshot of these observations is that in discussing the classical groups we need only concern ourselves with forms up to similarity.

Recall, from 19.16 that if n is even there is, up to equivalence, a unique symplectic form f on V. Write $Sp(V)$ for the isometry group $O(V, f)$. $Sp(V)$ is the *symplectic group* on V. As V is determined by n and F, I'll also write $Sp_n(F)$ for $Sp(V)$. $Sp_n(F)$ is the *n-dimensional symplectic group* over F.

If f is unitary then $O(V, f)$ is called a *unitary group*. Similarly if Q is a nondegenerate quadratic form then $O(V, Q)$ is an *orthogonal group*. In general there are a number of similarity classes of forms on V and hence more than one unitary group or orthogonal group on V. Lemma 21.6 gives precise information when F is finite; we will consider that case in a moment. In any event I'll write $GU(V)$ or $O(V)$ for a unitary or orthogonal group on V, respectively, even though there may be more than one such group. GU(V) is (the) *general unitary group*. Write SU(V) and SO(V) for SL(V) ∩ GU(V) and SL(V) ∩ O(V), respectively (recall the special linear group SL(V) is defined and discussed in section 13). SU(V) and SO(V) are the *special unitary group* and *special orthogonal group*, respectively. Write $\Omega(V)$ for the commutator group of O(V).

Suppose for the moment that $F = GF(q)$ is the finite field of order q. Then write $\mathrm{Sp}_n(q)$ for $\mathrm{Sp}_n(F)$. Also, from 21.6, there is a unitary form on V precisely when $q = r^2$ is a square, in which case the form is unique, and I write $\mathrm{GU}_n(r)$ and $\mathrm{SU}_n(r)$ for GU(V) and SU(V). Notice $r \neq |F|$, rather $r = |F|^{1/2}$ in the unitary case. If n is odd there is an orthogonal form on V only when q is odd, in which case all such forms are similar and I write $\mathrm{O}_n(q)$, $\mathrm{SO}_n(q)$, and $\Omega_n(q)$ for O(V), SO(V), and $\Omega(V)$. Finally if n is even then up to equivalence there are just two nondegenerate quadratic forms Q_ε on V, distinguished by the sign $\mathrm{sgn}(Q_\varepsilon) = \varepsilon = +1$ or -1 of the form. Write $\mathrm{O}_n^\varepsilon(q)$, $\mathrm{SO}_n^\varepsilon(q)$, and $\Omega_n^\varepsilon(q)$ for the corresponding groups.

For each group G we can restrict the representation $P: \mathrm{GL}(V) \to \mathrm{PGL}(V)$ of GL(V) on the projective space PG(V) to G and obtain the image PG of G which is a group of automorphisms of the projective space PG(V). Thus for example we obtain the groups $\mathrm{PSp}_n(q)$, $\mathrm{PGU}_n(r)$, $\mathrm{PO}_n^\varepsilon(q)$, $\mathrm{P}\Omega_n^\varepsilon(q)$, etc. It will develop much later that the groups $\mathrm{PSp}_n(q)$, $\mathrm{PSU}_n(r)$, and $\mathrm{P}\Omega_n^\varepsilon(q)$ are simple unless n and q are small. In this section we prove these groups are (usually) perfect (i.e. each group is its own commutator group). This fact together with Exercise 7.8 is used in 43.11 to establish the simplicity of the groups.

Recall from section 13 that subspaces of V of dimension 1, 2, and $n - 1$ are called points, lines, and hyperplanes, respectively, and in general subspaces of V are objects of the projective space PG(V). If V has a form f or Q, then from section 19 we have a notion of totally singular and nondegenerate subspace, and hence totally singular and nondegenerate points, lines, and hyperplanes.

(22.1) If $g \in \mathrm{O}(V, f)$ then $C_V(g) = [V, g]^\perp$.

Proof. Let $U = C_V(g)$. For $u \in U$ and $v \in V$, $(u, v) = (ug, vg) = (u, vg)$. Thus as

$$v + u^\perp = \{x \in V : (x, u) = (v, u)\}$$

we have $vg \in v + u^\perp$. Hence

$$[v, g] \in \bigcap_{u \in U} u^\perp = U^\perp.$$

Therefore $[V, g] \leq U^\perp$. But, by Exercise 4.2.3 and 19.2, $\dim([V, g]) = \dim(U^\perp)$, so the proof is complete.

Recall the definition of a transvection in section 13. I'll prove the next two lemmas together.

(22.2) $O(V, f)$ (or $O(V, Q)$) contains a transvection if and only if each of the following holds:

(1) V possesses isotropic points.
(2) If (V, Q) is orthogonal then char$(F) = 2$.

(22.3) Let $G = O(V, f)$ (or $O(V, Q)$) and assume t is a transvection in G. Then

(1) $U = [V, t]$ is an isotropic point and $C_V(t) = U^\perp$.
(2)   Let Δ_U be the set of transvections with center $U = \langle u \rangle$ and let $R = R_U = \langle \Delta_U \rangle$ be the *root group* of t. Then $R^\# = \Delta_U$, for each $r \in R$ and $y \in V$ we have $yr = y + a_r(y, u)u$ for some $a_r \in F$, and one of the following holds:

(i) (V, f) is symplectic and the map $r \mapsto a_r$ is an isomorphism of R with the additive group of F.
(ii) (V, f) is unitary and $r \mapsto a_r e$ is an isomorphism of R with Fix(θ), where $e \in F^\#$ with $e^\theta = -e$.
(iii) (V, Q) is orthogonal, char$(F) = 2$, $R \cong Z_2$, U is nonsingular, and $a_t = Q(u)^{-1}$.

(3)   If (V, f) is symplectic or unitary then each singular point is the center of a transvection and G is transitive on the root groups of transvections. If (V, Q) is orthogonal each nonsingular point is the center of a unique transvection.
(4)   Assume either (V, f) is symplectic and $H = G$ or (V, f) is unitary and $H = SU(V)$. Then one of the following holds:

(i) $R \le H^{(1)}$.
(ii) $n = 2$ and $|\text{Fix}(\theta)| \le 3$.
(iii) $n = 4$, (V, f) is symplectic, and $|F| = 2$.

Now to the proof of 22.2 and 22.3. First 22.3.1 follows from 22.1 and the definition of a transvection. In particular if G possesses a transvection then V possesses isotropic points, so we may assume $U = \langle u \rangle$ is an isotropic point of V. By 19.12 there is an isotropic vector $x \in V - U^\perp$ with x, u a hyperbolic pair in the hyperbolic hyperplane $W = \langle u, x \rangle$. Let $X = \{u, x\} \subseteq Y$ be a basis for V with $Y - X$ a basis for $W^\perp$. Let $t = t_a$ be the transvection in GL(V) with $C_V(t) = U^\perp$ and $xt = x + au$, where a is some fixed member of $F^\#$. Then, by 19.6, $t \in G$ if and only if $(v_1 t, v_2 t) = (v_1, v_2)$ for all $v_1, v_2 \in X$, and if (V, Q) is orthogonal also $Q(vt) = Q(v)$ for all $v \in X$. By construction if suffices to check these equalities when $v_1 = x = v_2$, and, if (V, Q) is orthogonal, for $v = x$. The check reduces to a verification that:

(*) $a + \varepsilon a^\theta = 0$, and if (V, Q) is orthogonal also $a = -Q(u)^{-1} \neq 0$, where $\varepsilon = -1$ if (V, f) is symplectic, and $\varepsilon = +1$ otherwise.

If (V, f) is symplectic then (*) holds for each $a \in F^\#$, so $\Delta_U = \{t_a : a \in F^\#\}$. Also the map $a \mapsto t_a$ is an isomorphism of R with the additive group of F. If (V, f) is unitary then (*) has a solution if and only if there exists $e \in F^\#$ with $e^\theta = -e$, in which case a is a solution to (*) precisely when $a = be$ with $b \in \text{Fix}(\theta)$. Observe there is $c \in F$ with $c \neq c^\theta$ and, setting $e = c - c^\theta$, $e^\theta = -e$. Finally if (V, Q) is orthogonal then a is a solution to (*) if and only if $a = Q(u) \neq 0$ and $\text{char}(F) = 2$. Observe in each case $t_a : y \mapsto y + a(y, u)u$ for each $y \in X$, and hence also for each $y \in V$.

So 22.2 and the first two parts of 22.3 are established. The transitivity statement in 22.3.3 follows from Witt's Lemma, so it remains to establish 22.3.4. Assume the hypothesis of 22.3.4. Let L be the group generated by the transvections with centers in W. $W^\perp \leq C_V(L)$, so L is faithful on W and hence $L \leq O(W, f)$. Now, by Exercise 7.1 and 13.7, $L \cong \text{SL}_2(\text{Fix}(\theta))$. Then, by 13.6.4, either $R \leq L^{(1)} \leq H^{(1)}$, or $|\text{Fix}(\theta)| \leq 3$, and we may assume the latter with $n > 2$.

If (V, f) is unitary let v be a nonsingular vector in $W^\perp$ and $Z = \langle W, v \rangle$, while if (V, f) is symplectic let Z be a nondegenerate subspace containing W of dimension 4 or 6, for $|F| = 3$ or 2, respectively. Let $K = C_H(Z^\perp)$, so that $K = \text{Sp}(Z)$ or $\text{SU}(Z)$. If $R \leq K^{(1)}$ then the proof is complete, so without loss $V = Z$. But now Exercise 7.1 completes the proof.

(22.4) Assume either (V, f) is symplectic and $G = O(V, f) = \text{Sp}(V)$, or (V, f) is unitary of dimension at least 2, $\text{Fix}(\theta) = \{aa^\theta : a \in F\}$, and $G = \text{SL}(V) \cap O(V, f) = \text{SU}(V)$. Then either

(1) G is generated by the transvections in $O(V, f)$, or
(2) (V, f) is unitary, $|F| = 4$, and $n = 3$.

Proof. If (V, f) is symplectic let $\Gamma = V^\#$ and Ω the set of hyperbolic bases of V. If (V, f) is unitary let

$$\Gamma = \{v \in V : (v, v) = 1\}.$$

Let T be the group generated by the transvections in $O(V, f)$. I'll show:

(i) T is transitive on Γ unless 22.4.2 holds, and
(ii) T is transitive on Ω if (V, f) is symplectic.

Observe that the stabilizer in G of $w \in \Omega$ is trivial, so (ii) implies $T = G$. Observe also that by Exercise 7.1 the lemma holds if $n = 2$, so we may assume $n > 2$.

Suppose (i) holds. If (V, f) is unitary and $x \in \Gamma$ then $G_x = \mathrm{SU}(x^\perp)$, and, as $n > 2$, induction on n implies $G_x \leq T$ unless $n = 4$ and $|F| = 4$, where Exercise 7.3 says the same thing. Hence, by (i) and 5.20, $G = T$. So take (V, f) symplectic and let $X = (x_i : 1 \leq i \leq n)$ and $Y = (y_i : 1 \leq i \leq n)$ be members of Ω. We need to prove $Y = Xs$ for some $s \in T$. By (i) we may take $x_1 = y_1$. Claim $y_2 \in x_2(T_{x_1})$. As $(x_1, x_2) = (x_1, y_2) = 1$, $y_2 = x_2 + v$ for some $v \in (x_1)^\perp$. If $x_2 \notin v^\perp$ then there is a transvection t with center $\langle v \rangle$ and $x_2 t = y_2$, and as $v \in (x_1)^\perp$, $t \in T_{x_1}$. If $x_2 \in v^\perp$ the same argument shows x_2 and y_2 are conjugate to $x_1 + x_2$ in T_{x_1}, and hence also to each other. So the claim holds and hence we may take $x_2 = y_2$. But now X is conjugate to Y in $\mathrm{Sp}(\langle x_1, x_2 \rangle^\perp) \cap T$ by induction on n, so the lemma holds.

So it remains to prove (i). Let $x, y \in \Gamma$ and $U = \langle x, y \rangle$. If U is nondegenerate then $x \in y(T \cap O(U, f))$ since (i) holds when $n = 2$. Thus we may take $0 \neq u \in \mathrm{Rad}(U)$ and it suffices to show there exists a point $\langle z \rangle$ (such that z is nonsingular if f is unitary) with $\langle x, z \rangle$ and $\langle y, z \rangle$ nondegenerate. If $n > 3$ just choose $z \in u^\perp$ with $\langle x, z \rangle$ nondegenerate. Thus we can assume $n = 3$, so that (V, f) is unitary. Further we may assume $|F| > 4$. Let u, v be a hyperbolic basis for $y^\perp$. We may take $x = y + u$. Then $x^\perp = \langle u, y - v \rangle$ and if $z = au + y - v$, then $\langle y, z \rangle \cap y^\perp = \langle au - v \rangle$. It suffices to choose z so that z and $au - v$ are nonsingular. Equivalently, if $T = T^F_{\mathrm{Fix}(\theta)}$ is the trace of F over $\mathrm{Fix}(\theta)$, then $T(a) \neq 0$ or 1. Hence as $|F| > 4$ and $T : F \to \mathrm{Fix}(\theta)$ is surjective, we can choose z as desired.

A field F is *perfect* if $\mathrm{char}(F) = 0$ or $\mathrm{char}(F) = p > 0$ and the p-power map is a surjection from F onto F (i.e. $F = F^p$). For example finite fields and algebraically closed fields are perfect.

(22.5) Let F be a perfect field of characteristic 2 and (V, Q) orthogonal of dimension at least 4. Let $\langle u \rangle$ be a nonsingular point of V, t the transvection with center $\langle u \rangle$, and $G = O(V, Q)$. Then

(1) $G_{\langle u \rangle} = C_G(t)$ is represented as $\mathrm{Sp}(u^\perp / \langle u \rangle)$ on $u^\perp / \langle u \rangle$, with $\langle t \rangle$ the kernel of this representation.

(2) $C_G(t)$ is transitive on the nonsingular points in $u^\perp$ distinct from $\langle u \rangle$.

(3) If $|F| > 2$ then $u^\perp = [u^\perp, C_G(t)]$.

(4) Either G is generated by, and is transitive on, its transvections, or $n = 4$, $|F| = 2$, and $\mathrm{sgn}(Q) = +1$.

Proof. Let $U = \langle u \rangle$ and $H = G_U$. As t is the unique transvection with center U, $H = C_G(t)$ and t is the kernel of the representation π of H on $U^\perp/U = \bar{M}$. Observe that, if f is the bilinear form on V defined by Q, then $(\bar{M}, \bar{f})$ is a symplectic space, where $\bar{f}(\bar{x}, \bar{y}) = f(x, y)$. Let $\bar{x} \in \bar{M}^\#$ and $W = \langle x, u \rangle$. As F is perfect we can choose u with $Q(u) = 1$, and W contains a unique singular point $\langle w \rangle$. For $a \in F^\#$, let t_a be the transvection with center $\langle aw + u \rangle$. Then $t_a \pi$ is a transvection on $\bar{M}$ with center $\bar{W}$ and, for $y \in M$, $\bar{y}(t_a \pi) = \bar{y} + a^2(\bar{y}, \bar{w})\bar{w}$ by 22.3.2. Thus, as $F = F^2$, $R\pi = \langle t_a \pi : a \in F^\# \rangle$ is the full root group of $t_a \pi$ in $\mathrm{Sp}(\bar{M})$ by 22.3.2. So, by 22.4, $H\pi = \mathrm{Sp}(\bar{M})$. Therefore (1) is established. Also $[y, t_a] = a(y, w)(aw + u)$ by 22.3.2, so, if $|F| > 2$, then, choosing $y \notin w^\perp$, we have $u \in [y, R]$, so, as $\bar{M} = [\bar{M}, H\pi]$, (3) holds. To prove (2) observe that $H\pi$ is transitive on $\bar{M}^\#$ by (1), so it suffices to show $N_H(W)$ is transitive on the set Γ of points of W distinct from $\langle u \rangle$ and $\langle w \rangle$. Let $\bar{w}, \bar{v}$ be a hyperbolic pair in $\bar{M}$. By (1), for $a \in F^\#$ there is $g_a \in H$ with $wg_a = aw$ and $vg_a = a^{-1}v$. Hence $\langle g_a : a \in F^\# \rangle$ is transitive on Γ.

It remains to prove (4). Let T be the group generated by the transvections in G. We've seen that $H \le T$, so to prove $T = G$ it suffices by 5.20 to show T is transitive on the set Δ of nonsingular points of V. This will also show G is transitive on its transvections. Let Z be a second nonsingular point. We must show $Z \in UT$. This follows from (2) if $(Z + U)^\perp$ contains a nonsingular point distinct from U and Z, so assume otherwise. Then $U + Z = (U + Z)^\perp$ and U and Z are the only nonsingular points in $U + Z$. This forces $n = 4$ and $|F| = 2$. Let $A = \mathrm{Rad}(U + Z)$. If B is a nonsingular point in $A^\perp - (U + Z)$ then $U + B$ and $Z + B$ are nondegenerate, so $U, Z \in BT$ and hence $Z \in UT$. Thus no such point exists, which forces $\mathrm{sgn}(Q) = +1$.

If $\mathrm{char}(F) \ne 2$ and (V, Q) is orthogonal, then a *reflection* on V is an element r in $O(V, Q)$ such that $[V, r]$ is a point of V. $[V, r]$ is called the *center* of r.

(22.6) Let $\mathrm{char}(F) \ne 2$ and (V, Q) orthogonal. Then
  (1) If r is a reflection on V then r is an involution, $[V, r]$ is nonsingular, and $C_V(r) = [V, r]^\perp$.
  (2) If $U = \langle u \rangle$ is a nonsingular point of V then there exists a unique reflection r_u on V with center U. Indeed $xr_u = x - (x, u)u/Q(u)$ for each $x \in V$.

Proof. Let r be a reflection on V. Then $[V, r] = \langle v \rangle$ is a point. By 22.1, $C_V(r) = v^\perp$, so $v^\perp$ is a hyperplane by 19.2. By 22.2, r is not a transvection, so $v \notin v^\perp$ and hence v is nonsingular. Next $vr = av$ for some $1 \ne a \in F^\#$ and $Q(v) = Q(vr) = Q(av) = a^2 Q(v)$, so $a = -1$. Hence r is an involution.

Conversely let $U = \langle u \rangle$ be a nonsingular point. By (1) there is at most one reflection with center U, while a straightforward calculation shows the map listed in (2) is such a reflection.

The proof of the following lemma comes essentially from I.5.1 on page 19 of Chevalley [Ch 2].

(22.7) Let (V, Q) be an orthogonal space. Then either

(1) $O(V, Q)$ is generated by its transvections or reflections, or
(2) $|F| = 2, n = 4$, and $sgn(Q) = +1$.

Proof. If $n = 2$ or $|F| = 2$ the result follows from Exercise 7.2 or 22.5.4, respectively. If $n = 1$, then $char(F) \neq 2$ and $O(V, Q)$ is generated by the reflection $-I$. Thus we may take $n > 2$ and $|F| > 2$.

Let T be the group generated by all transvections or reflections in $G = O(V, Q)$ and suppose $h \in G - T$. Pick h so that $dim(C_V(h))$ is maximal in the coset hT.

Suppose $y \in V$ with $z = [y, h]$ nonsingular. $yh = y+z$ and $Q(yh) = Q(y)$, so $Q(z) + (y, z) = 0$. In particular $y \notin z^{\perp}$ and $y+z = y - (y, z)z/Q(z) = yr_z$, where r_z is the transvection or reflection with center $\langle z \rangle$. Thus $yh = yr_z$, so $y \in C_V(hr_z)$. By 22.1, $C_V(h) = [V, h]^{\perp} \subseteq z^{\perp} = C_V(r_z)$, so $C_V(h) \leq C_V(hr_z)$. Hence $dim(C_V(hr_z)) \geq dim(\langle C_V(h), y \rangle) > dim(C_V(h))$, contrary to the choice of h.

Therefore $[V, h]$ is totally singular. I claim next that T is transitive on the maximal totally singular subspaces of V. Assume not and pick two such spaces M and N such that $M \notin NT$ and, subject to this constraint, with $dim(M \cap N)$ maximal. Then $M \neq N$ so $M < M + N$, and hence by maximality of M there is a nonsingular vector $x = m + n \in M + N$. As x is nonsingular, $(m, n) \neq 0$. Also $Q(x) = (m, n)$, so $mr_x = -n \in N$, while $M \cap N \leq x^{\perp} = C_V(r_x)$. Thus $M \cap N < \langle M \cap N, n \rangle \leq Mr_x \cap N$, and then, by maximality of $M \cap N$, $Mr_x \in NR$. But now $M \in NR$, contrary to the choice of M and N.

So the claim is established. Next there is a maximal totally singular subspace M with $[V, h] \leq M$. Then $M^{\perp} \leq [V, h]^{\perp} = C_V(h)$. Let $H = C_G(M^{\perp}) \cap C_G(V/M^{\perp})$, so that $h \in H$. As T is transitive on maximal totally singular subspaces, it follows that $G = N_G(M)T$ and each member of G has a T-coset representative in H^g for some $g \in G$. Then as $H \trianglelefteq N_G(M)$, $HT \trianglelefteq G$, so each member of G has a T-coset representative in HT, and hence $G = HT$.

By Exercise 4.8, H is abelian, so G/T is abelian. Thus $[h, g] \in T$ for each $g \in G$, so $[h, g] \notin h^G$. If h acts on a proper nondegenerate subspace U of

V then $h = h_1 h_2$ with $h_1 \in O(U, Q)$ and $h_2 \in O(U^\perp, Q)$. By induction on n, $h_i \in T$, so $h \in T$. Hence h acts on no such subspace. In particular $C_V(h)$ is totally isotropic, so, as $[V, h]$ is totally singular, 22.1 and 19.3.2 imply $C_V(h) = [V, h] = M$, and V is hyperbolic. Further, for $v \in V$, $[v, h] \in v^\perp$, since $\langle v, [v, h] \rangle$ is not nondegenerate.

By 19.14 there is a totally singular subspace N of V with $V = M \oplus N$. Let $x_1 \in M^\#$. As $M = [V, h] = [N, h]$, $\phi: n \mapsto [n, h]$ is a vector space isomorphism of N and M. Hence there is $y_2 \in N$ with $x_1 = [y_2, h]$. By the last remark of the previous paragraph, $(x_1, y_2) = 0$. As ϕ is an isomorphism, $(N - x_1^\perp)\phi \not\subseteq y_2^\perp$, so there is $y_1 \in N - x_1^\perp$ with $\phi(y_1) = x_2 \notin y_2^\perp$. Let $X = (x_1, x_2, y_1, y_2)$. Then $V_1 = \langle X \rangle$ is nondegenerate and h-invariant, so $V = V_1$.

For $a \in F^\#$ define $ah \in GL(V)$ by $v(ah) = v + a[v, h]$ for each $v \in V$. Notice $Q(v(ah)) = Q(v)$ as $[v, h] \in v^\perp \cap M$. Therefore $ah \in H$. Indeed setting $X_a = (ax_1, x_2, a^{-1}y_1, y_2)$ we have $M_X(h) = M_{X_a}(ah)$ and $J(X, f) = J(X_a, f)$. So the element $g \in GL(V)$ with $Xg = X_a$ is in G by 19.6 and $h^g = ah$. Now, as $|F| > 2$, we can choose a with $a - 1 \neq 0$. Then $[h, g] = h^{-1}h^g = ((-1)h)(ah) = (a - 1)h \in h^G$, contrary to an earlier remark. The proof is complete.

Let (V, Q) be an orthogonal space. We next construct the *Clifford algebra* $C = C(Q)$ of (V, Q). The treatment here will be abbreviated. For a more complete discussion see chapter 2 in Chevalley [Ch 1]. C is the tensor algebra (cf. Lang [La], chapter 16, section 5) of V, modulo the relations $x \otimes x - Q(x)1 = 0$, for $x \in V$. For our purposes it will suffice to know the following:

(22.8) Let (V, Q) be an orthogonal space with ordered basis X. Choose X to be orthogonal if $\mathrm{char}(F) \neq 2$ and choose X to be a hyperbolic basis of the underlying symplectic space (V, f) if $\mathrm{char}(F) = 2$. Let $C = C(Q)$ be the Clifford algebra of (V, Q). Then C is an F-algebra with the following properties:

(1) There is an injective F-linear map $\rho: V \to C$ such that C is generated as an F-algebra by $V\rho$. Write e_x for $x\rho$ if $x \in X$.
(2) For $S = \{x_1, \dots, x_m\} \subseteq X$ with $x_1 < \cdots < x_m$ write $e_S = e_{x_1} \dots e_{x_m}$. Then $(e_S: S \subseteq X)$ is a basis for C over F. In particular $e_\varnothing = 1 = 1_c$ and $\dim_F(C) = 2^n$.
(3) For $u, v \in V$, $(u\rho)^2 = Q(u) \cdot 1$ and $u\rho v\rho + v\rho u\rho = (u, v) \cdot 1$.
(4) Let C_i be the subspace of C spanned by the vectors e_S, $S \subseteq X$, $|S| \equiv i \bmod 2$, $i = 0, 1$. Then $\{C_0, C_1\}$ is a *grading* of C. That is $C = C_0 \oplus C_1$ and $C_i C_j \subseteq C_{i+j}$, for $i, j \in \{0, 1\}$, where $i + j$ is read mod 2.

(5) If u is a nonsingular vector of V then $u\rho$ is a unit in C with inverse $Q(u)^{-1}u\rho$, and, for $v \in V$, $(v\rho)^{u\rho} = -(vr_u)\rho$, where r_u is the reflection or transvection in $O(V, Q)$ with center $\langle u \rangle$.

(6) The *Clifford group* G of C is the subgroup of units in C which permute $V\rho$ via conjugation. The representation π of G on V (subject to the identification of V with $V\rho$ via ρ) is a representation of G on (V, Q) with $G\pi = O(V, Q)$ if n is even and $G\pi = SO(V, Q)$ if n is odd. If u is a nonsingular vector in V then $u\rho\pi = -r_u$.

(7) $\ker(\pi)$ is the set of units in $Z(C)$ and $Z(C) = F1$ or $F1 + Fe_X$ for n even or odd, respectively. If n is odd, no unit in C induces $-I$ on C by conjugation.

(8) There is an involutory algebra antiisomorphism t of C such that $e_S t = e_{x_m} \ldots e_{x_1}$ for each $S = \{x_1, \ldots, x_m\} \subseteq X$.

Proof. I'll sketch a proof. If $\mathrm{char}(F) \neq 2$ the multilinear algebra can be avoided as in chapter 5, section 4 of Artin [Ar]. A full treatment can be found in Chevalley [Ch 1].

By definition, $C = T/K$ where $T = \bigoplus_{i=0}^{\infty} T_i(V)$ is the tensor algebra and $K = \langle x \otimes x - Q(u)1 : x \in V \rangle$. In particular $T_0(V) = F1$ and there is a natural isomorphism $\rho_0 : V \to T_1(V)$. Then $\rho : V \to C$ is the map $v \mapsto v\rho_0 + K$ induced by ρ_0, and (1) will follow from (2), once that part is established. $e_S = e_{x_1} \ldots e_{x_m} = x_1 \otimes \cdots \otimes x_m + K$, so C_i is the image of $\bigoplus_{j \equiv i} T_j(V)$ in C. Hence (4) follows from (2) and the definition of multiplication in T. The universal property of T implies there is an involutory antiisomorphism t_0 of T with $(x_1 \otimes \cdots \otimes x_m)t_0 = x_m \otimes \cdots \otimes x_1$. As t_0 preserves K it induces t on C. Thus (8) holds. Part (3) is a direct consequence of the definition of C, since $x \otimes x - Q(u)1 \in K$. An easy induction argument using (3) shows $e_S e_T$ is a linear combination of the elements e_R, $R \subseteq X$, for each $S, T \subseteq X$, so $\langle e_R : R \subseteq X \rangle$ spans C. Using the universal property of the tensor algebra, Chevalley shows on page 39 of [Ch 1] that there is a homomorphic image of C of dimension 2^n, completing the proof of (2), and hence also of (1) and (4). I omit this demonstration.

Part (5) is a straightforward consequence of (3). If $\mathrm{char}(F) \neq 2$ then (3) shows $e_S^{e_x} = (-1)^m e_S$ for $x \in X$ and $S \subseteq X$, where $m = |S|$ if $x \notin S$ and $m = |S| - 1$ if $x \in S$. Since $X\rho$ generates C as an F-algebra, (7) follows in this case. If $\mathrm{char}(F) = 2$ then choose X so that each of its members is nonsingular. Then (3) shows $[e_x, e_S] = 0$ if $S \subseteq x^{\perp}$, while $[e_x, e_S] = Q(x)^{-1}e_{S+y}$ if S contains the unique y in $X - x^{\perp}$, where $S + y$ is the symmetric difference of S with $\{y\}$. It follows that $[e_x, C] = \langle e_S : S \subseteq x^{\perp} \rangle$ is of dimension 2^{n-1}. So, as $2^{n-1} = \dim(C)/2$ and e_x is an involution, $[e_x, C] = C_C(e_x)$. Thus (7) holds in this case too.

Let G and π be as in (6). For $g \in G$ and $v \in V$, $Q(vg\pi)1 = ((v\rho)^g)^2 = ((v\rho)^2)^g = Q(v)1$, so $Q(vg\pi) = Q(v)$. Hence $G\pi \leq O(V, Q)$. Let G_0 be the subgroup of G generated by the elements $u\rho$ as u varies over the nonsingular vectors of V. By (5), $u\rho\pi = -r_u$, so, by 22.7, $G\pi = G_0\pi = O(V, Q)$ if char$(F) = 2$. Further if char$(F) \neq 2$ and n is even then $-I$ is a product of elements $-r_u$, so $r_u \in G_0\pi$ and, again by 22.7, $G\pi = G_0\pi = O(V, Q)$. Finally if n is odd then $\det(-r_u) = +1$, so $G_0\pi \leq SO(V, Q)$, and then, as $O(V, Q) = \langle -I \rangle \times SO(V, Q)$, 22.7 says $G_0\pi = SO(V, Q)$. Then to complete the proof of (6) it remains only to observe that, by (7), $-I \notin G\pi$, so $G_0\pi = G\pi$.

(22.9) Let (V, Q) be an orthogonal space, $C = C(Q)$ its Clifford algebra, G the Clifford group of (V, Q), and $G^+ = G \cap C_0$ the *special Clifford group*. Let π be the representation of 22.8.6. Then
  (1) G^+ is a subgroup of G of index 2.
  (2) If char$(F) \neq 2$ then $G^+\pi = SO(V, Q)$.
  (3) If char$(F) = 2$ then $G^+\pi$ is of index 2 in $O(V, Q)$.
  (4) $G^+\pi$ contains no transvections or reflections.

Proof. Part (1) is a consequence of 22.8.4. If n is even then, by 22.8.7, $\ker(\pi) \leq G^+$ and, by 22.8.6, $G\pi = O(V, Q)$, so $G^+\pi$ is of index 2 in $O(V, Q)$. Also $-I \in G^+\pi$ by the proof of 22.8, so by 22.8.6 each transvection or reflection r_u is not in $G^+\pi$. Thus the lemma holds in this case as reflections are not in $SO(V, Q)$. If n is odd then $G\pi = SO(V, Q)$ by 22.8.6, while, by 22.8.7, $G = G^+\ker(\pi)$. So again the lemma holds.

(22.10) Let $v_1, \ldots, v_m$ be nonsingular vectors in the orthogonal space (V, Q) such that $r_{v_1} \ldots r_{v_m} = 1$. Then the product $Q(v_1) \ldots Q(v_m)$ is a square in F.

Proof. Let $c = v_1\rho \ldots v_m\rho$. As $r_{v_1} \ldots r_{v_m} = 1$, m is even, because by 22.9 there is a subgroup of $O(V, Q)$ of index 2 containing no transvection or reflection. Hence $c\pi = (-1)^m r_{v_1} \ldots r_{v_m} = 1$ by 22.8.6. So $c \in \ker(\pi)$. Indeed, as m is even, $c \in G^+ \cap \ker(\pi) = F^\# \cdot 1$ by 22.8.7. So $c = a \cdot 1$ for some $a \in F^\#$.

Let t be the antiisomorphism of 22.8.8. It follows that $c(ct) = c^2 = a^2 \cdot 1$. On the other hand $c(ct) = v_1\rho \ldots v_m\rho v_m\rho \ldots v_1\rho$ as t is an antiisomorphism. Further $(v_i\rho)^2 = Q(v_i)1$, so $c(ct) = (Q(v_1) \ldots Q(v_m))1$, completing the proof.

Let $F^2 = \{a^2 : a \in F^\#\}$ be the subgroup of squares in $F^\#$, and consider the factor group $F^\#/F^2$. For example if F is a finite field of odd order then $F^\#/F^2$ is of order 2, while if F is perfect of characteristic 2 then $F^\# = F^2$. Define a map $\theta : O(V, Q) \to F^\#/F^2$ as follows. For $g \in O(V, Q)$, $g = r_{x_1} \ldots r_{x_m}$ for suitable transvections or reflections r_{x_i} with center $\langle x_i \rangle$. (Except in the exceptional case of 22.7, which I'll ignore.) Define $\theta(g) = Q(x_1) \ldots Q(x_m)F^2$. Observe first that $Q(ax_i) = a^2 Q(x_i) \in Q(x_i)F^2$, so the definition of θ is independent of the choice of generator x_i of $\langle x_i \rangle$. Also if $g = r_{y_1} \ldots r_{y_k}$ then

$1 = r_{x_1} \ldots r_{x_m} r_{y_k} \ldots r_{y_1}$, so, by 22.10, $Q(x_1) \ldots Q(x_m)F^2 = Q(y_1) \ldots Q(y_k)F^2$. Thus θ is independent of the choice of transvections and reflections too. θ is called the *spinor norm* of $O(V, Q)$. From the preceding discussion it is evident that.

(22.11) The spinor norm θ is a group homomorphism of $O(V, Q)$ into $F^\#/F^2$.

(22.12) If Q is not definite then the spinor norm maps $G^+\pi$ surjectively onto $F^\#/F^2$.

Proof. V contains a hyperbolic plane U and for each $a \in F^\#$ there exist u, $v \in U$ with $Q(v) = 1$ and $Q(v) = a$. Now $\theta(r_v r_u) = a$.

(22.13) Let (V, Q) be a hyperbolic orthogonal space, let $\Gamma = \Gamma(V)$ be the set of maximal totally singular subspaces of V, and define a relation $\sim$ on Γ by $A \sim B$ if $\dim(A/(A \cap B))$ is even. Then $\sim$ is an equivalence relation with exactly two equivalence classes.

Proof. Given a triple A, B, C of members of Γ define

$$\delta(A, B, C) = \dim(A/(A \cap B)) + \dim(A/(A \cap C)) + \dim(B/(B \cap C)).$$

Observe that the lemma is equivalent to the assertion that $\delta(A, B, C)$ is even for each triple A, B, C from Γ.

Assume the lemma is false and pick a counterexample V with n minimal. As V is hyperbolic, $n = 2m$ is even. If $m = 1$ then $|\Gamma| = 2$, so the result holds. Thus $n > 1$. Let $A, B, C \in \Gamma$ with $\delta(A, B, C)$ odd.

Let $D = A \cap B \cap C$ and suppose $D \neq 0$. Let $U = D^\perp$ and $\bar{U} = U/D$. As we have seen several times already, Q induces a quadratic form $\bar{Q}$ on $\bar{U}$. Further $\bar{U}$ is hyperbolic with $\bar{A}, \bar{B}, \bar{C} \in \Gamma(\bar{U})$, so, by minimality of V, $\delta(\bar{A}, \bar{B}, \bar{C})$ is even. As $\delta(A, B, C) = \delta(\bar{A}, \bar{B}, \bar{C})$ we have a contradiction to the choice of A, B, C. Hence $D = 0$.

Suppose next $E = A \cap B \neq 0$ and let $C_0 = (C \cap E^\perp) + E$. By 19.2, $C_0 \in \Gamma$. $0 \neq E = A \cap B \cap C_0$, so, by a previous case, $\delta(A, B, C_0)$ is even. But $X \cap C_0 = (X \cap C) + E$ and $X \cap C \cap E = 0$ for $X = A$ and B, so $\delta(A, B, C) \equiv \delta(A, B, C_0) \bmod 2$, again a contradiction.

Thus $\delta(A, B, C) = 3m$. Hence m is odd. Therefore if $T, S \in \Gamma$ with $T \sim A \sim S$ then $A \cap T \neq 0$, so, by the last case, $T \sim S$. Hence $\sim$ is an equivalence relation. Finally let $R \in \Gamma$ with $A \cap R$ a hyperplane of A. Then $0 \neq A \cap R$, so, by the last case, $\delta(A, B, R)$ is even, and hence $B \sim R$. This shows $\sim$ has two classes and completes the proof.

(22.14) Let (V, Q) be a hyperbolic orthogonal space, let $G = O(V, Q)$, and let H be the subgroup of G preserving the equivalence relation of 22.13. Then
 (1) $|G : H| = 2$.
 (2) H is the image of the special Clifford group under the map of 22.8.6.
 (3) Reflections and transvections are in $G - H$.

Proof. By Witt's Lemma, G is transitive on the set Γ of maximal totally singular subspaces of V, so $|G : H|$ is the number of equivalence classes of Γ. That is $|G : H| = 2$ by 22.13. It's easy to check that (3) holds. Then, by (1), (3), and 22.7, H is the subgroup of G consisting of the elements which are the product of an even number of transvections or reflections, while by 22.9 this subgroup is the image of the special Clifford group under the map of 22.8.6.

I close this section with a brief discussion of some geometries associated to the classical groups. A few properties of these geometries are derived in Exercise 7.8, while chapter 14 investigates these geometries in great detail.

Assume the Witt index m of (V, f) or (V, Q) is positive. In this case there are some interesting geometries associated to the space and preserved by its isometry group. The reader may wish to refer to the discussion in section 3 on geometries.

The *polar geometry* Γ of (V, f) or (V, Q) is the geometry over $I = \{0, 1, \ldots, m - 1\}$ whose objects of type i are the totally singular subspaces of V of projective dimension i, with incidence defined by inclusion. Evidently $O(V, f)$ is represented as a group of automorphisms of Γ. Indeed the similarity group $\Delta(V, f)$ is also so represented.

If (V, Q) is a hyperbolic orthogonal space there is another geometry associated to (V, Q) which is in many ways nicer than the polar geometry. Assume the dimension of V is at least 6, so that the Witt index m of (V, Q) is at least 3. The *oriflamme geometry* Γ of (V, Q) is the geometry over $I = \{0, 1, \ldots, m - 1\}$ whose objects of type $i < m - 2$ are the totally singular subspaces of projective dimension i, and whose objects of types $m - 1$ and $m - 2$ are the two equivalence classes of maximal totally singular subspaces of (V, Q) defined by the equivalence relation of 22.13. Incidence is inclusion, except between objects U and W of type $m - 1$ and $m - 2$, which are incident if $U \cap W$ is a hyperplane of U and W. In this case the subgroup of $\Delta(V, Q)$ of index 2 preserving the equivalence relation of 22.13 is represented as a group of automorphisms of Γ.

Remarks. The standard reference for much of the material in this chapter is Dieudonné [Di]. In particular this is a good place to find out who first proved what in the subject.

We will encounter groups generated by reflections again in sections 29
and 30.

Observe that, by 13.8, 22.4.4, 22.4, and Exercise 7.6, almost all the finite
classical groups $SL_n(q)$, $Sp_n(q)$, $\Omega_n^\varepsilon(q)$, and $SU_n(q)$ are perfect. This fact is
used to prove in 43.12 that the projective groups $PSL_n(q)$, $PSp_n(q)$, $P\Omega_n^\varepsilon(q)$,
and $PSU_n(q)$ are simple unless n and q are small.

Since by some measure most finite simple groups are classical, the study
of the classical groups is certainly important. Moreover along with Lie theory
(cf. chapter 14 and section 47) the representations of the classical groups on
their associated spaces is the best tool for studying the classical groups. On the
other hand the study of the classical groups is a special topic and the material
in this chapter is technical. Thus the casual reader may wish to skip, or at least
postpone, this chapter.

Exercises for chapter 7

1. Let V be an n-dimensional vector space over F. Prove:
   (1) If $n = 2$ then $SL(V) = Sp(V)$.
   (2) If θ is an automorphism of F of order 2, $n = 2$, and (V, f) is a
       hyperbolic unitary space, then $SL(V) \cap O(V, f) \cong SL_2(\text{Fix}(\theta))$.
   (3) Let $|F| = q^2 < \infty$ and (V, f) a 3-dimensional unitary space over F.
       Then there exists a basis X of V such that

$$J(X, f) = \begin{bmatrix} 0 & 0 & 1 \\ 0 & 1 & 0 \\ 1 & 0 & 0 \end{bmatrix}.$$

   Moreover if P consists of those $g \in SU(V)$ with

$$M_X(g) = \begin{bmatrix} 1 & a & c \\ 0 & 1 & b \\ 0 & 0 & 1 \end{bmatrix}$$

   then $|P| = q^3$ and $[P, P]$ is a root group of $SU(V)$.
   (4) Let $|F| = q < \infty$ and (V, f) a $2m$-dimensional symplectic space
       over F with $m > 1$. Then there exists a basis $X = (x_i : 1 \leq i \leq 2m)$
       such that

$$J(X, f) = \begin{pmatrix} 0 & I_m \\ -I_m & 0 \end{pmatrix}$$

   where I_m is the m by m identity matrix. Moreover if P consists of those $g \in \mathrm{Sp}(V)$ with

$$M_X(g) = \begin{bmatrix} 1 & 0 & 0 \\ a & I_{2m-2} & 0 \\ b & c & 1 \end{bmatrix}$$

where a and c are column and row vectors, respectively, then $|P| = q^{2m-1}$. If q is odd, $[P, P]$ is a root group of $\mathrm{Sp}(V)$. If $q = 2$ and $m = 3$ then P contains a transvection and $P = [P, H]$ where $H = \mathrm{Sp}(U)$ and $U = \langle x_1, x_{2m} \rangle^{\perp}$.

2. Let (V, Q) be a 2-dimensional orthogonal space over F. Prove:
   (1) $O(V, Q)$ is the semidirect product of a subgroup H by $\langle r \rangle \cong \mathbb{Z}_2$, where r inverts H.
   (2) Each element of $O(V, Q) - H$ is a transvection or reflection. In particular $O(V, Q)$ is generated by such elements.
   (3) If $O(V, Q)$ is hyperbolic then H is isomorphic to the multiplicative group $F^{\#}$ of F.
   (4) If (V, Q) is definite then there exists a quadratic Galois extension K of F such that (V, Q) is similar to (K, N_F^K) and $H \cong \{a \in K : aa^{\theta} = 1\}$, where $\langle \theta \rangle = \mathrm{Gal}(K/F)$.

3. Let (V, f) be a 4-dimensional unitary space over a field F of order 4, X an orthonormal basis for V, $\Delta = \{\langle x \rangle : x \in X\}$, and $G = \mathrm{SU}(V)$. Prove $N_G(\Delta)^{\Delta} \cong S_4$, $G_{\Delta} \cong E_{27}$, and $N_G(\Delta)$ is generated by transvections. Let $D \in \Delta$, T the subgroup of G generated by the transvections in G, and Γ the set of conjugates of Δ under $N_G(D)$. Prove $N_G(D)^{\Gamma} \cong A_4$ and $|G_{\Gamma}| = 54$. Prove $N_G(D) \le T$.

4. Let q be a prime power. Prove
   (1) $Z(\mathrm{GU}_n(q))$ and $\mathrm{GU}_n(q)/\mathrm{SU}_n(q)$ are cyclic of order $q + 1$.
   (2) $Z(\mathrm{SU}_n(q))$ and $\mathrm{PGU}_n(q)/U_n(q)$ are cyclic of order $(q + 1, n)$.

5. Assume the hypothesis and notation of Exercise 4.7 with $\mathrm{char}(F) \ne 2$. Let $W = V_3$, $\alpha = \pi_3$, and define $Q: W \to F$ by

$$Q(ax^2 + bxy + cy^2) = b^2 - 4ac.$$

Prove
   (1) Q is a nondegenerate quadratic form on W with bilinear form $(ax^2 + bxy + cy^2, rx^2 + sxy + ty^2) = 2bs - 4(at + cr)$.
   (2) For each $g \in G$, $g\alpha$ is a similarity of (W, Q) with $\lambda(g\alpha) = \det(g)^2$.
   (3) $(G\alpha)S$ is the group $\Delta(W, Q)$ of all similarities of (W, Q), where S is the group of scalar maps on W.
   (4) Up to similarity, (W, Q) is the unique 3-dimensional nondefinite orthogonal space over F.

(5) If F is finite or algebraically closed every 3-dimensional orthogonal space over F is similar to (W, Q).

(6) $\langle rr^h : r \in R, h \in G\alpha \rangle = \Delta(W, Q)^{(1)} \cong L_2(F)$, where R is the set of reflections in $O(W, Q)$.

6. Let (V, Q) be an n-dimensional orthogonal space over a field F with $n \geq 3$.

   (1) Assume (V, Q) is not definite and if $|F| \leq 3$ and $n \leq 4$ assume $n = 4$ and $\text{sgn}(Q) = -1$. Prove the following subgroups are equal.

      (i) $\Omega(V, Q)$.

      (ii) The kernel in $G^+\pi$ of the spinor norm, where G^+ is the special Clifford group.

      (iii) $\langle rr^g : r$ reflection or transvection, $g \in O(V, Q)\rangle$.

   (2) If $\text{char}(F) \neq 2$ prove $\Omega(V, Q)$ is perfect unless $|F| = 3$ and either $n = 3$ or $n = 4$ and $\text{sgn}(Q) = +1$. If F is finite prove $O(V, Q)/\Omega(V, Q) \cong E_4$, and $-I \in \Omega(V, Q)$ if and only if n is even and $\text{sgn}(Q) \equiv |F|^{n/2}$ mod 4.

   (3) If $\text{char}(F) = 2$ and F is perfect, prove either $\Omega(V, Q)$ is perfect and $|O(V, Q) : \Omega(V, Q)| = 2$ or $|F| = 2, n = 4$, and $\text{sgn}(Q) = +1$.

(Hint: To prove $\Omega(V, Q)$ perfect use (1) and show rr^g is contained in a perfect subgroup of $O(V, Q)$ for each reflection or transvection r and each $g \in G$. Toward that end use Exercise 7.5 in (2) and 22.5 in (3).)

7. Let G be a permutation group on a set I of finite order n and V the permutation module for G over F with G-invariant basis $X = (x_i : i \in I)$. Define a bilinear form f on V by $f(x_i, x_j) = \delta_{ij}$ (the Kronecker delta) for $i, j \in I$. Let $z = \sum_{i \in I} x_i$, U the core of the permutation module V, and $\bar{V} = V/\langle z \rangle$. If $\text{char}(F) = 2$ define a quadratic form Q on U by $Q(\sum a_i x_i) = \sum a_i^2 + \sum_{i<j} a_i a_j$. Prove:

   (1) $U = z^\perp$.

   (2) $G \leq O(V, f)$.

   (3) If $\text{char}(F) \neq 2$ then (V, f) is an orthogonal space. If $\text{char}(F) = 2$ then $(\bar{U}, \bar{f})$ is a symplectic space preserved by G and if further $b \not\equiv 2$ mod 4 then $(\bar{U}, \bar{Q})$ is an orthogonal space preserved by G, where $\bar{f}(\bar{u}, \bar{v}) = f(u, v)$ and $\bar{Q}(\bar{u}) = Q(u)$ for $u, v \in U$.

   (4) If $G = S_6, n = 6$, and $|F| = 3$ then $(\bar{U}, \bar{f})$ is of sign -1, $O(\bar{U}, \bar{f}) = \langle -I \rangle \times G$, and $A_6 \cong \Omega_4^-(3)$.

   (5) If $G = S_5, n = 5$, and $|F| = 2$, then $(\bar{U}, \bar{Q})$ is of sign -1 and $O(\bar{U}, \bar{Q}) = G \cong O_4^-(2)$.

   (6) If $G = S_6, n = 6$, and $|F| = 2$, then $G = O(\bar{U}, \bar{f}) \cong Sp_4(2)$.

   (7) If $G = S_8, n = 8$, and $|F| = 2$, then $G = O(\bar{U}, \bar{Q}) \cong O_6^+(2)$.

8. Let (V, f) be a symplectic or unitary space over F, or (V, Q) an orthogonal space over F. Assume the Witt index m of the space is positive. Let Γ be the polar geometry of the space over $I = \{0, 1, \ldots, m-1\}$ and G the isometry

group of the space or (V, Q) a hyperbolic orthogonal space with $m \geq 3$, Γ the oriflamme geometry of (V, Q), and G the subgroup of $O(V, Q)$ preserving the equivalence relation of 22.13. Let Z be a maximal hyperbolic subspace of V, $X = (x_i: 1 \leq i \leq 2m)$ a hyperbolic basis for Z, $V_i = \langle x_{2j-1}: 1 \leq j \leq i \rangle$, and $Y = \{\langle x \rangle : x \in X\}$. Let $T = \{V_i: 1 \leq i \leq m\}$ if Γ is the polar geometry and $T = \{V_i, V'_{m-1}: 1 \leq i \leq m, i \neq m - 1\}$ if Γ is the oriflamme geometry, where $V'_{m-1} = \langle V_{m-2}, x_{2m-3}, x_{2m} \rangle$. Prove

(1) If $m = 1$ then G is 2-transitive on the points of Γ, while if $m > 1$ then G is rank 3 on these points.

(2) G is flag transitive on Γ.

(3) T is a flag of Γ of type I.

(4) $B = G_T$ is the semidirect product of U with $H = G_Y$, where U is the subgroup of G centralizing V_1, V_{i+1}/V_i, $1 \leq i < m - 1$, and
  (a) V_m/V_{m-1} and $(V_m)^\perp/V_m$ if Γ is a polar space, or
  (b) $(V_{m-1} \cap V_m)/V_{m-2}$ and $(V_{m-1} + V_m)/(V_{m-1} \cap V_m)$ if Γ is an oriflamme geometry.

(5) U is nilpotent and H is the direct product of m copies of $F^\#$ with $O(Z^\perp, f)$ (or $O(Z^\perp, Q)$).

(6) Let $i \in I$. Then either U fixes a unique object of type i in Γ or Γ is a polar geometry and V is a hyperbolic orthogonal space.

(7) $B = N_G(U)$.

(8) Assume F is finite of characteristic p and Γ is oriflamme if V is hyperbolic orthogonal. Then $U \in \mathrm{Syl}_p(G)$.

(9) $N_G(Y)^Y$ is $\mathbb{Z}_2 \mathrm{wr} S_m$ or of index 2 in that group, for Γ a polar space or oriflamme geometry, respectively.

(10) Let S be a flag of corank 1 in T. Then either the residue Γ_S of S is isomorphic to the projective line over F and $(G_S)^{\Gamma_s} \cong PGL_2(F)$ or $L_2(F)$, or Γ is a polar geometry, S is of type $\{0, \ldots, m - 2\}$, Γ_S is isomorphic to the set of singular points of $W = (V_{m-1})^\perp/V_{m-1}$, and either $(G_S)^{\Gamma_s} \cong PO(W, f)$ (or $PO(W, Q)$) or V is hyperbolic orthogonal and $|\Gamma_S| = 2$.

9. Let V be a finite dimensional vector space over a field F and f a nontrivial sesquilinear form on V. Then

(1) If $\mathrm{char}(F) \neq 2$ and f is bilinear then $f = g + h$ where g and h are symmetric and skew symmetric forms on V, respectively, and $O(V, f) \leq O(V, g) \cap O(V, h)$.

(2) If $\mathrm{char}(F) = 2$ and f is bilinear there exists a nontrivial symmetric bilinear form g on V with $O(V, f) \leq O(V, g)$.

(3) Let $\mathrm{char}(F) = 2$ and assume f is bilinear and symmetric. Let $U = \{x \in V: f(x, x) = 0\}$. Prove U is a subspace of V which is of codimension at most 1 if F is perfect.

(4) Assume f is sesquilinear with respect to the involution θ and f is *skew hermitian*; that is $f(x, y) = -f(y, x)^\theta$ for all $x, y \in V$. Prove f is similar to a hermitian form.

(5) If f is sesquilinear with respect to an involution θ, then there exists a nontrivial hermitian symmetric form g on V with $O(V, f) \leq O(V, g)$.

10. Let F be a field and f a sesquilinear form on V with respect to the automorphism θ of F, such that for all $x, y \in V$, $f(x, y) = 0$ if and only if $f(y, x) = 0$. Prove that either

(1) $f(x, x) = 0$ for all $x \in V, \theta = 1$, and f is skew symmetric, or

(2) there exists $x \in V$ with $f(x, x) \neq 0$ and one of the following holds:

    (a) $\theta = 1$ and f is symmetric.

    (b) $|\theta| = 2$ and f is similar to a hermitian symmetric form.

    (c) $|\theta| > 2$ and $\text{Rad}(V)$ is of codimension 1 in V.

8

p-groups

Chapter 8 investigates p-groups from two points of view: first through a study of p-groups which are extremal with respect to one of several parameters (usually connected with p-rank) and second through a study of the automorphism group of the p-group.

Recall that if p is a prime then the p-rank of a finite group is the maximum dimension of an elementary abelian p-subgroup, regarded as a vector space over GF(p). Section 23 determines p-groups of p-rank 1, p-groups in which each normal abelian subgroup is cyclic, and, for p odd, p-groups in which each normal abelian subgroup is of p-rank at most 2. Perhaps most important, the p-groups of symplectic type are determined (a p-group is of *symplectic type* if each of its characteristic abelian subgroups is cyclic).

The Frattini subgroup is introduced to study p-groups and their automorphisms. Most attention is focused on p'-groups of automorphisms of p-groups; a variety of results on the action of p'-groups on p-groups appear in section 24. One very useful result is the Thompson $A \times B$ Lemma. Also of importance is the concept of a critical subgroup.

23 Extremal p-groups

In this section p is a prime and G is a p-group.

The *Frattini subgroup* of a group H is defined to be the intersection of all maximal subgroups of H. $\Phi(H)$ denotes the Frattini subgroup of H.

(23.1) (1) $\Phi(H)$ char H.

(2) If $X \subseteq H$ with $H = \langle X, \Phi(H) \rangle$, then $H = \langle X \rangle$.

(3) If $H/\Phi(H)$ is cyclic, then H is cyclic.

(23.2) If G is a p-group then $\Phi(G)$ is the smallest normal subgroup H of G such that G/H is elementary abelian.

Proof. If M is a maximal subgroup of G then, by Exercise 3.2, $M \trianglelefteq G$ and $|G:M| = p$, so, by 8.8, $G^{(1)} \leq M$. Hence $G^{(1)} \leq \Phi(G)$, so, by 8.8, $G/\Phi(G)$ is abelian. Also, as $G/M \cong \mathbb{Z}_p$, $g^p \in M$ for each $g \in G$. so $g^p \in \Phi(G)$. Hence $G/\Phi(G)$ is elementary abelian.

Conversely let $H \trianglelefteq G$ with $G/H = G^*$ elementary abelian. Then $G^* = G_1^* \times \cdots \times G_n^*$ with $G_i^* \cong \mathbb{Z}_p$, so setting $H_i = \langle G_j : j \neq i \rangle$, $|G : H_i| = p$ and $H = \bigcap H_i$. Thus H_i is maximal in G so $H = \bigcap H_i \geq \Phi(G)$.

Observe that, as a consequence of 23.2, a p-group G is elementary abelian if and only if $\Phi(G) = 1$.

Recall that if n is a positive integer then $\Omega_n(G)$ is the subgroup of G generated by all elements of order at most p^n.

(23.3) Let $G = \langle x \rangle$ be cyclic of order $q = p^n > 1$ and let $A = \mathrm{Aut}(G)$. Then

(1)   The map $a \mapsto m(a)$ is an isomorphism of A with the group $U(q)$ of units of the integers modulo q, where $m(a)$ is defined by $xa = x^{m(a)}$ for $a \in A$. In particular A is abelian of order $\phi(q) = p^{n-1}(p-1)$.

(2)   The subgroup of A of order $p - 1$ is cyclic and faithful on $\Omega_1(G)$.

(3)   If p is odd then a Sylow p-group of A is cyclic and generated by the element b with $m(b) = p + 1$. In particular the subgroup of A of order p is generated by the element b_0 with $m(b_0) = p^{n-1} + 1$.

(4)   If $q = 2$ then $A = 1$ while if $q = 4$ then $A = \langle c \rangle \cong \mathbb{Z}_2$, where $m(c) = -1$.

(5)   If $p = 2$ and $q > 4$ then $A = \langle b \rangle \times \langle c \rangle$ where b is of order 2^{n-2} with $m(b) = 5$, and c is of order 2 with $m(c) = -1$. The involution b_0 in $\langle b \rangle$ satisfies $m(b_0) = 2^{n-1} + 1$ and $m(cb_0) = 2^{n-1} - 1$.

Proof. I leave part (1) as an exercise and observe also that $\alpha : a \mapsto m(a) \bmod p$ is a surjective homomorphism of A onto $U(p)$ with kernel $C_A(\Omega_1(G))$. So, as $|U(p)|_{p'} = p - 1 = |U(q)|_{p'}$, the subgroup of A of order $p - 1$ is isomorphic to $U(p)$ and faithful on $\Omega_1(G)$, while $\ker(\alpha) = \{a \in A : m(a) \equiv 1 \bmod p\} \in \mathrm{Syl}_p(A)$. Next $U(p)$ is the multiplicative group of the field of order p, and hence cyclic, so (2) holds. Thus we may take $q > p$. Evidently if $m(c) = -1$ then c is of order 2. So, as $|A| = 2$ if $q = 4$, (4) holds. Thus we can assume $n > 1$, and $n > 2$ if $p = 2$. Choose b as in (3) or (5). Then $b \in \ker(\alpha)$, so $b^{p^{n-1}} = 1$. Thus if p is odd it remains to show $b^{p^{n-2}} = b_0$ and if $p = 2$ show $b^{2^{n-3}} = b_0$.

Observe:

$$(kp^m + 1)^p \equiv (1 + kp^{m+1} + k^2 p^{2m+1}(p-1)/2) \bmod p^{m+2}$$

$$\equiv 1 + kp^{m+1}$$

if $m > 1$ or p is odd. Hence as $m(b) = 1 + s$ with $s = p$ if p is odd and $s = 4$ if $p = 2$, it follows that $m(b^{p^{n-2}}) = 1 + p^{n-1} = m(b_0)$ if p is odd, while $m(b^{2^{n-3}}) = 2^{n-1} + 1 = m(b_0)$ if $p = 2$. So the proof is complete.

Next the definition of four extremal classes of p-groups. The *modular p-group* Mod_{p^n} of order p^n is the split extension of a cyclic group $X = \langle x \rangle$ of order p^{n-1} by a subgroup $Y = \langle y \rangle$ of order p with $x^y = x^{p^{n-2}+1}$. Mod_{p^n} is defined only when $n \geq 3$, where, by 23.3 and 10.3, Mod_{p^n} is well defined and determined up to isomorphism. Similar comments hold for the other classes. If $p = 2$ and $n \geq 2$ the *dihedral group* D_{2^n} is the split extension of X by Y with $x^y = x^{-1}$ and if $n \geq 4$ the *semidihedral group* SD_{2^n} is the split extension with $x^y = x^{2^{n-2}-1}$.

The fourth class is a class of nonsplit extensions. Let G be the split extension of $X = \langle x \rangle$ of order $2^{n-1} \geq 4$ by $Y = \langle y \rangle$ of order 4 with $x^y = x^{-1}$. Notice $\langle x^{2^{n-2}}, y^2 \rangle = Z(G)$. Define the *quaternion group* Q_{2^n} of order 2^n to be the group $G/\langle x^{2^{n-2}} y^2 \rangle$.

The modular, dihedral, semidihedral, and quaternion groups are discussed in Exercises 8.2 and 8.3. Observe $\text{Mod}_8 = D_8$.

(23.4) Let G be a nonabelian group of order p^n with a cyclic subgroup of index p. Then $G \cong \text{Mod}_{p^n}$, D_{2^n}, SD_{2^n}, or Q_{2^n}.

Proof. Notice that, as G is nonabelian, $n \geq 3$ by Exercise 2.4. Let $X = \langle x \rangle$ be of index p in G. By Exercise 3.2, $X \trianglelefteq G$. As X is abelian but G is not, $X = C_G(X)$ by Exercise 2.4. So $y \in G - X$ acts nontrivially on X. As $y^p \in X$, y induces an automorphism of X of order p. By 23.3, $\text{Aut}(X)$ has a unique subgroup of order p unless $p = 2$ and $n \geq 4$, where $\text{Aut}(X)$ has three involutions. In the first case by 23.3, $x^y = xz$ for some z of order p in X. In the remaining case $p = 2$ and $x^y = x^{-1}z^\varepsilon$, where $\varepsilon = 1$ or 0 and z is the involution in X.

Now if the extension splits we may choose y of order p and by definition $G \cong \text{Mod}_{p^n}$, D_{2^n}, or SD_{2^n}. So assume the extension does not split. Observe $C_X(y) = \langle x^p \rangle$ if $x^y = xz$, while $C_X(y) = \langle z \rangle$ otherwise. Also $y^p \in C_X(y)$. As G does not split over X, $\langle y, C_X(y) \rangle$ does not split over $C_X(y)$, so, as $\langle y, C_X(y) \rangle$ is abelian, it is cyclic. Thus $C_X(y) = \langle y^p \rangle$. Hence we may take $y^p = x^p$ if $x^y = xz$ and $y^2 = z$ otherwise.

Suppose $x^y = xz$. Then $z = [x, y]$ centralizes x and y, so, by 8.6, $(yx^{-1})^p = y^p x^{-p} z^{p(p-1)/2} = z^{p(p-1)/2}$, while $z^{p(p-1)/2} = 1$ unless $p = 2$. So, as G does not split, $p = 2$. Here $z = x^{2^{n-2}}$ and if $n \geq 4$ then, setting $i = 2^{n-3} - 1$, $(yx^i)^2 = 1$. If $n = 3$ then $x^y = x^{-1}$, which we handle below.

So $p = 2$, $x^y = x^{-1}z^\varepsilon$, and $y^2 = z$. If $\varepsilon = 0$, then by definition $G \cong Q_{2^n}$, so take $\varepsilon = 1$. Then, as $z \in Z(G)$, $(yx)^2 = y^2 x^y x = zx^{-1}zx = 1$, so the extension does indeed split.

(23.5) Let G be a nonabelian p-group containing a cyclic normal subgroup U of order p^n with $C_G(U) = U$. Then either

(1) $G \cong D_{2^{n+1}}$, $Q_{2^{n+1}}$, or $SD_{2^{n+1}}$, or
(2) $M = C_G(\mho^1(U)) \cong \text{Mod}_{p^{n+1}}$ and $E_{p^2} \cong \Omega_1(M)$ char G.

Proof. Let $G^* = G/U$. As $U = C_G(U)$, $G^* = \text{Aut}_G(U) \le \text{Aut}(U)$. As G is non-abelian, $G^* \ne 1$ and $n \ge 2$. If G^* is of order p then the lemma holds by 23.4 and Exercise 8.2, so assume $|G^*| > p$. Then by 23.3 there exists $y^* \in G^*$ of order p with $u^y = u^{p^{n-1}+1}$, where $U = \langle u \rangle$. Let $M = \langle y, U \rangle$. By 23.4, $M \cong \text{Mod}_{p^{n+1}}$ and, by Exercise 8.2, $E = \Omega_1(M) \cong E_{p^2}$. It remains to show E char G. By 23.3, G^* is abelian and either G^* is cyclic, or $p = 2$ and there exists $g^* \in G^*$ with $u^g = u^{-1}$. In the first case $\Omega_1(G^*) = M^*$, so $E = \Omega_1(M) = \Omega_1(G)$ char G. In the second $\mho^1(U) = \langle u^2 \rangle = \langle [u, g] \rangle$ and as G^* is abelian, $G^{(1)} \le U$. Hence $G^{(1)} = \mho^1(U)$ or U, and in either case $\mho^1(U)$ char $G^{(1)}$, so $\mho^1(U)$ char G. Therefore $E = \Omega_1(C_G(\Omega^1(U)))$ char G.

A *critical subgroup* of G is a characteristic subgroup H of G such that $\Phi(H) \le Z(H) \ge [G, H]$ and $C_G(H) = Z(H)$. Observe that in particular a critical subgroup is of class at most 2.

(23.6) Each p-group possesses a critical subgroup.

Proof. Let S be the set of characteristic subgroups H of G with $\Phi(H) \le Z(H) \ge [G, H]$. Let H be a maximal member of S; I claim H is a critical subgroup of G. Assume not and let $K = C_G(H)$, $Z = Z(H)$, and define X by $X/Z = \Omega_1(Z(G/Z)) \cap K/Z$. Then $K \not\le H$ and $Z = H \cap K$, so, as $K \trianglelefteq G$, $X \ne Z$ by 5.15. But notice $XH \in S$, contradicting the maximality of H.

A p-group G is *special* if $\Phi(G) = Z(G) = G^{(1)}$. A special p-group is said to be *extraspecial* if its center is cyclic.

(23.7) The center of a special p-group is elementary abelian.

Proof. Let G be special and $g, h \in G$. Then $g^p \in \Phi(G) = Z(G)$, so, by 8.6.1, $1 = [g^p, h] = [g, h]^p$. Hence $G^{(1)}$ is elementary, so, as $Z(G) = G^{(1)}$, the lemma holds.

(23.8) Let E be an extraspecial subgroup of G with $[G, E] \le Z(E)$. Then $G = E C_G(E)$.

Proof. Let $Z = \langle z \rangle = Z(E)$. As $E/Z \le \text{Aut}_G(E) \le C = C_{\text{Aut}(E)}(E/Z)$, it suffices to show $E/Z = C$. Let $\alpha \in C$ and $(x_i Z: 1 \le i \le n)$ a basis for E/Z. Then

$[x_i, \alpha] = z^{m_i}$ for some $0 \le m_i < p$, and, as $E = \langle x_i : 1 \le i \le n \rangle$ by 23.1, α is determined by the integers $(m_i : 1 \le i \le n)$. Thus $|C| \le p^n = |E/Z|$, and the lemma holds.

A p-group is said to be of *symplectic type* if it has no noncyclic characteristic abelian subgroups.

(23.9) If G is of symplectic type then $G = E * R$ where

(1) Either E is extraspecial or $E = 1$, and
(2) Either R is cyclic, or R is dihedral, semidihedral, or quaternion, and of order at least 16.

Proof. By 23.6, G possesses a critical subgroup H. Let $U = Z(H)$. By hypothesis U is cyclic. Let Z be the subgroup of U of order p and $G^* = G/Z$. For $h, k \in H$, $h^p \in U$, so $[h, k]^p = [h^p, k] = 1$, by 8.6. Thus $H^{(1)} \le Z$ so H^* is abelian. Let $K^* = \Omega_1(H^*)$ and E^* a complement to $Z(K)^*$ in K^*. K char G, so $Z(K)$ is cyclic. Hence if K is abelian then H^* is cyclic, so, by Exercise 2.4, $U = H$. Now 23.5 applies and says the lemma holds with $E = 1$ and $R = G$ or $E = G \cong D_8$ or Q_8 and $R = 1$.

So K is nonabelian and then E is extraspecial. For $g \in G$ and $e \in E$, $e^g = eu$ for some $u \in U \cap K$, and $e^p \in Z \le Z(G)$, so $e^p = e^{pg} = (eu)^p = e^p u^p$. Hence $u^p = 1$, so that $[G, E] = Z$. Therefore, by 23.8, $G = E * R$, where $R = C_G(E)$. Thus $H = E * C_H(E)$. Recall H^* is abelian and $\Omega_1(H^*) = K^* = E^* \times Z(K)^*$ with $Z(K)$ cyclic. We conclude $C_H(E)^*$ is cyclic and hence $C_H(E)$ is abelian by Exercise 2.4. Thus as $H = EC_H(E)$, $C_H(E) = U$. Also $C_R(U) = C_R(H) \le R \cap H = U$, and we may assume $R \ne U$. So $|U| \ge p^2$ and 23.5 applies to R. If $|U| = p^2$ then $\mho^1(U) = Z \le Z(G)$. So, by 23.5, R is dihedral, quaternion or modular of order p^3, and in particular R is extraspecial. But then $G = E^* R$ is also extraspecial, so the lemma holds. Thus we may take $|U| > p^2$ and assume R satisfies 23.5.2. Let $M = C_R(\mho^1(U))$. Then $N = C_G(\mho^1(U))$ char G and $N = EM$. By Exercise 8.2, $\Omega_2(N) = E\Omega_2(M) \cong \mathbb{Z}_p \times (E * \mathbb{Z}_{p^2})$ so $Z(\Omega_2(N))$ is noncyclic, a contradiction.

(23.10) Let E be an extraspecial p-group, $Z = Z(E)$, and $\tilde{E} = E/Z$.

(1) Regard Z as the field of integers modulo p and $\tilde{E}$ as a vector space over Z. Define $f : \tilde{E} \times \tilde{E} \to Z$ by $f(\tilde{x}, \tilde{y}) = [x, y]$. Then f is a symplectic form on $\tilde{E}$, so $(\tilde{E}, f)$ is a symplectic space over Z.

(2) $m(\tilde{E}) = 2n$ is even.

(3) If $p = 2$ define $Q : \tilde{E} \to Z$ by $Q(\tilde{x}) = x^2$. Then Q is a quadratic form on $\tilde{E}$ associated to f, so $(\tilde{E}, Q)$ is an orthogonal space over Z.

(4)   Let $Z \leq U \leq E$. Then U is extraspecial or abelian if and only if $\tilde{U}$ is nondegenerate or totally isotropic, respectively. If $p = 2$ then U is elementary abelian if and only if $\tilde{U}$ is totally singular.

Proof. As $Z = \Phi(E)$, $\tilde{E}$ is elementary abelian, so by 12.1 we can regard $\tilde{E}$ as a vector space over Z in a natural way. Notice that under this convention the group operations on $\tilde{E}$ and Z are written additively. By 8.5.4, $[xy, z] = [x, z]^y [y, z] = [x, z][y, z]$, with the latter equality holding as E is of class 2. This says f is linear in its first variable and a similar argument gives linearity in the second variable. As $Z = Z(E)$, f is nondegenerate. $[x, y] = [y, x]^{-1}$, or, in additive notation, $f(\tilde{x}, \tilde{y}) = -f(\tilde{y}, \tilde{x})$. Thus (1) holds.

Notice (1) and 19.16 imply (2). Let $p = 2$. By 8.6, $(xy)^2 = x^2 y^2 [x, y]$, or, in additive notation, $Q(\tilde{x} + \tilde{y}) = Q(\tilde{x}) + Q(\tilde{y}) + f(\tilde{x}, \tilde{y})$. Thus (3) holds. The proof of (4) is straightforward.

(23.11) Assume p is odd and G is of class at most 2. Then $\Omega_1(G)$ is of exponent p.

Proof. Let x and y be elements of G of order p. Then $[x, y] = z \in Z(G)$. By 8.6.1, $z^p = [x^p, y] = 1$, so, by 8.6.2, $(xy)^p = x^p y^p z^{p(p-1)/2} = 1$.

(23.12) Let p be odd and E an extraspecial p-group. Then $\Omega_1(E)$ is of exponent p and index at most p in E. If $\Omega_1(E) \neq E$ then $\Omega_1(E) = X \times E_0$ where $X \cong \mathbb{Z}_p$ and E_0 is of order p or extraspecial, and $E = E_1 * E_0$ with $E_1 \cong \text{Mod}_{p^3}$.

Proof. Let $Y = \Omega_1(E)$. By 23.11, Y is of exponent p. Suppose $|E : Y| = p$. Then, in the notation of 23.10, $\tilde{Y}$ is a hyperplane of $\tilde{E}$, and hence of odd dimension, so, as all symplectic spaces are of even dimension, $\tilde{Y}$ is degenerate. Let $\tilde{R}$ be a point in $\text{Rad}(\tilde{Y})$. As $\tilde{R}^\perp$ is a hyperplane of $\tilde{E}$, $\tilde{R}^\perp = \tilde{Y}$. Hence, by 23.10.4, $Y = C_E(R)$. As Y is of exponent p, $R = X \times Z$ for some X of order p. Let $\tilde{E}_0$ be a complement to $\tilde{R}$ in $\tilde{Y}$. By 19.3 and 23.10.4, E_0 is extraspecial or $E_0 = Z$; of course $Y = X \times E_0$. Let $E_1 = C_E(E_0)$. By 19.3 and 23.10, E_1 is extraspecial. As $Y \neq E$, $E_1 > \Omega_1(E_1)$ so, by 23.4, $E_1 \cong \text{Mod}_{p^3}$.

It remains to show $|E : Y| \leq p$. Let $u, v \in E$ and $U = \langle u, v \rangle$. It suffices to show $|U : \Omega_1(U)| \leq p$. If U is abelian this holds because $Z = \Phi(E)$. If U is nonabelian appeal to 23.4.

By 23.10 an extraspecial p-group is of order p^{1+2n} for some positive integer p. If p is odd, denote by p^{1+2n} an extraspecial p-group of exponent p and order p^{1+2n}. 2^{1+2n} denotes any extraspecial 2-group of order 2^{1+2n}; by 1.13

there are no such groups of exponent 2. Write $D_8^n Q_8^m$ for a central product of n copies of D_8 with m copies of Q_8, and all centers identified.

(23.13) Let p be an odd prime and n a positive integer. Then up to isomorphism there is a unique extraspecial p-group E of order p^{1+2n} and exponent p. E is the central product of n copies of p^{1+2}.

Proof. By 23.10 and 19.16, E is a central product of n extraspecial subgroups E_i, $1 \le i \le n$, of order p^3 and center $Z = Z(E)$. Now Exercise 8.7 completes the proof.

(23.14) Let n be a positive integer. Then up to isomorphism D_8^n and $D_8^{n-1} Q_8$ are the unique extraspecial groups of order 2^{2n+1}. D_8^n has 2-rank $n + 1$ while $D_8^{n-1} Q_8$ has 2-rank n, so the groups are not isomorphic.

Proof. By 23.10 and 21.2, E is a central product of n extraspecial groups E_i, $1 \le i \le n$, of order 8 with $Z(E_i) = Z(E) = Z$. Z is of order 2 so $\mathrm{Aut}(Z) = 1$. Hence 11.2 says E is determined up to isomorphism by the groups E_i. Again by 23.10 and 21.2, we can choose $(\tilde{E}_i, Q)$ hyperbolic for $i < n$ and $(\tilde{E}_n, Q)$ either hyperbolic or definite. By 1.13, E_i is not of exponent 2, so, by 23.4 $E_i \cong D_8$ or Q_8. By 23.10.4, $(\tilde{E}_i, Q)$ is hyperbolic in the first case and definite in the second. Finally 23.10 and 21.2 imply the remark about the 2-rank and complete the proof.

(23.15) Let A be a maximal abelian normal subgroup of G and $Z = \Omega_1(A)$. Then
  (1) $A = C_G(A)$.
  (2) $(C_G(A/Z) \cap C(Z)^{(1)} \le A$.
  (3)   If p is odd then $\Omega_1(C_G(Z)) \le C_G(A/Z)$.

Proof. Let $C = C_G(A)$. $A \le C \trianglelefteq G$, so if $C \ne A$ there is D/A of order p in $Z(G/A) \cap C/A$. Then $D \trianglelefteq G$ and D is abelian by Exercise 2.4, contradicting the maximality of A. A straightforward calculation shows $(C_G(A/Z) \cap C(Z))^{(1)} \le C(A)$, so (1) implies (2). Let p be odd, x of order p in $C_G(Z)$ and $X = \langle x, A \rangle$. Let $Y = \langle x, C_A(\langle x, Z \rangle / Z) \rangle$. Then Y is of class at most 2, so, by 23.11, $W = \Omega_1(Y)$ is of exponent p. Thus $W = \langle x, Z \rangle$. But W char Y so $N_X(Y) \le N_X(W) = Y$, so $Y = X$ and (3) holds.

(23.16) Let p be an odd prime and Z a maximal elementary abelian normal subgroup of G. Then $Z = \Omega_1(C_G(Z))$.

Proof. Let $X = \Omega_1(C_G(Z))$. I'll show X is of exponent p. Hence if $X \neq Z$ then there is D/Z of order p in $Z(G/Z) \cap X/Z$ and, by Exercise 2.4, D is elementary abelian, contradicting the maximality of Z.

Let A be a maximal abelian normal subgroup of G containing Z. Then $Z = \Omega_1(A)$ by maximality of Z. By 23.15.3, $[X, A] \leq Z$, so by 23.15.2, $X^{(1)} \leq A$. Choose $U \leq X$ of minimal order subject to $U = \Omega_1(U)$ and U not of exponent p. Then there exist x and y in U of order p with xy not of order p. By minimality of U, $U = \langle x, y \rangle$. By 7.2, $V = \langle x^U \rangle \neq U$, so V is of exponent p. Hence $[x, y] \in V$ is of order at most p, so, as $X^{(1)} \leq A$, $[x, y] \in Z$. As $X \leq C(Z)$, U is of exponent p by 23.11, contrary to the choice of U.

(23.17) Let p be an odd prime and assume G contains no normal abelian subgroup of rank 3. Then G is of p-rank at most 2.

Proof. By Exercise 8.4 we may assume $E_{p^2} \cong Z \trianglelefteq G$. Let $H = C_G(Z)$ and $E_{p^3} \cong A \leq G$. Then $|A: A \cap H| \leq p$ and hence $m((A \cap H)Z) \geq 3$. Thus $m(H) \geq 3$. However by hypothesis Z is a maximal elementary abelian normal subgroup of G, so $Z = \Omega_1(H)$ by 23.16.

24 Coprime action on *p*-groups

In this section p is a prime, G is a p-group, and A is a p'-group of automorphisms of G, unless the conditions are explicitly relaxed as in the Thompson $A \times B$ Lemma.

(24.1) A is faithful on $G/\Phi(G)$.

Proof. Suppose $b \in A$ centralizes $G/\Phi(G)$. We wish to show $b = 1$. If not there is a prime q and a nontrivial power of b which is a q-element and centralizes $G/\Phi(G)$, so without loss b is a q-element. Let $B = \langle b \rangle$ and $g \in G$. Then B acts on the coset $X = g\Phi(G)$. By 5.14, $m \equiv |X| \bmod q$, where m is the number of fixed points of B on X, and, as $|X| = |\Phi(G)|$ is a power of p, $|X| \not\equiv 0 \bmod q$, so B centralizes some $x \in X$. Hence B centralizes a set Y of coset representatives for $\Phi(G)$ in G, so, by 23.1.2, $G = \langle Y \rangle \leq C_G(B)$. Hence $B = 1$, completing the proof.

(24.2) (Thompson $A \times B$ Lemma). Let AB be a finite group represented as a group of automorphisms of a p-group G, with $[A, B] = 1 = [A, C_G(B)]$, B a p-group, and $A = O^p(A)$. Then $[A, G] = 1$.

Proof. Form the semidirect product H of G by AB and identify AB and G with subgroups of H. We may assume $[A, G] \neq 1$ so, as $A = O^p(A)$, $[X, G] \neq 1$

for some p'-subgroup X of A, and replacing A by X we may assume A is a p'-group.

$G \trianglelefteq H = GBA$ with $A \leq N_H(B)$, so GB is a normal p-subgroup of H, and, replacing G by GB, we may assume $B \leq G$. Then $B \leq Q = C_G(A)$, so $C_G(Q) \leq C_G(B)$. Also $C_G(B) \leq Q$ by hypothesis, so $C_G(Q) \leq Q$.

As $[A, G] \neq 1$, $Q \neq G$, so, by Exercise 2.2, Q is properly contained in $N_G(Q)$. So, by definition of Q, $[A, N_G(Q)] \neq 1$, and hence we may assume $Q \trianglelefteq G$.

Let $G^* = G/Q$. As A is a p'-group, $C_{G^*}(A) = C_G(A)^* = Q^* = 1$, by 18.7.4. Thus $[G, A] \not\leq Q$.

As $Q \trianglelefteq G$, $[G, Q, A] \leq [Q, A] = 1$, so $[G, Q, A] = [Q, A, G] = 1$. Hence by the Three-Subgroup Lemma, 8.7, $[A, G, Q] = 1$. Thus $[A, G] \leq C_G(Q) \leq Q$, by paragraph 2, contrary to the last paragraph.

(24.3) If G is abelian then A is faithful on $\Omega_1(G)$.

Proof. Without loss, A centralizes $\Omega_1(G)$. Let X be of order p in G and $G^* = G/X$. By Exercise 3.1, A is faithful on G^*, so, by induction on the order of G, A is faithful on $\Omega_1(G^*)$, and hence without loss $G^* = \Omega_1(G^*)$. Now, by 12.1 and Exercise 4.1.1, we may take $C_{G^*}(A) = 1$. Thus, by 18.7.4, $X = C_G(A)$, so $X = \Omega_1(G)$. Hence, as G is abelian, 1.11 implies G is cyclic. Now 23.3 supplies a contradiction.

(24.4) $G = [G, A]C_G(A)$.

Proof. Let $G^* = G/\Phi(G)$. By 23.2, G^* is an elementary abelian p-group, so, by Exercise 4.1, $G^* = [G^*, A] \times C_{G^*}(A)$. By 8.5.3, $[G^*, A] = [G, A]^*$ and, by 18.7.4, $C_{G^*}(A) = C_G(A)^*$. Hence $G = \langle [G, A], C_G(A), \Phi(G) \rangle$, so, by 23.1, $G = \langle [G, A], C_G(A) \rangle$. Finally, by 8.5.6, $[G, A] \trianglelefteq G$, so $\langle [G, A], C_G(A) \rangle = [G, A]C_G(A)$ by 1.7.2.

(24.5) $[G, A] = [G, A, A]$.

Proof. Let $H = [G, A]$. By 8.5.6, $H \trianglelefteq G$ and $[H, A] \trianglelefteq H$. Thus $C_G(A)$ acts on $[H, A]$, so $[H, A] \trianglelefteq HC_G(A) = G$. Next $H = [H, A]C_H(A)$ so $[G, A] \leq [H, A]$ by 8.5. But of course $[H, A] \leq [G, A]$ as $H \leq G$.

(24.6) If G is abelian then $G = [G, A] \times C_G(A)$.

Proof. Let G be a minimal counterexample and $X = [\Omega_1(G), A]$. By 24.3, $X \neq 1$ and, by 12.1 and Exercise 4.1.1, $C_X(A) = 1$. By minimality of G, $C_{([G,A]/X)}(A) = 1$, so $C_{[G,A]}(A) = 1$. Now 24.4 completes the proof.

(24.7) If $G = [G, A]$ and A centralizes every characteristic abelian subgroup of G, then G is special and $Z(G) = C_G(A)$.

Proof. As A centralizes each characteristic abelian subgroup of G, so does $G = [G, A]$ by Exercise 3.6. Thus $Z = Z(G)$ is the unique maximal characteristic abelian subgroup. $[Z_2(G), G, G] = 1$, so, by the Three-Subgroup Lemma, $Z_2(G)$ centralizes $G^{(1)}$. Hence $Z_2(G) \cap G^{(1)}$ is abelian, and therefore contained in Z, so $G^{(1)} \leq Z$. By 24.6, $G/G^{(1)} = (Z/G^{(1)}) \times [G/G^{(1)}, A]$ so, as $G = [G, A]$, $Z = G^{(1)}$. Finally suppose G has exponent $p^n > p$. Let $g, h \in G$. By 8.6, $[g^{p^{n-1}}, h^{p^{n-1}}] = [g^{p^n}, h^{p^{n-2}}] = 1$, so $\mho^{n-1}(G)$ is abelian and hence $\mho^{n-1}(G) \leq Z$. But then G/Z is of exponent p. So $Z = \Phi(G)$.

(24.8) If p is odd then A is faithful on $\Omega_1(G)$.

Proof. Choose G to be a minimal counterexample and let $a \in A^{\#}$ centralize $\Omega_1(G)$. By 24.5 and minimality of G, $G = [G, a]$. By 24.3, a centralizes each characteristic abelian subgroup of G, so, by 24.7, G is special with $Z = Z(G) = C_G(a)$. By 23.7, $Z = \Omega_1(G)$. Let $g \in G - Z$, $z = g^p$ and $v = [g, g^{-a}]$. Then $z, v \in Z = \Omega_1(G)$, so $v^p = 1$. Notice that, as $Z = C_G(a)$, $(g^{-a})^p = z^{-1}$, and $gg^{-a} = h \notin Z$ by 18.7.4. Now, by 8.6, $h^p = zz^{-1}v^{p(p-1)/2} = 1$, contradicting $Z = \Omega_1(G)$.

(24.9) Let H be a critical subgroup of G. Then
   (1) A is faithful on H.
   (2)   If p is odd then A is faithful on $\Omega_1(H)$, and there exists a critical subgroup H of G such that $\Omega_1(H)$ contains each element of order p in $C_G(\Omega_1(H))$.

Proof. By definition of H, $C_G(H) \leq H$, so by the Thompson $A \times B$ Lemma (applied to 'A' $= C_A(H)$ and 'B' $= H$), $C_A(H) = 1$. Thus (1) holds. Part (1) and 24.8 imply the first statement in (2). To prove the second, choose H with $L = \Omega_1(H)$ maximal. It suffices to show $Y = \Omega_1(C_G(L)) \leq L$. Assume not and let V be a maximal elementary abelian normal subgroup of Y. By 23.16, $V = \Omega_1(C_Y(V))$, so, as $Y \not\leq L$, $V \not\leq L$. Thus $V \cap Z_2(Y) \not\leq L$, so $\Omega_1(Z_2(Y)) = K \not\leq L$. By 23.11, K is of exponent p, so $X \not\leq L$, where $X/Z(L) = Z(G/Z(L)) \cap (K/Z(L))$. Now define S as in the proof of 23.6. Then $XL \in S$ so, by the proof of 23.6, XL is contained in a critical subgroup C of G. But $L < XL \leq \Omega_1(C)$, contradicting the choice of H.

Remarks. The discussion of p-groups in this chapter is essentially the same as Gorenstein's treatment of p-groups [Gor 4], which was influenced in turn by lecture notes of Phillip Hall.

P. Hall originally classified the p-groups of symplectic type. The notion of a 'critical subgroup' is due to J. Thompson as is of course the Thompson $A \times B$ Lemma.

Almost all of the material in this chapter is basic and belongs in the repertoire of any finite group theorist. For the simple group theorist it represents an important part of the foundation of the local group theory involved in the classification. For example the importance of p-groups of symplectic type is reflected in the second case of Theorem 48.3. More generally the results of this chapter will be used repeatedly in chapters 10 through 16.

Exercises for chapter 8

1. Let q be a prime and A an elementary abelian q-group acting on a q'-group G. Prove $G = \langle C_G(B): |A:B| = q \rangle$. (Hint: Use 18.7 to reduce to the case G a p-group. Then use Exercise 4.1 and 23.1.)

2. Let $G \cong \text{Mod}_{p^n}$, $n \geq 3$, with $\mathbb{Z}_{p^{n-1}} \cong X = \langle x \rangle \trianglelefteq G$, y of order p in $G - X$, and $x^y = x^{p-1}$. Prove
   (1) G is of class 2 with $Z(G) = \Phi(G) = \langle x^p \rangle \cong \mathbb{Z}_{p^{n-2}}$.
   (2) $G^{(1)} = \langle x^{p^{n-2}} \rangle \cong \mathbb{Z}_p$.
   (3) $\Omega_m(G) = \langle x^{p^{n-m}}, y \rangle \cong \mathbb{Z}_{p^m} \times \mathbb{Z}_p$ for $0 < m < n - 1$, unless $p^n = 8$.

3. Let $G \cong D_{2^n}$, $n \geq 2$, Q_{2^n}, $n \geq 3$, or SD_{2^n}, $n \geq 4$. Let $\mathbb{Z}_{2^{n-1}} \cong X = \langle x \rangle \trianglelefteq G$ and $y \in G - X$ with y an involution if G is dihedral or semidihedral and y of order 4 if G is quaternion. Prove
   (1) $G^{(1)} = \Phi(G) = \langle x^2 \rangle \cong \mathbb{Z}_{2^{n-2}}$.
   (2) Either G is dihedral of order 4 or $Z(G) = \langle x^{2^{n-2}} \rangle$ is of order 2.
   (3) G is of class $n - 1$.
   (4) X is the unique cyclic subgroup of G of index p, unless G is dihedral of order 4 or quaternion of order 8.
   (5) $G - X$ is the union of two conjugacy classes of G with representatives y and yx. Each member of $G - X$ is an involution if G is dihedral, each is of order 4 if G is quaternion, while if G is semidihedral then y is of order 2 and xy of order 4.
   (6) G has two maximal subgroups distinct from X. If G is dihedral of order at least 8, both are dihedral. If G is quaternion of order at least 16, both are quaternion. If G is semidihedral then one is dihedral and the other quaternion.
   (7) Quaternion groups have a unique involution.

4. Let G be a p-group with no noncyclic normal abelian subgroups. Prove G is cyclic, quaternion, semidihedral, or dihedral, and in the last case $|G| > 8$.

If H is a p-group with just one subgroup of order p, prove H is cyclic or quaternion.

5. Let G be a nonabelian p-group of symplectic type and exponent p or 4. Set $Z = Z(G)$, $\tilde{G}, = G/Z$, $A = \mathrm{Aut}(G)$, and $A^* = \mathrm{Out}(G)$. Prove
   (1) $\mathrm{Inn}(G) = C_A(\tilde{G})$.
   (2) $C_A(Z)^* = \mathrm{Sp}(\tilde{G})$ and A^* is the group of all similarities of some symplectic form on $\tilde{G}$ if p is odd.
   (3) If $p = 2$ then $G \cong D^n$, $D^{n-1}Q$, or $\mathbb{Z}_4 * D^n$, and $A^* \cong O_{2n}^+(2)$, $O_{2n}^-(2)$, or $\mathrm{Sp}_{2n}(2)$, respectively.
6. Let G be a 2-group containing an involution x with $C_G(x) \cong E_4$. Then G is dihedral or semidihedral.
7. Let p be an odd prime. Prove
   (1) Up to isomorphism there is a unique extraspecial group E of order p^3 and exponent p.
   (2) $\mathrm{Aut}_{\mathrm{Aut}(E)}(Z(E)) \cong \mathrm{Aut}(Z(E))$.
   (3) Up to isomorphism there is a unique central product of n copies of E with identified centers.
8. Let A be a π'-group acting on a π-group G. Prove
   (1) $G = [G, A]C_G(A)$, and
   (2) $[G, A] = [G, A, A]$.
9. Let r be a prime, A an elementary abelian r-group acting on a solvable r'-group G, $D \leq A$, and B a noncyclic subgroup of A. Prove $[G, D] = \langle [C_G(b), D] : b \in B^\# \rangle$.
10. Let A be a p'-group with a unique minimal normal subgroup B, assume A acts faithfully on a nontrivial p-group P, and assume A is faithful on no proper subgroup of P. Prove that either
    (1) P is elementary abelian and A is irreducible on P, or
    (2) $P = [P, B]$ is special, $[B, Z(P)] = 1$, and A is irreducible on $P/Z(P)$. If $[A, Z(P)] = 1$ and AP possesses a faithful irreducible representation over some field, then P is extraspecial.
11. Let p be an odd prime and G a p-group with $m(G) > 3$. Prove G has a normal abelian subgroup of p-rank at least 4. (Hint: Let G be a counterexample, V an elementary abelian normal subgroup of G of maximal rank, $H = C_G(V)$, and $E_{p^4} \cong A \leq G$. Show $V = \Omega_1(H) \cong E_{p^3}$, $m(A \cap H) = 2$, and A is the unique E_{p^4}-subgroup of AV. Let $K = N_G(A)$ and $g \in N_G(K) - K$. Show AA^g is of class at most 2 and $AA^g \leq AV$.)
12. Let G be a p-group and $(G_i : 1 \leq i \leq n)$ a family of subgroups of G which generates G. Then, for any family $(x_i : 1 \leq i \leq n)$ of elements of G, G is generated by $((G_i)^{x_i} : 1 \leq i \leq n)$.

9

Change of field of a linear representation

Let $\pi: G \to \mathrm{GL}(V, F)$ be an FG-representation, E a subfield of F, and K an extension field of F. Then V is also a vector space over E with $\mathrm{GL}(V, F) \leq \mathrm{GL}(V, E)$, so π also defines an EG-representation. Further, by a tensoring process discussed in section 25, π induces a KG-representation π^K on a K-space V^K. This chapter investigates the relationship among these representations. It will often be very useful to extend F to K by passing from π to π^K. For example several results at the end of chapter 9 are established in this way.

π is said to be *absolutely irreducible* if π^K is irreducible for each extension K of F, and F is said to be a *splitting field* for G if every irreducible FG-representation is absolutely irreducible. It develops in section 25 that π is absolutely irreducible precisely when $F = \mathrm{End}_{FG}(V)$ and in section 27 that if G is finite then a splitting field is obtained by adjoining a suitable root of unity to F. It's particularly nice to work over a splitting field. For example in section 27 it is shown that, over a splitting field, the irreducible representations of the direct product of groups are just the tensor products of irreducible representations of the factors.

Section 26 investigates representations over finite fields, where change of field goes very smoothly. Lemma 26.6 summarizes many of the relationships involved.

Section 27 introduces the minimal polynomial of a linear transformation. Semisimple and unipotent elements are discussed and it is shown that if F is perfect then each member g of $\mathrm{GL}(V)$ admits a *Jordan decomposition*; that is g can be written uniquely as the commuting product of a semisimple element and a unipotent element.

25 Tensor products

In this section G is a group, F is a field, and V a finite dimensional vector space over F.

Let $(V_i: 0 \leq i \leq m)$ be vector spaces over F, and denote by $L(V_1, \ldots, V_m; V_0)$ the set of all maps $\alpha: V_1 \times \cdots \times V_m \to V_0$ such that for each $i, 1 \leq i \leq m$, and each choice of $v_j \in V_j, j \neq i$, the map $\theta: V_i \to V_0$ defined by $v_i \theta = (v_1, \ldots, v_m)\alpha$ is an F-linear transformation. Such maps are called *m-linear.* $L(V_1, \ldots, V_m; V_0)$ is a vector space under F via

$$v(\alpha + \beta) = v\alpha + v\beta \quad a \in F, v \in V_1 \times \cdots \times V_m,$$

$$v(a\alpha) = a(v\alpha) \quad \alpha, \beta \in L(V_1, \ldots, V_m; V_0).$$

A *tensor product* of $V_1, \ldots, V_m$ is an F-space T together with $\pi \in L(V_1, \ldots, V_m; T)$ with the following universal property: whenever U is an F-space and $\alpha \in L(V_1, \ldots, V_m; U)$, there exists a unique $\beta \in \mathrm{Hom}_F(T, U)$ with $\pi\beta = \alpha$.

(25.1) Tensor products exist and are unique up to isomorphism.

Proof. See for example page 408 in Lang [La].

Because of 25.1 there is a unique tensor product of $V_1, \ldots, V_m$ which is denoted by $V_1 \otimes \cdots \otimes V_m$ or $\bigotimes_{i=1}^{m} V_i$. Write $v_1 \otimes \cdots \otimes v_m$ for the image of $(v_1, \ldots, v_m)$ under the map (denoted by π above) associated to the tensor product. The elements $v_1 \otimes \cdots \otimes v_m$, $v_i \in V_i$, are called *fundamental tensors*. It is easy to verify from the universal property that:

(25.2) $V_1 \otimes \cdots \otimes V_m$ is generated as an F-space by the fundamental tensors.

Here are some elementary properties of the tensor product; they can be found for example in Lang, Chapter 16.

(25.3) Let $(V_i : 1 \le i \le m)$ be F-spaces. Then
  (1) $V_1 \otimes V_2 \cong V_2 \otimes V_1$.
  (2) $(V_1 \otimes V_2) \otimes V_3 \cong V_1 \otimes (V_2 \otimes V_3)$.
  (3) $(\bigoplus_{U \in I} U) \otimes V \cong \bigoplus_{U \in I}(U \otimes V)$ for any direct sum $\bigoplus_{U \in I} U$ of F-spaces.
  (4) Let X_i be a basis of V_i, $i = 1, 2$. Then

$$X_1 \otimes X_2 = \{x_1 \otimes x_2 : x_i \in X_i\}$$

is a basis for $V_1 \otimes V_2$.
  (5) Let $\alpha_i \in \mathrm{End}_F(V_i)$. Then there exists a unique map $\alpha_1 \otimes \cdots \otimes \alpha_n \in \mathrm{End}_F(V_1 \otimes \cdots \otimes V_m)$ with $(v_1 \otimes \cdots \otimes v_m)(\alpha_1 \otimes \cdots \otimes \alpha_m) = v_1\alpha_1 \otimes \cdots \otimes v_m\alpha_m$.
  (6) For $v_i, u_i \in V_i$, $i = 1, 2$, and $a \in F$:

$$(v_1 + u_1) \otimes v_2 = (v_1 \otimes v_2) + (u_1 \otimes v_2),$$

and

$$v_1 \otimes (v_2 + u_2) = (v_1 \otimes v_2) + (v_1 \otimes u_2),$$

$$a(v_1 \otimes v_2) = av_1 \otimes v_2 = v_1 \otimes av_2.$$

If $\pi_i\colon G \to \mathrm{GL}(V_i)$, $1 \leq i \leq m$, are FG-representations, then by 25.3.5 there is an FG-representation $\pi_1 \otimes \cdots \otimes \pi_m$ of G on $V_1 \otimes \cdots \otimes V_m$ defined by $g(\pi_1 \otimes \cdots \otimes \pi_m) = g\pi_1 \otimes \cdots \otimes g\pi_m$, for $g \in G$. $\pi_1 \otimes \cdots \otimes \pi_m$ is the *tensor product* of the representations $\pi_1, \ldots, \pi_m$.

A special case of these constructions is of particular interest. Let K be an extension field of F. Then K is a vector space over F, so the tensor product $K \otimes U$ can be formed for any F-space U. Let X and B be bases for U and K over F, respectively. By 25.3 each member of $K \otimes U$ can be written uniquely as $\sum_{(b,x) \in B \times X} a_{b,x}(b \otimes x)$, with $a_{b,x} \in F$. As $a_{b,x}(b \otimes x) = ((a_{b,x})b) \otimes x$ with $(a_{b,x})b \in K$, it follows that each member of $K \otimes U$ is of the form $\sum_{x \in X}(c_x \otimes x)$, $c_x \in K$. Indeed it turns out $K \otimes U$ can be made into a vector space $K \otimes_F U = U^K$ over K by defining scalar multiplication via:

$$a\left(\sum_{x \in X}(c_x \otimes x)\right) = \sum_{x \in X}(ac_x \otimes x) \quad a, c_x \in K, x \in X.$$

These remarks are summarized in the following lemma; see chapter 16, section 3 in Lang [La] for example.

(25.4) Let K be an extension field of F and X a basis for a vector space U over F. Then $U^K = K \otimes_F U$ is a vector space over K with $1 \otimes X = \{1 \otimes x : x \in X\}$ a basis for U^K.

It will be useful to have the following well known property of this construction, which can be found on page 419 in Lang [La].

(25.5) If $L \geq K \geq F$ is a tower of fields and U an F-space then $L \otimes_F U \cong L \otimes_K (K \otimes_F U)$.

Notice that, for $g \in \mathrm{End}_F(U)$, $1 \otimes g \in \mathrm{End}_K(U^K)$, where $1 \otimes g$ is the map defined in 25.3.5 with respect to the identity map 1 on K. That is $1 \otimes g\colon a \otimes x \mapsto a \otimes xg$. In this way $\mathrm{End}_F(U)$ is identified with a subalgebra of $\mathrm{End}_K(U^K)$. Further if $\pi\colon G \to \mathrm{GL}(U)$ is an FG-representation then we obtain a KG-representation $\pi^K\colon G \to \mathrm{GL}(U^K)$ defined by $\pi^K = 1 \otimes \pi$, where 1 is the trivial representation of G on K. Equivalently $(1 \otimes x)(g\pi^K) = 1 \otimes xg\pi$ for each $x \in X$, $g \in G$. Observe that if U is finite dimensional then $M_X(g\pi) = M_{(1 \otimes X)}(g\pi^K)$

Recall the definition of enveloping algebra in section 12.

(25.6) Let $\pi\colon G \to \mathrm{End}_F(V) = E$ be an FG-representation, A the enveloping algebra of π in E, and K an extension of F. Then the enveloping algebra of π^K in $\mathrm{End}_K(V^K)$ is isomorphic to A^K as a K-space.

Proof. We may regard E and $\text{End}_K(V^K) = E^K$ as the rings $F^{n \times n}$ and $K^{n \times n}$, respectively, with E the set of matrices in E^K whose entries are in F. Now

$$A = \left\{ \sum_{g \in G} a_g(g\pi) : a_g \in F \right\}, \quad \text{so} \quad A^K = \left\{ \sum_{g \in G} b_g(g\pi) : b_g \in K \right\}.$$

But, as matrices, $g\pi$ and $g\pi^K$ are the same, so, as G is a group, A^K is the subalgebra of E^K generated by $G\pi^K$. That is A^K is the enveloping algebra of π^K.

Let G_1 and G_2 be groups and π_i an FG_i-representation. Denote by $\pi_1 \otimes \pi_2$ the tensor product $\bar{\pi}_1 \otimes \bar{\pi}_2$ where $\bar{\pi}_i$ is the representation of $G_1 \times G_2$ with $\bar{\pi}_i$ restricted to G_i equal to π_i and $\bar{\pi}_i$ restricted to G_{3-i} trivial. This is a small abuse of notation which will hopefully cause no problem. The convention is used in the proof of the next lemma.

Notice that if $G_1 \cong G_2 \cong G$ then G is diagonally embedded in $G_1 \times G_2$ via the map $g \mapsto (g, g)$, and if we identify G with this diagonal subgroup via the isomorphism, then the tensor product representation $\pi_1 \otimes \pi_2 : G \to \text{GL}(V_1 \otimes V_2)$ of G is the restriction of the tensor product representation of $G_1 \times G_2$ to this diagonal subgroup.

(25.7) Let K be a Galois extension of F, Γ the Galois group of K over F, and V an FG-module. Then

(1) There is a unique $F(\Gamma \times G)$-representation on V^K with $(\gamma, g) : a \otimes v \mapsto a\gamma \otimes vg$ for each $a \in K, v \in V, \gamma \in \Gamma, g \in G$.

(2) If W is a KG-submodule of V^K with $W\Gamma = W$, then $W = U^K$ for some FG-submodule U of V.

Proof. The representation in (1) is just the tensor product representation $\alpha \otimes \beta$ of $\Gamma \times G$ where α is the action of Γ on K, and β is the representation of G on V.

Assume W is as in (2), and extend a basis $Z = \{z_i : 1 \le i \le m\}$ for W inside of $X' \cup Z$ to a basis $Y = \{z_1, \ldots, z_m, x'_{m+1}, \ldots, x'_n\}$ of V^K, where $x'_i = 1 \otimes x_i$, and $X = \{x_i : 1 \le i \le n\}$ is an F-basis for V. For $i \le m$, $x'_i = \sum_{j>m}(a_{ij}x'_j + w_i)$, for $w_i \in W$ and $a_{ij} \in K$. Let $\gamma \in \Gamma$. Then

$$w_i\gamma = x'_i - \sum_{i>m}(a_{ij})\gamma x'_j = \sum_{j>m}(a_{ij} - a_{ij}\gamma)x'_j + w_i.$$

But by hypothesis $w_i\gamma \in W$, so, as Y is a basis for V^K, $a_{ij} = a_{ij}\gamma$ for all i, j.

Hence $a_{ij} \in \text{Fix}(\Gamma) = F$. As X' is a basis for V^K, $\{w_i: 1 \le i \le m\}$ is a basis for W and we have shown $w_i = 1 \otimes v_i$, where

$$v_i = x_i - \sum_{j>m} a_{ij} x_j \in V.$$

Thus $W = U^K$ where U is the subspace of V generated by $(v_i: 1 \le i \le m)$.

An irreducible FG-module V is *absolutely irreducible* if V^K remains irreducible for each extension K of F.

(25.8) Let V be an irreducible FG-module. Then V is absolutely irreducible if and only if $F = \text{End}_{FG}(V)$.

Proof. Assume $F = \text{End}_{FG}(V)$. Then, by 12.16, $E = \text{End}_F(V)$ is the enveloping algebra for G on V. So, if K is an extension of F and A' the enveloping algebra of G in $E' = \text{End}_K(V^K)$, then, by 25.4 and 25.6, $\dim_K(A') = \dim_F(A) = n^2 = \dim_K(E')$, so $A' = E'$. In particular A', and hence also G, is irreducible on V^K.

Conversely assume V is absolutely irreducible. Then V^K is irreducible where K is the algebraic closure of F. By 12.17, $E' = \text{End}_K(V^K)$ is the enveloping algebra for G in E', so, by 25.6, $n^2 = \dim_K(E') = \dim_F(A)$. Let $D = \text{End}_{FG}(V)$. By 12.16, $A \cong D^{m \times m}$ where $m = \dim_D(V)$. Then $n = \dim_F(V) = mk$, $k = \dim_F(D)$ and $\dim_F(A) = m^2 k$. So $m^2 k^2 = n^2 = m^2 k$ and hence $k = 1$. That is $F = D$.

If $\pi: G \to \text{GL}(V)$ is an FG-representation, X is a basis for V, and $\sigma \in \text{Aut}(F)$, then $\pi^\sigma: G \to \text{GL}(V)$ is the FG-representation with $M_X(g\pi^\sigma) = M_X(g\pi)^\sigma$. Here if $A = (a_{ij}) \in F^{n \times n}$ then $A^\sigma = (a_{ij}^\sigma)$. Notice that if $\tilde{\pi}^\sigma$ is the representation defined with respect to a different basis of V, then $\tilde{\pi}^\sigma$ is equivalent to π^σ by a remark after 13.1. So π^σ is independent of X, up to equivalence. I'll sometimes write V^σ for V regarded as an FG-module with respect to π^σ.

Recall the *character* of an FG-representation π is the function $\chi: G \to F$ defined by $\chi(g) = \text{Tr}(g\pi)$.

Let V be an FG-module and k a field with $F \le k \le \text{End}_{FG}(V)$. Then the action of k on V makes V into a k-space. Further that k-space structure extends the F-space structure and is preserved by G, so we can regard V as a kG-module. Similarly if K is a subfield of F then V is certainly a K-space and G preserves that K-structure. So V is also a KG-module.

(25.9) Let $\alpha: G \to \mathrm{GL}(V)$ be an irreducible FG-representation such that $K = \mathrm{End}_{FG}(V)$ is a field, and let β be the representation of G on V regarded as a KG-module, and χ the character of β. Then

(1) $K = F[\chi]$, where $F[\chi]$ is the F-subalgebra of K generated by the elements $\chi(g)$, $g \in G$.

(2) Assume L is a normal extension of K and $\sigma \in \mathrm{Gal}(K/F)^{\#}$. Then $L \otimes_K V^{\sigma}$ is not LG-isomorphic to $L \otimes_K V$.

Proof. Let X be a basis for V over K, $m = |X|$, and M the enveloping algebra of α in $\mathrm{End}_F(V)$. By definition $g\alpha = g\beta$ for each $g \in G$ and M is the F-subalgebra generated by $G\alpha$. Thus M consists of the elements $\sum a_g(g\alpha)$, $a_g \in F$. By 12.16, the map $\sum a_g(g\alpha) \mapsto \sum a_g M_X(g\beta)$ is an isomorphism of M with the ring $K^{m \times m}$ of all m by m matrices over K. Hence for each $x \in K$ there is $A \in M$ with $\mathrm{Tr}(A) = x$, and $A = \sum_{g \in G} a_g M_X(g\beta)$ for some $a_g \in F$. Now $x = \sum_{g \in G} a_g \chi(g) \in F[\chi]$. So (1) is established.

Assume the hypothesis of (2) and let E be the fixed field of σ. As $\sigma \neq 1$, $E \neq K$. As $K = \mathrm{End}_{EG}(V)$ we may assume $E = F$. Let γ be an extension of σ to L and $U = L \otimes_K V$. Then $L \otimes_K V^{\sigma} = U^{\gamma}$ and the character of G on U is still the character χ of β. Thus if U is LG-isomorphic to U^{γ} then $\chi = \chi^{\gamma}$, so $\chi(g)$ is contained in the fixed field k of γ. But $k \cap K = F$, so, by (1), $K = F$, a contradiction.

(25.10) Let V be an irreducible FG-module, k a Galois extension of F, and $K = \mathrm{End}_{FG}(V)$. Then

(1) $V^K = \bigoplus_{a \in A} W^a$ for some irreducible kG-module W and some $A \subseteq \mathrm{Gal}(k/F) = \Gamma$ with $\Gamma = AN_{\Gamma}(W)$, where $N_{\Gamma}(W) = \{\gamma \in \Gamma : W^{\gamma} \cong W\}$.

(2) Let U be an irreducible kG-module. Then V is an FG-submodule of U (regarded as an FG-module) precisely when U is kG-isomorphic to W^{σ} for some $\sigma \in \Gamma$.

(3) If $k \leq K$ then $A = \Gamma$ and W^{σ} is kG-isomorphic to V for some $\sigma \in \Gamma$.

(4) If K is a Galois extension of F then A is a set of coset representatives for $N_{\Gamma}(W)$ in Γ.

Proof. Let $\Gamma = \mathrm{Gal}(k/F)$. By 25.7, $\Gamma \times G$ is represented on V^K. Let W be an irreducible kG-submodule of V^k and $M = \langle W\gamma : \gamma \in \Gamma \rangle$. If $M \neq V^k$ then, by 25.7.2, $M = U^k$, for some FG-submodule U of V. As $0 \neq M \neq V^k$, $0 \neq U \neq V$ contradicting the irreducibility of V. Hence $V^k = M$ and then (1) follows from 12.5. Also V^k is generated as an F-module by the copies $(aV : a \in k^{\#})$ of V, so V^k is a homogeneous FG-module and hence each summand W^a is the sum of copies of V, as an FG-module. This gives half of (2).

Assume U is an irreducible kG-module and V an FG-submodule of U. Let X be an F-basis of V. As U is irreducible and V an FG-submodule of U, X generates U as a kG-module. Also $1 \otimes X$ is a basis for V^k over k, so we can define:

$$\alpha : V^k \to U$$
$$\sum_{x \in X} a_x(1 \otimes x) \mapsto \sum_{x \in X} a_x x \quad a_x \in k.$$

Then α is a surjective kG-homomorphism, so $V^k / \ker(\alpha) \cong U$ as a kG-module. Hence (1) and 12.5 imply $U \cong W^\sigma$ for some $\sigma \in A$. So (2) holds.

In (3) we may regard V as an irreducible kG-module, so, by (2), $V \cong W^\sigma$ for some $\sigma \in A$. Then, by (1), $|A| = \dim_F(V)/\dim_k(V) = |k : F| = |\Gamma|$ so $A = \Gamma$ and (3) is established.

To prove (4) let L be the extension generated by k and K and assume K is Galois over F. Then L is Galois over K, and, by (3), 25.5, and 25.3.3:

$$V^L \cong L \otimes_K V^K = \bigoplus_{\sigma \in \mathrm{Gal}(K/F)} L \otimes_K U^\sigma$$

with U KG-isomorphic to V. Now $L \otimes_K U^\sigma$ is an irreducible LG-module for each $\sigma \in \mathrm{Gal}(K/F)$, since, by 25.8, V is absolutely irreducible as a KG-module. Further if $\sigma \neq \tau$ then $L \otimes_K U^\sigma \not\cong L \otimes_K U^\tau$ by 25.9.2. On the other hand if a and b are distinct members of A with $W^a \cong W^b$ then $L \otimes_k W^a \cong L \otimes_k W^b$ so some irreducible occurs in V^L with multiplicity greater than 1, a contradiction. This completes the proof of (4).

A *splitting field* for a finite group G is a field F with the property that every irreducible FG-representation is absolutely irreducible. Notice that by 25.8 an irreducible FG-module V is absolutely irreducible precisely when $F = \mathrm{End}_{FG}(V)$. Hence, by 12.17:

(25.11) If F is algebraically closed then F is a splitting field for each finite group.

It will turn out in section 27 that if G is a finite group then a splitting field for G is obtained by adjoining a suitable root of unity to $\mathrm{GF}(p)$ for any prime p.

26 Representations over finite fields

The hypotheses of section 25 are continued in this section. In addition assume F is of finite order.

The following observations make things go particularly smoothly when F is finite.

(26.1) (1) Each finite dimensional division algebra over F is a finite field, and hence a Galois extension of F.

(2) If V is an irreducible FG-module then $\text{End}_{FG}(V)$ is a finite Galois extension of F.

The first remark follows from the well known facts that finite division rings are fields, and that every finite field is Galois over each of its subfields. The second remark is a consequence of the first and the hypothesis that V is of finite dimension over F.

(26.2) Let V be an irreducible FG-module, k a finite extension of F, and $\Gamma = \text{Gal}(k/F)$. Then

(1) $V^k = \bigoplus_{a \in A} W^a$ for some irreducible kG-module W and any set A of coset representatives for $N_\Gamma(W)$ in Γ.

(2) Let U be an irreducible kG-module. Then V is an FG-submodule of U precisely when U is kG-isomorphic to W^σ for some $\sigma \in \Gamma$.

Proof. This is a direct consequence of 26.1 and 25.10.

Let $\pi: G \to \text{GL}(V)$ be an FG-representation and K a subfield of F. We say π can be *written over* K if there exists an F-basis X of V such that each entry of $M_X(g\pi)$ is in K for each $g \in G$.

(26.3) Let $\pi: G \to \text{GL}(V)$ be an irreducible FG-representation, K a subfield of F, and $\langle \sigma \rangle = \text{Gal}(F/K)$. Then the following are equivalent:

(1) π can be written over K.

(2) $V = F \otimes_K U$ for some irreducible KG-submodule U of V.

(3) V is FG-isomorphic to V^σ.

Proof. The equivalence of (1) and (2) is trivial as is the implication (1) implies (3). Assume (3) and let U be an irreducible KG-submodule of V. Then $\Gamma = \text{Gal}(F/K) = N_\Gamma(V)$ as $\Gamma = \langle \sigma \rangle$ and $V \cong V^\sigma$. Hence, by 26.2, $U^F = V$. That is (3) implies (2).

(26.4) Let V be an irreducible FG-module, K a subfield of F, U an irreducible KG-submodule of V, and $E = N_F(U)$. Then $V = F \otimes_E U$.

Proof. Let $U^F = F \otimes_E U$ and $\Gamma = \text{Gal}(F/E)$. By 26.2, $U^F = \bigoplus_{a \in A} V^a$, where A is a set of coset representatives for $\Delta = N_\Gamma(V)$ in Γ. Let L be the fixed field of Δ and W an irreducible LG-submodule of V; as V is a homogeneous EG-module we may assume $U \leq W$. If $L \neq F$ then, by induction on $|F : E|$, $L \otimes_E U \cong W$, while, by 26.3, $V \cong F \otimes_L W$, So $V \cong F \otimes_E U$. Thus we may take $L = F$. But then $\Delta = 1$ so

$$\dim_E(U) = \dim_F(U^F) = |F : E| \dim_F(V) = \dim_E(V),$$

so $U = V$. Hence $F = N_F(U) = E$, and the lemma holds.

(26.5) Let V be an irreducible FG-module. Then the following are equivalent:
(1) V can be written over no proper subfield of F.
(2) V is an irreducible KG-module for each subfield K of F.
(3) $N_{\mathrm{Aut}(F)}(V) = 1$.

Proof. This follows from 26.3 and 26.4.

An FG-module V is *condensed* if V is absolutely irreducible and can be written over no proper subfield of F.

Theorem 26.6. Let p be a prime, F_p the field of order p, $\bar{F}_p$ its algebraic closure, Δ the set of finite subfields of $\bar{F}_p$, and Ω the set of pairs (F, V) where $F \in \Delta$ and V is an (isomorphism type of an) irreducible finite dimensional FG-module. Define a relation $\nearrow$ on Ω by $(F, V) \nearrow (K, U)$ if $F \leq K$ and V is an FG-submodule of U. Let $\sim$ be the equivalence relation on Ω generated by $\nearrow$. Then

(1) $(F, V) \nearrow (K, U)$ if and only if $F \leq K$ and U is a summand of $K \otimes_F V$.

(2) Let Λ be an equivalence class of $\sim$. Then, for each $F \in \Delta$, $\Lambda_F = \{(F, V) \in \Lambda\}$ is nonempty and $\mathrm{Aut}(F)$ is transitive on Λ_F. In particular $|\Lambda_{F_p}| = 1$ and the map $\Lambda \mapsto \Lambda_{F_p}$ is a bijection between the set of equivalence classes of $\sim$ and the isomorphism classes of irreducible finite dimensional $F_p G$ modules.

(3) If $(F, V), (K, U) \in \Omega$ with $F \leq K$ then $(F, V) \sim (K, U)$ if and only if $(F, V^\sigma) \nearrow (K, U)$, for some $\sigma \in \mathrm{Aut}(F)$.

(4) In each equivalence class Λ of $\sim$ there exists a unique $F \in \Delta$ such that the members of Λ_F are condensed. Indeed for $(F_p, V) \in \Lambda$, $F = \mathrm{End}_{F_p G}(V)$ and $(F, V) \in \Lambda$ with V a condensed FG-module.

Proof. Part (1) follows from 26.2. Let $(F, V) \in \Omega$. If $E \leq F$ then there is an irreducible EG-module V_E of V, and we saw during the proof of 25.10 that V is a homogeneous EG-module, so V_E is determined up to isomorphism. Write V_p for V_{F_p}.

Let $(K, U) \in \Omega$. Claim $(F, V) \sim (K, U)$ if and only if $V_p = U_p$. The sufficiency of $V_p = U_p$ is immediate from the definition of $\sim$; to prove necessity it suffices to take $(F, V) \nearrow (K, U)$ and to show $V_p \cong U_p$. But this follows from the last paragraph.

Let Λ be an equivalence class of $\sim$ and $(F_p, V) \in \Lambda$. By 26.2, for each $F \in \Delta$, $V^F = \bigoplus_{a \in A} W^a$, for some set A of coset representatives for $N_{\text{Aut}(F)}(W)$ in Aut(F). By (1) and the claim,

$$\Lambda_F = \{(F, W^a): a \in \text{Aut}(F)\}.$$

That is (2) holds.

To prove (3) observe that if $(K, U) \in \Omega$ and $F \leq K$ then $(F, U_F) \nearrow (K, U)$. Then, by (2), $(F, W) \sim (K, U)$ if and only if $W^\sigma = U_F$ for some $\sigma \in \text{Aut}(F)$. So (3) holds.

Let $F = \text{End}_{F_p G}(V)$. Then $(F, V) \in \Lambda$ and, by 25.8, V is an absolutely irreducible FG-module. By 25.10.3, 25.10.4, and 26.5, V can be written over no proper subfield of F, so V is condensed.

Finally suppose (K, U) is another condensed member of Λ. To complete the proof of (4) we must show $K = F$. Let k be the subfield of $\bar{F}_p$ generated by K and F. Use 26.5 and the fact that V is condensed as an FG-module to conclude $V \ncong V^\sigma$ for $\sigma \in \text{Aut}(F)^\#$; then by 26.2,

$$F \otimes_{\text{GF}(p)} V = \bigoplus_{\sigma \in \text{Aut}(F)} V^\sigma.$$

Then by 25.3 and 25.5,

$$k \otimes_{\text{GF}(p)} V = k \otimes_{\text{GF}(p)} \left(F \otimes_{\text{GF}(p)} V\right) = k \otimes_F \left(\bigoplus_{\gamma \in \text{Aut}(F)} V^\gamma\right)$$

$$= \bigoplus_{\gamma \in \text{Aut}(F)} (k \otimes_F V^\gamma).$$

As V is absolutely irreducible as an FG-module, $k \otimes_F V^\gamma$ is irreducible for each γ. Hence $k \otimes_{F_p} V$ has exactly $|F : F_p|$ irreducible summands. But, by symmetry between K and F, $k \otimes_{F_p} V$ also has $|K : F_p|$ irreducible summands, so $K = F$.

Theorem 26.6 defines an equivalence relation on the finite dimensional representations of G over finite fields of characteristic p. This equivalence relation has the property that each class contains a representative over each finite field of characteristic p. Hence we can think of such a representation as written over any finite field of characteristic p. However the lemma suggests that to each class there is associated a field F over which the representation is best written: namely the unique field over which the representation is condensed. I will refer to F as the *field of definition* of the representation. F can also be described as the smallest field over which the representation is absolutely irreducible. Equivalently $F = \mathrm{End}_{F_p G}(V)$, where V is the unique $F_p G$-module in the class.

27 Minimal polynomials

In section 27 F is a field, V is a finite dimensional vector space over F, and G is a group.

Suppose for the moment that A is a finite dimensional algebra over F and $F[x]$ is a polynomial ring. For $a \in A$ and $f(x) = \sum_{i=0}^{m} b_i x^i \in F[x]$, define $f(a) = \sum_{i=0}^{m} b_i a^i \in A$. Then the map $\alpha_a : F[x] \to A$ such that $\alpha_a : f \mapsto f(a)$, is an F-algebra homomorphism. The assumption that A is finite dimensional forces $\ker(\alpha_a) \neq 0$ since $F[x]$ is of infinite dimension. As $F[x]$ is a principal ideal domain (PID), $\ker(\alpha_a)$ is a principal ideal, and indeed there is a unique monic polynomial $f_a(x)$ with $\ker(\alpha_a) = (f_a)$. By definition f_a is the *minimal polynomial* of a. By construction f_a divides each polynomial which annihilates a. Further f_a is monic of degree at most $\dim_F(A)$, since $\dim_F(A) \geq \dim_F(F[x]/(f_a)) = \deg(f_a)$.

Applying these observations to the n^2-dimensional F-algebra $\mathrm{End}_F(V)$, it follows that each $g \in \mathrm{End}_F(V)$ has a minimal polynomial $\min(g) = \min(g, F, V)$. Observe that if $\pi : A \to B$ is an F-algebra isomorphism then a and $a\pi$ have the same minimal polynomial. In particular if X is a basis for V then $M_X : \mathrm{End}_F(V) \to F^{n \times n}$ is such an isomorphism, so $\min(g) = \min(M_X(g))$. If K is an extension of F, then $M_{1 \otimes X}(1 \otimes g) = M_X(g)$, so the minimal polynomial of $1 \otimes g$ over K divides the minimal polynomial of g over F. Indeed an easy application of rational canonical form shows the two minimal polynomials are equal. (Reduce to the case where V is a cyclic FG-module. Then there is a basis $X = (x_i : 1 \leq i \leq n)$ for V in which $x_i g = x_{i+1}$ for $i < n$ and $x_n g = -\sum_{i=0}^{n-1} a_{i+1} x_i$, where $f(x) = x^n + \sum_{i=0}^{n-1} a_i x^i$ is the minimal polynomial for g over F. Then $f(1 \otimes g) = 0$ and if $h(x) \in K[x]$ properly divides f then $h(1 \otimes g) \neq 0$ as

$$\{1 \otimes x_i : 1 \leq i \leq \deg(h) + 1\}$$

is linearly independent and $1 \otimes x_{i+1} = (1 \otimes x_1)g^i$.) Thus we have shown:

(27.1) The minimal polynomial of a linear transformation is unchanged by extension of the base field. That is, if K is an extension of F and $g \in \mathrm{End}_F(V)$, then $\min(g, F, V) = \min(1 \otimes g, K, V^K)$.

Let $g \in \mathrm{End}_F(V)$ and $a \in F$. We say a is a *characteristic value for g* if there exists $v \in V^{\#}$ with $vg = av$. We call v a *characteristic vector for a*.

(27.2) Let $g \in \mathrm{End}_F(V)$ and $a \in F$. Then a is a characteristic value for g if and only if a is a root of the minimal polynomial of g.

Proof. Let $f(x) = \min(g)$. If a is not a root of f then $(f, x - a) = 1$ so there exist $r, s \in F[x]$ with $fr + (x - a)s = 1$. Now if $v \in V^{\#}$ with $vg = av$ then $v = v1 = v(f(g)r(g) + (g - aI)s(g)) = 0$, since $f(g) = 0 = v(g - aI)$. This contradiction shows characteristic values of g are roots of f.

Conversely if a is a root of f, then $f = (x - a)h$, for some $h \in F[x]$. If a is not a characteristic value of g then $\ker(g - aI) = 0$ so $(g - aI)^{-1}$ exists. Hence, as $0 = f(g) = (g - aI)h(g)$, we also have $h(g) = 0$. But then f divides h, a contradiction.

$g \in \operatorname{End}_F(V)$ is *semisimple* if the minimal polynomial of g has no repeated roots.

(27.3) Let $g \in \operatorname{End}_F(V)$ and assume $\min(g)$ splits over F. Then g is semisimple if and only if g is diagonalizable.

Proof. This is Exercise 9.2.

(27.4) Let S be a finite subset of commuting elements of $\operatorname{End}_F(V)$, and assume the minimal polynomial of each number of S splits over F. Then there exists a basis X of V such that for each $s \in S$ the following hold:

(1) $M_X(s)$ is lower triangular.
(2) The entries on the main diagonal of $M_X(s)$ are the eigenvalues of $\min(s)$.
(3) If s is semisimple then $M_X(s)$ is diagonal.

Proof. Induct on $n + |S|$. If $n = 1$ the result is trivial, so take $n > 1$. If some $g \in S$ is a scalar transformation then by induction on $|S|$, the result holds for $S - \{g\}$, and then also for S. So no member of S is a scalar transformation.

Let $g \in S$. By 27.2, g possesses a characteristic value a_1. Let V_1 be the eigenspace of a_1 for g. Then $S \subseteq C(g) \leq N(V_1)$. As g is not a scalar, $V_1 \neq V$, so, by induction on n there is a basis X_1 for V_1 with $M_{X_1}(S|_{V_1})$ as claimed in the lemma. In particular there is a 1-dimensional subspace U of V_1 fixed by S.

If g is semisimple, then, by 27.3, $V = \bigoplus_{i=1}^m V_i$, where V_i is the eigenspace of the characteristic value a_i of g. Now, choosing a basis X_i for V_i as in the last paragraph, we see that $X = \bigcup_{i=1}^m X_i$ is a basis with the desired properties.

So we can assume no member of S is semisimple. Thus (3) is established. Finally S acts on $V/U = \bar{V}$ and, by induction on n, there is a basis $\bar{X}$ for $\bar{V}$ with $M_{\bar{X}}(S|_{\bar{V}})$ triangular. Now pick $X = (x_i : 1 \leq i \leq n)$ with $U = \langle x_1 \rangle$ and $(\bar{x}_i : 1 \leq i \leq n) = \bar{X}$; then $M_X(s)$ is triangular for each $s \in S$.

$g \in \operatorname{End}_F(V)$ is *nilpotent* if $g^m = 0$ for some positive integer m, and g is *unipotent* if $g = I + h$ for some nilpotent $h \in \operatorname{End}_F(V)$.

(27.5) Let $g \in \text{End}_F(V)$ and $n = \dim_F(V)$. Then

(1) The following are equivalent:

    (i) g is nilpotent, unipotent, respectively.

    (ii) $\min(g) = x^m$, $(x - 1)^m$, respectively, for some positive integer m.

    (iii) There exists a basis X for V such that $M_X(g)$ is lower triangular and all entries on the main diagonal are 0, 1, respectively.

(2) Unipotent elements are of determinant 1, and hence nonsingular.

(3) If g is nilpotent and semisimple then $g = 0$.

(4) If g is unipotent and semisimple then $g = I$.

(5) Let $\text{char}(F) = p > 0$. Then $g^{p^n} = 1$ if g is unipotent. Conversely if $g^{p^m} = 1$ for some positive integer m, then g is unipotent.

(6) If $\text{char}(F) = 0$ and $g \in \text{GL}(V)$ is of finite order then g is semisimple.

(7) If $\text{char}(F) = p > 0$ and $g \in \text{GL}(V)$ is of finite order m then g is semisimple if and only if $(m, p) = 1$.

Proof. All parts of the lemma are reasonably straightforward, but I'll make a couple of remarks anyway. If $g \in \text{GL}(V)$ is of finite order m the minimal polynomial of g divides $x^m - 1$, which has no multiple roots if $(m, \text{char}(F)) = 1$. Hence (6) and half of (7) hold. Parts (4) and (5) imply the remaining half of (7), since any power of a semisimple element is semisimple.

Recall a field F is perfect if $\text{char}(F) = 0$ or $\text{char}(F) = p > 0$ and $F = F^p$, where F^p is the image of F under the p-power map $a \mapsto a^p$. For example finite fields are perfect as are algebraically closed fields. We need the following elementary fact, which appears for example as the Corollary on page 190 of Lang [La].

(27.6) If F is perfect then every polynomial in $F[x]$ is separable.

(27.7) If F is perfect and $\alpha \in \text{End}_F(V)$, then there exists $f \in F[x]$ and a positive integer e with $(f, f') = 1$ (where f' is the derivative of f) and $f^e(\alpha) = 0$. If $\beta \in \text{End}_F(V)$ with $f(\beta) = 0$ then β is semisimple.

Proof. Let $\min(\alpha) = \prod_{i=1}^m f_i^{e_i}$ with f_i irreducible. Let

$$e = \max\{e_i : 1 \le i \le m\} \quad \text{and} \quad f = \prod_{i=1}^m f_i.$$

Then $f^e(\alpha) = 0$ and, by 27.6, f_i has no repeated roots. Hence f has no repeated roots, so $(f, f') = 1$. Also, if $f(\beta) = 0$ then $\min(\beta)$ divides f, and hence has no repeated roots, so β is semisimple.

The proof of the following lemma comes from page 71 of [Ch 2].

(27.8) Let F be perfect and $\alpha \in \operatorname{End}_F(V)$. Then there exist $\beta, \gamma \in \operatorname{End}_F(V)$ with $\alpha = \beta + \gamma$, β semisimple, and γ nilpotent. Further $\beta = t(\alpha)$ for some $t(x) \in F[x]$.

Proof. Choose f and e as in the last lemma. Then there exist h, $h_1 \in F[x]$ with $1 = f'h + fh_1$. Define an F-algebra homomorphism $\phi: F[x] \to F[x]$ by $(g\phi)(x) = g(x - f(x)h(x))$. Observe $g\phi \equiv (g - g'fh) \bmod f^2$, so in particular $f\phi \equiv f - f'fh = f - f(1 - fh_1) \equiv 0 \bmod f^2$. Then proceeding by induction on m and using the fact that ϕ is a homomorphism, $f\phi^m \equiv 0 \bmod f^{2m}$. Choose m with $2^m \geq e$. Then:

$$(27.8.1) \qquad f\phi^m \equiv 0 \bmod f^e, \quad \text{so} \quad (f\phi^m)(\alpha) = 0.$$

Next, for $g(x) = \sum_i a_i x^i \in F[x]$,

$$g\phi^k = \left(\sum_i a_i x^i \right) \phi^k = \sum_i a_i (x\phi^k)^i = g(x\phi^k),$$

as ϕ^k is a homomorphism, so

$$(27.8.2) \qquad g\phi^k = g(x\phi^k) \quad \text{for each } g \in F[x].$$

Also $x\phi = x - fh \equiv x \bmod f$, so proceeding by induction on k, and using the fact that ϕ is a homomorphism with $f\phi \equiv 0 \bmod f$, we conclude:

$$(27.8.3) \qquad x\phi^k \equiv x \bmod f.$$

We can now complete the proof of the lemma. Let $t = x\phi^m$, $\beta = t(\alpha)$, and $\gamma = \alpha - \beta$. Then $f(\beta) = f(t)(\alpha) = f(x\phi^m)(\alpha) = (f\phi^m)(\alpha) = 0$, by 27.8.2 and 27.8.1, respectively. Hence, by 27.7, β is semisimple. Also, by 27.8.3, $(x - t) \equiv 0 \bmod f$, so $\gamma^e = (\alpha - \beta)^e = (x - t)^e(\alpha) = 0$. Thus γ is nilpotent.

(27.9) Let F be perfect and $\alpha \in \operatorname{GL}(V)$. Then
    (1) There exist $\alpha_s, \alpha_u \in \operatorname{GL}(V)$ with α_s semisimple, α_u unipotent, and $\alpha = \alpha_s \alpha_u = \alpha_u \alpha_s$.
    (2) If $\alpha = \xi\mu = \mu\xi$ with $\xi, \mu \in \operatorname{GL}(V)$, ξ semisimple, and μ unipotent, then $\xi = \alpha_s$ and $\mu = \alpha_u$.
    (3) There exist polynomials $t(x), v(x) \in F[x]$ with $\alpha_s = t(\alpha)$ and $\alpha_u = v(\alpha)$.

Proof. By 27.8, $\alpha = \beta + \gamma$, with $\beta, \gamma \in \operatorname{End}_F(V)$, β semisimple, γ nilpotent, and $\beta = t(\alpha)$, for some $t(x) \in F[x]$. As $\beta = t(\alpha)$, $\beta \in Z(C(\alpha) \cap \operatorname{End}_F(V))$. Let $\bar{F}$ be the algebraic closure of F. By 27.4 and 27.5 there is a basis X of $\bar{F}$ with $M_X(\beta)$ diagonal and $M_X(\gamma)$ strictly lower triangular. From this it is evident that $\det(\beta) = \det(\alpha)$. Thus, as α is nonsingular, so is β.

Let $\alpha_s = \beta$ and $\alpha_u = I + \beta^{-1}\gamma$. As β^{-1} and γ commute and γ is nilpotent, $\beta^{-1}\gamma$ is nilpotent, so α_u is unipotent. By construction $\alpha = \alpha_s\alpha_u = \alpha_u\alpha_s$. So (1) and (3) hold.

Suppose ξ and μ are as in (2). As $\beta \in Z(C(\alpha))$, ξ and μ commute with β and α_u. By 27.4, β and ξ can be simultaneously diagonalized over $\bar{F}$, and hence $\beta^{-1}\xi$ is diagonalizable over $\bar{F}$, so $\beta^{-1}\xi$ is semisimple by 27.3. Similarly $\alpha_u\mu^{-1}$ is unipotent. Finally as $\beta\alpha_u = \alpha = \xi\mu$, $\beta^{-1}\xi = \alpha_\mu\mu^{-1}$ is both semisimple and unipotent, so, by 27.5.4, $\beta = \xi$ and $\alpha_u = \mu$.

α_s and α_u are called the *semisimple part* of α and the *unipotent part* of α, respectively, and the decomposition $\alpha = \alpha_s\alpha_u = \alpha_u\alpha_s$ is called the *Jordan decomposition* of α. As a consequence of 27.5 and 27.9 we have:

(27.10) Let $\alpha \in GL(V)$ be of finite order. Then α_s and α_u are powers of α. If char$(F) = 0$ then $\alpha = \alpha_s$, while if char$(F) = p > 0$ then $|\alpha_s| = |\alpha|_{p'}$ and $|\alpha_u| = |\alpha|_p$.

(27.11) Let F be perfect and $\alpha \in \text{End}_F(V)$, b a characteristic value of α in F, U the eigenspace of b for α in V, and K an extension of F. Then U^K is the eigenspace of b for $1 \otimes \alpha$ in V^K.

Proof. This is essentially an application of Jordan Form; I sketch a proof. Recall the map $f \mapsto f(\alpha)$ is an F-algebra representation of $F[x]$ on V with kernel $(M(x))$, where $M(x) = \min(\alpha)$. Let $M(x) = \prod_{i=1}^{r} p_i(x)^{e_i}$ be the prime factorization of M. From the theory of modules over a principal ideal domain (cf. Theorems 3 and 6 on pages 390 and 397 of Lang [La]) we know $V = \bigoplus_{i=1}^{r} V(i)$, where $V(i) = \ker(p_i(\alpha)^{e_i})$. Indeed as the polynomials $p_i(x)^{e_i}$ are relatively prime the same holds in V^K, so as $V(i)^K$ is contained in the kernel $(V^K)(i)$ of $p_i(x)^{e_i}$ on V^K, we conclude $V(i)^K = (V^K(i))$. Thus without loss $M = (x - b)^e$.

Again from the theory of modules over a PID, $V = \bigoplus_{i=1}^{s} V_i$, where $V_i = v_iF[x]$ is a cyclic module for $F[x]$ with annihilator $(x - b)^{e_i}$, $e = e_1 \geq e_2 \geq \cdots \geq e_s \geq 1$, and the invariants e_i are uniquely determined. As $V^K = \bigoplus_{i=1}^{s} V_i^K$ is a module with invariants e_i, we conclude the e_i are also the invariants of $F[x]$ on V^K, and it remains to observe that $U = \bigoplus_{i=1}^{s} U_i$, with $U_i = U \cap V_i$ of dimension 1.

(27.12) Let $\pi: G \to GL(V)$ be an FG-representation and K an extension of F. Then $\dim_K(C_{V^K}(g\pi^K)) = \dim_F(C_V(g\pi))$ for each $g \in G$.

Proof. This is a direct consequence of 27.11.

Recall the definition of a splitting field in section 25.

(27.13) Let G be a finite group of exponent m and $\pi: G \to \mathrm{GL}(V)$ an FG-representation. Let $k = m$ if $\mathrm{char}(F) = 0$ and $k = m_{p'}$ if $\mathrm{char}(F) = p > 0$. Let $n = \dim_F(V)$. Then

(1) Let χ be the character of π and $g \in G$. Then $\chi(g)$ is a sum of n kth roots of unity.

(2) F is a splitting field for G if F is finite and contains a primitive kth root of unity.

Proof. Let $\bar{F}$ be the algebraic closure of F. By 27.4 there is a basis X for $V^{\bar{F}}$ such that $M_X(g)$ is triangular. By 27.10, $g = g_s g_u$ with g_s semisimple and $|g_s|$ dividing k, with g_u unipotent, and with g_s and g_u powers of g. Thus the entries on the main diagonal of $M_X(g_u)$ are 1 and so the entries of $M_X(g)$ are the same as those of $M_X(g_s)$. In particular if a is such an entry then $a^k = 1$ as $(g_s)^k = 1$. So (1) holds.

Assume F is finite, F contains ω, a primitive kth root of 1, V is an irreducible FG-module, and let $K = \mathrm{End}_{FG}(V)$. By 26.1.2, K is a finite field extending F and hence containing ω. Let ψ be the character of V regarded as a KG-module. The argument of (1) shows $\psi(g) \in F$ for each $g \in G$. As $\psi(g) \in F$ for each $g \in G$, we conclude from 25.9.1 that $K = F$. Thus F is a splitting field for G by 25.8, completing the proof.

Recall that given representations α and β of groups G and H, respectively, there is a representation $\alpha \otimes \beta$ of $G \times H$. The definition of $\alpha \otimes \beta$ appears just before the statement of Lemma 25.7. This representation appears in the statement of the next two lemmas.

(27.14) Let $G \leq \mathrm{GL}(V)$, $M = C_{\mathrm{GL}(V)}(G)$, and assume V is a homogeneous FG-module. Write $I = \mathrm{Irr}(G, V, F)$ for the set of irreducible FG-submodules of V and choose $V_i \in I$, $1 \leq i \leq m$, with $V = \bigoplus_{i=1}^{m} V_i$. Let $K = \mathrm{End}_{FG}(V_1)$ be a field and $A = \mathrm{Hom}_{FG}(V_1, V)$. Then

(1) There exists $Y = (\alpha_i: 1 \leq i \leq m) \subseteq A$ with $V_1 \alpha_i = V_i$ and $\alpha_1 = 1$.

(2) A is a KG-module and Y is a K-basis for A. Y induces a unique K-space structure on V extending the F-structure such that $\alpha_i \in \mathrm{Hom}_{KG}(V_1, V_i)$ for each i. This structure is preserved by G.

(3) The map $\pi: M \to \mathrm{GL}(A; K)$ defined by $x\pi: \beta \mapsto \beta x$, $x \in M$, $\beta \in A$, is an isomorphism. M preserves the K-space structure on V.

(4) The map $\psi: A^{\#} \to I$ defined by $\psi: \beta \mapsto V_1\beta$ is a surjection and defines a bijection

$$\phi: S(A) \to S_G(V)$$

$$B \mapsto \langle b\psi: b \in B \rangle$$

between the set $S(A)$ of all K-subspaces of A and the set $S_G(V)$ of all FG-submodules of V. ϕ is a permutation equivalence of the actions of M on $S(A)$ and $S_G(V)$.

(5) The map $\theta: M \times G \to GL(V, K)$ defined by

$$(x, g)\theta: v \mapsto vxg$$

is a $K(M \times G)$-representation whose image is GM and which is equivalent to the tensor product of the representations of M on A and G on V_1 over K.

Proof. As V is homogeneous there exists an isomorphism $\alpha_1: V_1 \to V_i$ of FG-modules. Composing α_i with the inclusion $V_i \subseteq V$ we may regard α_i as a member of A. Choose $\alpha_1 = 1$. Then (1) holds. As $K = \text{End}_{FG}(V_1)$ and α_i is an FG-isomorphism, $K = \text{End}_{FG}(V_i)$ for each i and α_i is also a KG-isomorphism. Thus we have a unique scalar multiplication of K on V_i extending that of F such that $\alpha_i \in \text{Hom}_{KG}(V_1, V_i)$, so there is a K-space structure on V extending that on F. It is defined by:

$$a\left(\sum_{i=1}^{m} u\alpha_i\right) = \sum_{i=1}^{m} a(u\alpha_i), u \in V_1, a \in K.$$

Next $A = \text{Hom}_{FG}(V_1, \bigoplus_{i=1}^{m} V_i) = \bigoplus_{i=1}^{m} \text{Hom}_{FG}(V_i, V_i)$ is isomorphic to K^m as an F-space. Similarly $\text{Hom}_{KG}(V_1, V)$ is isomorphic to K^m as a K-space and is an F-subspace of A, so $A = \text{Hom}_{KG}(V_1, V)$ is also a K-space. Also, as Y is a K-linearly independent subset of A of order m, Y is a K-basis for A, so (2) holds.

Evidently, for $\beta \in A$ and $x \in M$, the composition βx is also in A, so the map π in (3) is a well-defined KM-representation. If $\alpha_i x = \alpha_i$ then, as G is irreducible on V_i, $V_i \leq C_V(x)$. Hence π is faithful. Let $(v_j: 1 \leq j \leq d)$ be a K-basis for V_1. Then

$$X = (v_j\alpha_i: 1 \leq j \leq d, 1 \leq i \leq m)$$

is a K-basis for V. An element of the general linear group $GL(A, K)$ on A (regarded as a K-space) may be regarded as an m by m matrix (a_{ij}) with respect to the basis Y of A. Given such an element define $x \in M$ by $v_j\alpha_i x = \sum_k v_j a_{ik}\alpha_k$. Then $x \in M$ with $x\pi = (a_{ij})$, and x preserves the K-structure on V. Hence π is an isomorphism and (3) holds.

Evidently ψ maps $A^{\#}$ into I and the induced map ϕ takes $S(A)$ into $S_G(V)$ and preserves inclusion. It is also clear that ϕ is an injection from the set $S_1(A)$ of 1-dimensional subspaces of A into I. Let $W \in I$ and let $\pi_i \colon W \to V_i$ be the ith projection. π_i is trivial or an isomorphism by Schur's Lemma, and there exists an isomorphism $\beta \colon V_1 \to W$. Then $a_i = \beta\pi_i\alpha_i^{-1} \in K$ and $a = \sum a_i\alpha_i \in A$ with $a\psi = V_1 a = W$, so $\phi \colon S_1(A) \to I$ is a bijection. For each $B, D \in S(A)$, $(B + D)\phi = B\phi + D\phi$; from this remark and its predecessor it is not difficult to complete the proof of (4).

Finally, by (3), M preserves the K-space structure on V. Hence the map θ in (5) is indeed a well-defined $K(M \times G)$-representation whose image is GM. The map $v_j\alpha_i \mapsto v_j \otimes \alpha_i$ induces an equivalence of θ with the tensor product representation

$$(x, g) \colon \alpha_i \otimes v_i \mapsto \alpha_i x \otimes v_j g$$

of $M \times G$ on $A \otimes V_1$, so (5) holds.

(27.15) Let $G_i, i = 1, 2$, be groups, F a splitting field for G_1 and G_2, and Δ_i a collection of representatives for the equivalence classes of finite dimensional irreducible FG_i-representations. Then the map $(\pi_1, \pi_2) \mapsto \pi_1 \otimes \pi_2$ is a bijection between $\Delta_1 \times \Delta_2$ and the set of equivalence classes of finite dimensional irreducible $F(G_1 \times G_2)$-modules.

Proof. Let $\pi_i \in \Delta_i$ with module V_i. By 27.14.4 and 27.14.5 there is a bijection between the $F(G_1 \times G_2)$-submodules of $V_1 \otimes V_2$ and the FG_2-submodules of V_2, so, as π_2 is irreducible, so is $\pi_1 \otimes \pi_2$. Conversely let $\pi \colon G_1 \times G_2 \to GL(V)$ be an irreducible FG-representation. By Clifford's Theorem, 12.13, V is a homogeneous FG_i-module, so, by 27.14.5, π is equivalent to $\pi_i \otimes \pi_2$ for some $\pi_i \in \Delta_i$. Indeed π_i is determined up to equivalence by the equivalence class of irreducible FG_i-submodules of V. So the lemma holds.

(27.16) Let G be a finite group and $\pi \colon G \to GL(V)$ an irreducible FG-representation. Then $Z(G\pi) = \langle z \rangle$ is a cyclic group of order relatively prime to the characteristic of F and, if F contains a primitive $|z|$th root of unity ω, then z acts on V by scalar multiplication via a power ω^k of ω with $(|z|, k) = 1$.

Proof. By 12.15, $Z(G\pi)$ is cyclic, say $Z(G\pi) = \langle z \rangle$. By Exercise 4.3, $n = |z|$ is relatively prime to char(F). So we can assume ω is a primitive nth root of 1 in F. Now z satisfies the polynomial $f(x) = x^n - 1$ so its minimal polynomial divides f and hence has roots powers of ω. So by 27.2 ω^k is a characteristic value for z for some $0 < k < n$, and then by Clifford's Theorem z acts by scalar multiplication via ω^k on V. Thus $n = |z| = |\omega^k|$, so $(k, n) = 1$.

(27.17) Let V be an irreducible FG-module. Assume G is finite and a semi-direct product of $H \unlhd G$ by X of prime order p. Assume $\dim(V) \neq p(\dim(C_V(X)))$. Then V is a homogeneous FH-module and if F is finite of order prime to p then V is an irreducible FH-module.

Proof. By Clifford's Theorem, 12.13, V is the direct sum of the homogeneous components $(V_i : 1 \leq i \leq r)$ of H on V, and X permutes these components transitively. So as $X = \langle x \rangle$ is of prime order p either V is a homogeneous FH-module or $p = r$ and X is regular on the components. But in the latter case $C_V(X) = \{\sum_{i=1}^{p} vx^i : v \in V_1\}$, so $\dim(V) = p \dim(C_V(X))$ contrary to hypothesis.

So assume F is finite of order q prime to p. By 27.13 there exists a finite extension K of F which is a splitting field for H and contains a primitive pth root of 1. By 26.2, $V^K = \bigoplus_{a \in A} W^a$ for some $A \subseteq \mathrm{Gal}(K/F)$. By 27.12, $\dim(C_{V^K}(X)) = \dim(C_V(X))$. So $\dim(W) \neq p \dim(C_W(X))$. Further if H is irreducible on W then, by 26.2, $\{W^a : a \in A\}$ is the set of H-homogeneous components of V^K. But if $0 \neq U$ is an FH-submodule of V then $0 \neq U^K$ is a $\mathrm{Gal}(K/F)$-invariant KH-submodule of V^K, so $W^a \leq U^K$ for some a, and then $V^K = U^K$. That is, H is irreducible on V. So, replacing (V, F) by (W, K), we may assume F is a splitting field for H and contains a primitive pth root of 1. The existence of the pth root forces $q \equiv 1 \bmod p$. Then by 27.14.4 the set I of irreducible H-submodules of V is of order $(q^m - 1)/(q - 1) \equiv m \bmod p$, where $\dim(V) = m \dim(U)$ for $U \in I$. As V is an irreducible FG-module, $V = \langle U^X \rangle$ is the sum of at most p conjugates of U, so $m \leq p$, with equality only if $V = \bigoplus_{i=1}^{p} U^{x^i}$. This last case is out by an argument in the last paragraph, so $m < p$. Further as V is an irreducible FG-module either $\{V\} = I$ or X is fixed point free on I, and we may assume the latter. But then, by 5.14, $m \equiv |I| \equiv 0 \bmod p$, contradicting $m < p$.

(27.18) Let p and q be primes with $q > p$, G a group of order pq, $X \in \mathrm{Syl}_p(G)$, and V a faithful FG-module with $(pq, \mathrm{char}(F)) = 1$ and $C_V(X) = 0$. Then G is cyclic.

Proof. Extending F if necessary, we may assume with 27.12 that F contains a primitive qth root of 1. By Exercise 2.5, G has a normal Sylow q-group H. As $(pq, \mathrm{char}(F)) = 1$, V is the direct sum of irreducible FG-modules by Maschke's Theorem, so H is faithful on one of these irreducibles, and hence we may assume V is an irreducible FG-module. So, by 27.17, V is a homogeneous FH-module. Hence 27.16 says H acts by scalar multiplication on V, so $H \leq Z(G)$. Thus $G = HX$ is cyclic.

Remarks. The classical theory of linear representations of finite groups considers representations over the complex numbers where things go relatively smoothly. Unfortunately many questions about finite groups require consideration of representations over fields of prime characteristic, particularly finite fields. For example we've seen that the study of representations in the category of groups of a group G on an elementary abelian p-group E is equivalent to the study of $GF(p)G$-representations on E regarded as a $GF(p)$-space. Representation theory over such less well behaved fields requires the kind of techniques introduced in this chapter. Lemma 27.18 provides one application of these techniques, and we will encounter others in section 36.

Little use is made in this book of the Jordan decomposition studied in section 27. It is however fundamental to the study of groups of Lie type as linear groups or algebraic groups.

Exercises for chapter 9

1. Let U, V, and W be FG-modules.
   (1) Define $\phi: L(U, V; W) \to \operatorname{Hom}_F((U, \operatorname{Hom}_F(V, W)))$ by $v(u(\alpha\phi)) = (u, v)\alpha$ for $u \in U$, $v \in V$, and $\alpha \in L(U, V; W)$. Prove ϕ is an isomorphism of F-spaces.
   (2) Prove ϕ is an isomorphism of $L(U, V; F)$ with $\operatorname{Hom}_F(U, V^*)$, where V^* is the dual of V.
   (3) G preserves $f \in L(U, V; F)$ if $f(ug, vg) = f(u, v)$ for each $g \in G$. Prove G preserves f if and only if $f\phi \in \operatorname{Hom}_{FG}(U, V^*)$.
   (4) Let θ be an automorphism of F of order at most 2 and $L_G(V, V^\theta)$ the set of sesquilinear forms on V with respect to θ which are preserved by G. Assume V is an irreducible FG-module. Prove $L_G(V, V^\theta) \neq 0$ if and only if V is isomorphic to $(V^\theta)^*$ as an FG-module, in which case each member of $L_G(V, V^\theta)^\#$ is nondegenerate. If V is absolutely irreducible prove the members of $L_G(V, V^\theta)$ are similar, if $\theta = 1$ each member is symmetric or each is skew symmetric, and if $|\theta| = 2$ some member is hermitian symmetric.
2. Prove Lemma 27.3. (Hint: Use the theory of modules over a PID as in the proof of 27.11.)
3. Let π_1 and π_2 be FG-representations, χ_i the character of π_i, and χ the character of $\pi_1 \otimes \pi_2$. Prove $\chi = \chi_1 \chi_2$; that is, for each $g \in G$, $\chi(g) = \chi_1(g)\chi_2(g)$.
4. Let G be a finite group and χ the character of a complex G-representation. Prove $\bar{\chi}(g) = \chi^*(g) = \chi(g^{-1})$ for each $g \in G$, where χ^* is the character of the dual representation and $\bar{\chi}(g)$ is the complex conjugate of $\chi(g)$.
5. (Spectral Theorem) Let V be a finite dimensional vector space over the complex numbers and f a positive definite unitary form on V. Then for

each $g \in O(V, f)$ there exists an orthonormal basis for (V, f) consisting of characteristic vectors for g. In particular every element of $O(V, f)$ is semisimple.

6. Let (V_i, f_i), $i = 1, 2$, be 2-dimensional symplectic spaces over a field F and let $V = V_1 \otimes V_2$. Let $\Delta(V_i)$ be the group of similarities g of V_i; that is $g \in GL(V_i)$ with $f_i(xg, yg) = \lambda(g)f_i(x, y)$ for all $x, y \in V_i$, and some $\lambda(g) \in F^{\#}$. Prove
   (1) There exists a unique nondegenerate symmetric bilinear form $f = f_1 \otimes f_2$ on V such that

   $$f(v_1 \otimes v_2, u_1 \otimes u_2) = f_1(v_1, u_1)f_2(v_2, u_2), u_i, v_i \in V_i.$$

   (2) There is a unique quadratic form Q on V associated to f with $Q(v_1 \otimes v_2) = 0$ for all $v_i \in V_i$.
   (3) (V, Q) is a 4-dimensional hyperbolic orthogonal space.
   (4) Let $\Delta_i = \Delta(V_i, f_i)$, $G_i = O(V_i, f_i)$, and π the tensor product representation of $\Delta = \Delta_1 \times \Delta_2$ on V (cf. the convention before 25.7). Prove $\Delta\pi \leq \Delta(V, Q)$ with $(g_1, g_2)\pi \in O(V, Q)$ if and only if $\lambda(g_1) = \lambda(g_2)^{-1}$. $\ker(\pi) = \{(\lambda I, \lambda^{-1} I): \lambda \in F^{\#}\}$.
   (5) Let $\alpha: (V_1, f_1) \to (V_2, f_2)$ be an isometry. Prove there is a unique $t \in GL(V)$ with $(u \otimes v\alpha)t = v \otimes u\alpha$. Prove t is a transvection or reflection in $O(V, Q)$, $(\Delta_1\pi)^t = \Delta_2\pi$, and $(G_1\pi)^t = G_2\pi$.
   (6) $\Delta(V, Q) = (\Delta\pi)\langle t \rangle$.
   (7) $\Omega(V, Q) = (G_1 G_2)\pi \cong SL_2(F) * SL_2(F)$, unless $|F| = 2$.

7. If π is an irreducible FG-representation and $\sigma \in \operatorname{Aut}(F)$, then π^{σ} is an irreducible FG-representation. If χ is the character of π then χ^{σ} is the character of π^{σ}, where $\chi^{\sigma}(g) = (\chi(g))^{\sigma}$ for $g \in G$.

8. Let V be an n-dimensional vector space over a field F of prime characteristic p and x an element of order p in $GL(V)$. Assume $n > p$. Prove $\dim(C_V(x)) > 1$.

9. Let V be a finite dimensional vector space over a field F, f a nontrivial sesquilinear form on V with respect to an automorphism θ of finite order m, and $G = O(V, f)$. Assume G is irreducible on V. Prove that either
   (1) V is FG-isomorphic to V^{θ} and G preserves a nondegenerate bilinear form on V, or
   (2) m is even and V is FG-isomorphic to V^{θ^2} but not to V^{θ}. Further $V = F \otimes_K U$ and G preserves a nondegenerate hermitian symmetric form on U, where K is the fixed field of θ^2 and U is an irreducible KG-submodule of V.
   (Hint: Use Exercise 9.1 and the fact that $(V^*)^*$ is FG-isomorphic to V.)

10. Let π be an irreducible CG-representation and σ a 1-dimensional CG-representation. Prove $\pi \otimes \sigma$ is an irreducible CG-representation.

10

Presentations of groups

A group F is *free* with free generating set X if it possesses the following universal property: each function $\alpha: X \to H$ of X into a group H extends uniquely to a homomorphism of F into H. We find in section 28 that for each cardinal C there exists (up to isomorphism) a unique free group F with free generating set of cardinality C. Less precisely: F is the largest group generated by X.

If W is a set of words in the alphabet $X \cup X^{-1}$, it develops that there is also a largest group G generated by X with $w = 1$ in G for each $w \in W$. This is the group $\mathrm{Grp}(X : W)$ generated by X subject to the relations $w = 1$ for $w \in W$.

In section 29 we investigate $\mathrm{Grp}(X : W)$ when $X = \{x_1, \ldots, x_n\}$ is finite and W consists of the words $(x_i x_j)^{m_{ij}} = 1$, for suitable integral matrices (m_{ij}). Such a group is called a *Coxeter group*. For example finite symmetric groups are Coxeter groups. We find that Coxeter groups admit a representation $\pi: G \to O(V, Q)$ where (V, Q) is an orthogonal space over the reals and $X\pi$ consists of reflections. If G is finite (V, Q) turns out to be Euclidean space. Finite Coxeter groups are investigated via this representation in section 30, which develops the elementary theory of root systems.

The theory of Coxeter groups will be used extensively in chapter 14 to study the classical groups from a geometric point of view.

28 Free groups

An object G in an algebraic category A is said to be *free* with *free generating set* X if X is a subset of G and, whenever H is an object in A and $\alpha: X \to H$ is a function from X into H, there exists a unique morphism $\beta: G \to H$ of G into H extending α. This section discusses free groups.

But first recall that a *monoid* is a set G together with an associative binary operation on G possessing an identity 1. Here's an example of a monoid. Let X be a set. A *word* in X is a finite sequence $x_1 x_2 \ldots x_n$ with x_i in X; n is the length of the word. The empty sequence is allowed and denoted by 1. Let M be the set of words in X and define the product of two words $x_1 \ldots x_n$ and $y_1 \ldots y_m$ to be the word $x_1 \ldots x_n y_1 \ldots y_m$ of length $n + m$. Observe that M is a monoid with identity the empty sequence 1. Indeed

(28.1) M is a free monoid with free generating set X.

For if H is a monoid and $\alpha: X \to H$ is a function then α can be extended to a morphism $\beta: M \to H$ defined by $(x_1 \ldots x_n)\beta = x_1\alpha \ldots x_n\alpha$. β is well defined as each word has a unique representation as a product of members of X. Evidently β is the unique extension of α. Indeed in general in any algebraic category if X is a generating set for an object G and $\alpha: X \to H$ is a function then there is at most one extension β of α to a morphism of G into H. This is because if β' is another extension then $K = \{g \in G: g\beta = g\beta'\}$ is a subobject of G containing X.

Next assume $X = Y \cup Y^{-1}$ with $Y \cap Y^{-1} = \varnothing$ and $y \mapsto y^{-1}$ is a bijection of Y with Y^{-1}. Set $(y^{-1})^{-1} = y$ for each $y \in Y$; thus $x \mapsto x^{-1}$ is a permutation of X of order 2. Define two words u and w to be *adjacent* if there exist words $a, b \in M$ and $x \in X$ such that $u = axx^{-1}b$ and $w = ab$, or vice versa. Thus adjacency is a reflexive and symmetric relation. Define an equivalence relation $\sim$ on M by $u \sim w$ if there exists a sequence $u = u_1, \ldots, u_n = w$ of words such that u_i and u_{i+1} are adjacent for each i, $1 \le i < n$. That is $\sim$ is the transitive extension of the adjacency relation. Write $\bar{w}$ for the equivalence class of a word w under $\sim$ and let F be the set of equivalence classes.

(28.2) If $u, v, w \in M$ with $u \sim v$ then $uw \sim vw$ and $wu \sim wv$.

Proof. There is a sequence $u = u_1, \ldots, u_n = v$ of words with u_i adjacent to u_{i+1}. Observe $u_iw = w_i$ is adjacent to w_{i+1} and $uw = w_1, \ldots, w_n = vw$, so $uw \sim vw$.

Now define a product on F by $\bar{u}\bar{v} = \overline{uv}$. By 28.2 this product is well defined. Further the product of the equivalence classes of the elements $x_n^{-1}, \ldots, x_1^{-1}$ is an inverse for $\bar{x}_1 \ldots \bar{x}_n$, so F is a group. Hence

(28.3) F is a group and $w \mapsto \bar{w}$ is a surjective monoid homomorphism of M onto F.

(28.4) F is a free group with free generating set $\bar{Y}$.

Proof. Observe first that $\bar{Y}$ generates F. This follows from 28.3 together with the fact that X generates M and $\bar{X} = \bar{Y} \cup \bar{Y}^{-1}$.

Now let H be a group and $\alpha: \bar{Y} \to H$ a function. Define $\beta: X \to H$ by $y\beta = \bar{y}\alpha$ and $y^{-1}\beta = (\bar{y}\alpha)^{-1}$ for $y \in Y$. As M is a free monoid on X, β extends to a morphism $\gamma: M \to H$. Define $\delta: F \to H$ by $\bar{w}\delta = w\gamma$. I must show δ is well defined; that is if $u \sim v$ then $u\gamma = v\gamma$. It suffices to assume u is adjacent to v, say $u = axx^{-1}b$ and $v = ab$. Then $u\gamma = (axx^{-1}b)\gamma = a\gamma x\beta(x\beta)^{-1}b\gamma = a\gamma b\gamma$, as desired. Evidently δ is a homomorphism extending α. As $\bar{Y}$ generates F, an earlier remark shows δ is the unique extension of α.

Lemma 24.8 shows that for each set S there exists a free group with free generating set S. The universal property implies:

(28.5) Up to isomorphism there exists a unique free group with free generating set of cardinality C for each cardinal C.

If $W \subseteq M$ is a set of words in X, write $\text{Grp}(Y : W)$ for the group F/N, where $N = \langle \bar{W}^F \rangle$ is the normal subgroup of F generated by the subset $\bar{W}$ of F. $\text{Grp}(Y : W)$ is the *group generated by Y subject to the relations $w = 1$ for $w \in W$*. That is $\text{Grp}(Y : W)$ is the largest group generated by the set Y in which $w = 1$ for each $w \in W$. To be more precise $\text{Grp}(Y : W) = F/N = G$ with Y and W identified with

$$(\bar{y}N : y \in Y) \quad \text{and} \quad (\bar{w}N : w \in W),$$

respectively. As $\bar{Y}$ generates F, Y generates G. As $\bar{w} \in N, w = 1$ in G for each $w \in W$. So G is generated by Y and each of the words in W is trivial. I'll also say *the relation $w = 1$ is satisfied* in G to indicate that $w = 1$ in G. G is the largest group with these properties in the following sense:

(28.6) Let $\alpha : Y \to Y\alpha$ be a function of Y onto a set $Y\alpha$, H a group generated by $Y\alpha$, and W a set of words $w = y_1^{\delta_1} \dots y_n^{\delta_n}$ in $Y \cup Y^{-1}$ with $w\alpha = (y_1\alpha)^{\delta_1} \dots (y_n\alpha)^{\delta_n} = 1$ in H for each $w \in W$. (That is H is generated by $Y\alpha$ and satisfies the relations $w = 1$ for $w \in W$.) Then α extends uniquely to a surjective homomorphism of $\text{Grp}(Y : W)$ onto H.

Proof. Let F be the free group on Y. Then there exists a unique homomorphism $\beta : F \to H$ of F onto H extending α. Let $N = \langle W^F \rangle$ and $\delta : v \mapsto vN$ the natural map of F onto $G = F/N$. Then $N = \ker(\delta)$. For $w = x_1 \dots x_n \in W$ with $x_i \in X$, $1 = w\alpha = x_1\alpha \dots x_n\alpha = x_1\beta \dots x_n\beta = (x_1, \dots x_n)\beta = w\beta$, so $w \in \ker(\beta)$. Thus $N \leq \ker(\beta)$, as N is the smallest normal subgroup of F containing W. But as $N \leq \ker(\beta)$, β induces a homomorphism $\gamma : G \to H$ with $\beta = \delta\gamma$. As β is surjective so is γ. Also, for $y \in Y$, $y\alpha = y\beta = y\delta\gamma = y\gamma$.

A *presentation* for a group G is a set Y of generators of G together with a set W of words in $Y \cup Y^{-1}$ such that the relation $w = 1$ is satisfied in G for each $w \in W$ and the homomorphism of $\text{Grp}(Y : W)$ onto G described in lemma 28.6 is an isomorphism. I'll summarize this setup with the statement $G = \text{Grp}(Y : W)$. Every group has at least one presentation; namely:

(28.7) For each group G,

$$G = \mathrm{Grp}(G \colon xy(xy)^{-1} = 1, x, y \in G)$$

is a presentation for G.

Proof. Let $g \mapsto \bar{g}$ be a bijection of G with a set $\bar{G}$, let F be the free group on $\bar{G}$, let W be the set of words $\bar{x}\bar{y}(\overline{xy})^{-1}$, $x, y \in G$, and let $N = \langle W^F \rangle$. Evidently G satisfies the relations defined by W so the map $\bar{g} \mapsto g$ extends to a homomorphism α of F onto G with $N \le \ker(\alpha)$. It remains to show $N = \ker(\alpha)$. Assume otherwise and let $v = \bar{x}_1 \ldots \bar{x}_n$ be a word in $\ker(\alpha) - N$ of minimal length n. As $1 \in N$, $n > 0$. If $n = 1$ then $x_1 = 1$ and $v = \bar{x}_1 = \bar{x}_1 \bar{x}_1 \bar{x}_1^{-1} \in W$, contrary to the choice of v. Hence $n \ge 2$ so $v = \bar{x}_1 \bar{x}_2 u$ for some word u of length $n - 2$. Now $w = \bar{x}_1 \bar{x}_2 (\overline{x_1 x_2})^{-1} \in W \subseteq N$, so $w^{-1} v = (\overline{x_1 x_2}) u \in \ker(\alpha) - N$. As $(\overline{x_1 x_2}) u$ is of length at most $n - 1$, the choice of v of minimal length is contradicted.

Here's a slightly more nontrivial example. The *dihedral group* of order $2n$ is defined to be the semidirect product of a cyclic group $X = \langle x \rangle$ of order n by a group $Y = \langle y \rangle$ of order 2, with respect to the automorphism $x^y = x^{-1}$. The case where $n = \infty$ and x is the infinite cyclic group is also allowed. Denote the dihedral group of order $2n$ by D_{2n}. Dihedral 2-groups have already been discussed in the chapter on p-groups.

(28.8) $D_{2n} = \mathrm{Grp}(x, y \colon x^n = y^2 = 1 = x^y x)$.

If $n = \infty$ the relation $x^n = 1$ is to be ignored. The proof of 28.8 is easy. Let $D = XY = D_{2n}$ and

$$G = \mathrm{Grp}(\bar{x}, \bar{y} \colon \bar{x}^n = \bar{y}^2 = 1 = \bar{x}^{\bar{y}} \bar{x}).$$

By 28.6 there is a homomorphism α of G onto D with $\bar{x}\alpha = x$ and $\bar{y}\alpha = y$. Then $n = |x|$ divides $|\bar{x}|$, so, as $\bar{x}^n = 1$, $|\bar{x}| = n$ and $\alpha \colon \bar{X} \to X$ is an isomorphism, where $\bar{X} = \langle \bar{x} \rangle$. Similarly, setting $\bar{Y} = \langle \bar{y} \rangle$, $\alpha \colon \bar{Y} \to Y$ is an isomorphism. $1 = \bar{x}\bar{x}^{\bar{y}}$ so $\bar{x}^{\bar{y}} = (\bar{x})^{-1}$. Thus $\bar{X} \trianglelefteq G = \langle \bar{x}, \bar{y} \rangle$, so $G = \bar{X}\bar{Y}$. Hence, as α restricted to $\bar{X}$ and $\bar{Y}$ is an isomorphism, α itself is an isomorphism.

29 Coxeter groups

Define a *Coxeter matrix* of size n to be an n by n symmetric matrix with 1s on the main diagonal and integers of size at least 2 off the main diagonal. To each Coxeter matrix $M = (m_{ij})$ of rank n there is associated a *Coxeter diagram*: this diagram consists of n nodes, indexed by integers $1 \le i \le n$, together with an edge of weight $m_{ij} - 2$ joining distinct nodes i and j, $1 \le i < j \le n$. We will be most concerned with Coxeter matrices with the following diagrams:

Thus a diagram of type A_n defines a Coxeter matrix of size n with $m_{ij} = 3$ if $|i - j| = 1$ and $m_{ij} = 2$ if $|i - j| > 1$. Similarly a diagram of type C_n defines a matrix with $m_{ij} = 3$ if $|i - j| = 1$ and $i, j < n$, $m_{n-1,n} = m_{n,n-1} = 4$, and $m_{ij} = 2$ if $|i - j| > 1$.

A *Coxeter system* with Coxeter matrix $M = (m_{ij})$ of size n is a pair (G, S) where G is a group, $S = (s_i: 1 \leq i \leq n)$ a family of elements of G, and

$$G = \mathrm{Grp}(S: (s_i s_j)^{m_{ij}} = 1, 1 \leq i \leq n, 1 \leq j \leq n).$$

G is a *Coxeter group* if there exists a family S such that (G, S) is a Coxeter system.

In the remainder of this section let (G, S) be a Coxeter system with matrix $M = (m_{ij})$ of size n. Let $S = (s_i: 1 \leq i \leq n)$. Notice

$$s_i^2 = (s_i s_i)^{m_{ii}} = 1.$$

(29.1) Let T be the set of conjugates of members of S under G and for each word $\mathbf{r} = r_1 \ldots r_m$ in the alphabet S and each $t \in T$ define

$$N(\mathbf{r}, t) = \{i: t = r_1^{(r_{i-1} \cdots r_1)}, 1 \leq i \leq m\}.$$

Then, if $r_1 \ldots r_m = r_1' \cdots r_k'$ in G, we have $|N(\mathbf{r}, t)| \equiv |N(\mathbf{r}', t)| \bmod 2$ for each $t \in T$.

Proof. Let Δ be the set product $\{\pm 1\} \times T$ and for $s \in S$ define $s\pi \in \mathrm{Sym}(\Delta)$ by $(\varepsilon, t)s\pi = (\varepsilon\delta(s, t), t^s)$, where $\delta(s, t) = -1$ if $s = t$ and $\delta(s, t) = +1$ if $s \neq t$. Observe that $s\pi$ is an involution. I'll show $(s_i\pi s_j\pi)^{m_{ij}} = 1$ for all i, j; hence, by 28.6, π extends to a permutation representation π of G on Δ. In particular π is a homomorphism so, if $g = r_1 \ldots r_m \in G$ and $r_i \in S$, then

$$(\varepsilon, t)g\pi = (\varepsilon, t)r_1\pi \ldots r_m\pi = \left(\varepsilon\left(\prod_{i=1}^{m} \delta(r_i, t^{r_1 \cdots r_{i-1}})\right), t^g\right).$$

Further $\delta(r_i, t^{r_1 \cdots r_{i-1}}) = -1$ exactly when $i \in N(\mathbf{r}, t)$. So $(\varepsilon, t)g\pi = (\varepsilon(-1)^{|N(\mathbf{r},t)|}, t^g)$ and hence $|N(\mathbf{r}, t)|$ mod 2 depends only on g and not on $\mathbf{r}$.

It remains to show $(s_i \pi s_j \pi)^{m_{ij}} = 1$; equivalently $\alpha = (s_i \pi s_j \pi)^{m_{ij}}$ fixes each $(\varepsilon, t) \in \Delta$. Now $(\varepsilon, t)\alpha = (\varepsilon\delta, t^{(s_i s_j)^{m_{ij}}}) = (\varepsilon\delta, t)$, as $(s_i s_j)^{m_{ij}} = 1$, where of course

$$\delta = \delta(s_i, t)\delta(s_j, t^{s_i})\ldots.$$

So we must show $\delta = 1$. But $\delta(s_i, t^{(s_i s_j)^k}) = -1$ for some $0 \le k < m_{ij}$, precisely when $t = (s_i s_j)^{2k} s_i$. Also $\delta(s_j, t^{(s_i s_j)^k s_i}) = -1$ precisely when $t = (s_i s_j)^{2k+1} s_i$. Further, by Exercise 10.1, $\langle s_i, s_j \rangle$ is either dihedral of order $2m$, where m divides m_{ij}, or of order at most 2. Hence if $t \notin \langle (s_i s_j) \rangle s_i$ then all terms in δ are $+1$, while if $t \in \langle (s_i s_j) \rangle s_i$ exactly $(2m_{ij})/|s_i s_j|$ terms are -1, as $t = (s_i s_j)^{d|s_i s_j|} s_i$ for $0 \le d < (2m_{ij})/|s_i s_j|$.

(29.2) The members of S are involutions.

Proof. In the proof of 29.1 a homomorphism π of G into $\mathrm{Sym}(\Delta)$ was constructed for which $s\pi$ was an involution for each $s \in S$. So $2 = |s\pi|$ divides $|s|$. But of course, as (G, S) is a Coxeter system, $s_i^2 = 1$.

If H is a group with generating set R then the *length* of $h \in H$ with respect to R is the minimal length of a word w in the alphabet $R \cup R^{-1}$ such that $w = h$ in H. Denote this length by $l(h) = l_R(h)$.

(29.3) Let $g \in G$ and $\mathbf{r} = r_1 \ldots r_m$ a word in the alphabet S with $g = \mathbf{r}$ in G. Define

$$\eta(\mathbf{r}) = |\{t \in T : |N(\mathbf{r}, t)| \equiv 1 \bmod 2\}|$$

in the notation of 29.1. Then $\eta(\mathbf{r}) = l(g)$.

Proof. By 29.1, if $\mathbf{r}' = g$ then $\eta(\mathbf{r}) = \eta(\mathbf{r}')$, while by definition of $l(g)$ there is $\mathbf{r}' = r_1' \ldots r_k'$ with $k = l(g)$ and $\mathbf{r}' = g$ in G. So, without loss, $m = l(g)$. So evidently $\eta(\mathbf{r}) \le m = l(g)$. If $\eta(\mathbf{r}) < m$ there are $i, j, 1 \le i \le j \le m$, with $r_i^{r_{i-1}\cdots r_1} = r_j^{r_{j-1}\cdots r_1}$, so $r_i r_{i+1} \ldots r_{j-1} = r_{i+1} \ldots r_j$ and hence $g = r_1 \ldots r_{i-1} r_i \ldots r_{j-1} r_j \ldots r_m = r_1 \ldots r_{i-1} r_{i+1} \ldots r_{j-1} r_{j+1} \ldots r_m$ is of length at most $m - 2$, contrary to the choice of $\mathbf{r}$.

(29.4) Let H be a group generated by a set R of involutions. Then (H, R) is a Coxeter system precisely when the following Exchange Condition is satisfied:

Exchange Condition: If $r_i \in R, 0 \le i \le n$, and $h = r_1 \ldots r_n \in H$ with $l(h) = n$ and $l(r_0 h) \le n$, then there exists $1 \le k \le n$ such that $r_0 r_1 \ldots r_{k-1} = r_1 \ldots r_k$ in H.

Proof. Suppose first that (H, S) is a Coxeter system and let r_i, h, satisfy the hypothesis of the Exchange Condition. Then, setting $\mathbf{r} = r_0 r_1 \dots r_n$, $\eta(\mathbf{r}) \leq l(r_0 h) < n+1$ by 29.3, so there are $i, j, 0 \leq i < j \leq n$ with $r_i^{r_{i-1} \cdots r_0} = r_j^{r_{j-1} \cdots r_0}$ and hence $r_i \dots r_{j-1} = r_{i+1} \dots r_j$. By 29.3, $\eta(r_1 \dots r_n) = l(h) = n$, so $i = 0$. Thus the Exchange Condition is satisfied.

Conversely suppose (H, R) satisfies the Exchange Condition. Let $\alpha: R \to X$ be a function into a group X such that, for each $r, s \in R$, $(r\alpha s\alpha)^{|rs|} = 1$. It will suffice to show α extends to a homomorphism of H into X.

Let $h \in H$, $n = l(h)$, and $r_1 \dots r_n = h = s_1 \dots s_n$ with $r_i, s_i \in R$. Claim

$$r_1\alpha \dots r_n\alpha = s_1\alpha \dots s_n\alpha \quad \text{and} \quad \{r_i: 1 \leq i \leq n\} = \{s_i: 1 \leq i \leq n\}.$$

Assume not and pick a counterexample with n minimal. Then $l(s_1 h) = n - 1 < l(h)$, so, by the Exchange Condition, $s_1 r_1 \dots r_{k-1} = r_1 \dots r_k$ for some k. Hence $s_1 r_1 \dots r_{k-1} r_{k+1} \dots r_n = h = s_1 \dots s_n$ so $r_1 \dots r_{k-1} r_{k+1} \dots r_n = s_2 \dots s_n$. Thus, by minimality of $n, r_1\alpha \dots r_{k-1}\alpha r_{k+1}\alpha \dots r_n\alpha = s_2\alpha \dots s_n\alpha$ and $\{s_2, \dots, s_n\} = \{r_i: i \neq k\}$. Also if $k < n$ then, by minimality of n, $s_1\alpha r_1\alpha \dots r_{k-1}\alpha = r_1\alpha \dots r_k\alpha$ and $\{s_1, r_1, \dots, r_{k-1}\} = \{r_1 \dots, r_k\}$, which combined with the last set of equalities establishes the claim. So $k = n$, $s_1 r_1 \dots r_{n-1} = h$, and $\{r_1, \dots, r_{n-1}\} = \{s_2, \dots, s_n\}$. Similarly $r_1 s_1 \dots s_{n-1} = h$ and $\{s_1, \dots, s_{n-1}\} = \{r_2 \dots, r_n\}$. In particular $\{r_1 \dots, r_n\} = \{s_1 \dots, s_n\}$, establishing half the claim. Replacing $r_1 \dots, r_n$, and $s_1, \dots, s_n$, by $s_1, r_1, \dots, r_{n-1}$ and $r_1, s_1, \dots, s_{n-1}$, and continuing in this manner, we obtain $(s_1 r_1)^{n/2} = (r_1 s_1)^{n/2}$ or $r_1(s_1 r_1)^{(n-1)/2} = s_1(r_1 s_1)^{(n-1)/2}$, with equality of images under α failing in the respective case. It follows that $(s_1 r_1)^n = 1$, so the order m of $s_1 r_1$ in H divides n. But by hypothesis the order of $s_1\alpha r_1\alpha$ divides m, so equality of images under α does hold, a contradiction.

So the claim is established. Since Coxeter systems satisfy the Exchange Condition we can record:

(29.5) Let $g \in G$ with $l(g) = m$ and $r_i, t_i \in S$ with $r_1 \dots r_m = t_1 \dots t_m = g$. Then $\{r_i: 1 \leq i \leq m\} = \{t_i: 1 \leq i \leq m\}$.

Now back to the proof of 29.4. Define $\alpha: H \to X$ by $h\alpha = r_1\alpha \dots r_n\alpha$, for $h = r_1 \dots r_n$ with $n = l(h)$ and $r_i \in R$. The claim shows α to be well defined. Let's see next that $(rh)\alpha = r\alpha h\alpha$ for $r \in R$. If $l(rh) = l(h) + 1$ this is clear, so assume not. Then, by the Exchange Condition, $rr_1 \dots r_{k-1} = r_1 \dots r_k$ for some $k \leq n$. By the claim, $r\alpha r_1\alpha \dots r_{k-1}\alpha = r_1\alpha \dots r_k\alpha$. Also $rh = r_1 \dots r_{k-1} r_{k+1} \dots r_n$ is of length at most $n - 1$. As $l(h) = n$, we conclude $l(rh) = n - 1$. So $(rh)\alpha = r_1\alpha \dots (r_{k-1})\alpha(r_{k+1})\alpha \dots r_n\alpha = r\alpha r_1\alpha \dots r_n\alpha = r\alpha h\alpha$, establishing the second claim.

It remains to show $g\alpha h\alpha = (gh)\alpha$ for $g, h \in H$. Assume not and choose a counter example with $l(g)$ minimal. By the last paragraph, $l(g) > 1$, so $g = rk$,

$r \in R, k \in H$ with $l(k) = l(g) - 1$. Then $(gh)\alpha = r\alpha(kh)\alpha = r\alpha k\alpha h\alpha = g\alpha h\alpha$ by minimality of $l(g)$, completing the proof.

Let V be an n-dimensional vector space over the reals $\mathbb{R}$ with basis $X = (x_i : 1 \leq i \leq n)$ and let Q be the quadratic form on V with $(x_i, x_j) = -\cos(\pi/m_{ij})$, where $(\ ,\ )$ is the bilinear form determined by Q as in section 19. Thus, for $u, v \in V$, $Q(v) = (v, v)/2$ and

$$(u, v) = Q(u) + Q(v) - Q(u + v).$$

(29.6) (1) $Q(x_i) = 1/2$.
    (2) $(x_i, x_j) \leq 0$ for $i \neq j$, with $(x_i, x_j) = 0$ if and only if $m_{ij} = 2$.

Proof. $m_{ii} = 1$, so $Q(x_i) = (x_i, x_i)/2 = -\cos(\pi)/2 = 1/2$. Similarly if $i \neq j$ then $m_{ij} \geq 2$, so $(x_i, x_j) = -\cos(\pi/m_{ij}) \leq 0$ with equality if and only if $m_{ij} = 2$.

By 29.6 and 22.6.2 there is a unique reflection r_i on V with center $\langle x_i \rangle$; moreover $v r_i = v - 2(v, x_i)x_i$ for $v \in V$.

(29.7) For $i \neq j$, $r_i r_j$ is of order m_{ij}, $\langle r_i, r_j \rangle \cong D_{2m_{ij}}$, and if $m_{ij} > 2$ then $\langle r_i, r_j \rangle$ is irreducible on $\langle x_i, x_j \rangle$.

Proof. Let $U = \langle x_i, x_j \rangle$, $D = \langle r_i, r_j \rangle$, $m = m_{ij}$, and $\theta = \pi/m$. Observe that, for $a, b \in \mathbb{R}$,

$$2Q(ax_1 + bx_2) = a^2 - 2ab\cos\theta + b^2 = (a - b\cos\theta)^2 + b^2(\sin\theta)^2 \geq 0,$$

with equality precisely when $a = b = 0$. Thus Q is a positive definite quadratic form on U, so in particular U is a nondegenerate subspace of V and hence $V = U \oplus U^\perp$. But $U^\perp \leq x_k^\perp \leq C_V(r_k)$ for $k = i$ and j, so $U^\perp \leq C_V(D)$. Hence D is faithful on U.

As Q is positive definite on U, $(x_k, x_k) = 1$, and $(x_i, x_j) = -\cos\theta$, (U, Q) is isometric to 2-dimensional Euclidean space $\mathbb{R}^2$ with the standard inner product and with $x_i = (1, 0)$ and $x_j = (\cos(\pi-\theta), \sin(\pi-\theta))$ in the standard coordinate system. Thus r_i and r_j are the reflections on $\mathbb{R}^2$ through the vertical axis and the axis determined by $\pi/2 - \theta$, respectively. Hence $r_i r_j$ is the rotation through the angle $-2\pi/m$, and therefore is of order m as desired.

Thus the first claim of 29.7 is established and the second is a consequence of the first and Exercise 10.1. $\langle x_k \rangle$ and $x_k^\perp \cap U$ are the only nontrivial proper subspaces of U fixed by r_k and if $m > 2$ then, by 29.6.2, $\langle x_i \rangle \neq x_j^\perp \cap U$, so D is irreducible on U.

(29.8) Let $W = \langle r_i : 1 \leq i \leq n \rangle$ be the subgroup of $O(V, Q)$ generated by the reflections $(r_i : 1 \leq i \leq n)$. Then there exists a surjective homomorphism $\alpha : G \to W$ with $s_i \alpha = r_i$ for each i. In particular α is an $\mathbb{R}G$-representation which identifies S with a set of reflections in $O(V, Q)$.

Proof. This is immediate from 29.7 and 28.6.

(29.9) (1) S is of order n.
  (2) For each $i \neq j$, $|s_i s_j| = m_{ij}$ and $\langle s_i, s_j \rangle \cong D_{2m_{ij}}$.

Proof. The map $\alpha : \langle s_i, s_j \rangle \to \langle r_i, r_j \rangle$ induced by the map of 29.8 is a surjective homomorphism. By Exercise 10.1 and 29.7, $\langle r_i, r_j \rangle = \mathrm{Grp}(r_i, r_j : r_i^2 = r_j^2 = (r_i r_j)^{m_{ij}})$ so, as $\langle s_i, s_j \rangle$ satisfies these relations, 28.6 says there is a homomorphism β of $\langle r_i, r_j \rangle$ onto $\langle s_i, s_j \rangle$ with $r_k \beta = s_k$. Then $\beta = \alpha^{-1}$ so α is an isomorphism and 29.7 implies (2). As $r_i \neq r_j$ for $i \neq j$, (1) holds.

Let $\Delta = \{1, \ldots, n\}$ be the set of nodes of the Coxeter diagram of (G, S). The graph of the diagram is the graph on Δ obtained by joining i to j if the edge between i and j in the Coxeter diagram is of weight at least 1, or equivalently if $m_{ij} \geq 3$.

(29.10) Let $(\Delta_k : 1 \leq k \leq r)$ be the connected components of the graph of the Coxeter diagram Δ of (G, S) and let $G_k = \langle s_i : i \in \Delta_k \rangle$. Then
  (1) $G = G_1 \times \cdots \times G_r$ is the direct product of the subgroups G_k, $1 \leq k \leq r$.
  (2) V is the orthogonal direct sum of the subspaces $V_k = [G_k \alpha, V]$, $1 \leq k \leq r$.

Proof. If i and j are in distinct components of Δ then $|s_i s_j| = m_{ij} = 2$, so $[s_i, s_j] = 1$. Thus G is the central product of the subgroups G_k, $1 \leq k \leq r$, and hence there is a surjective homomorphism β of $G_1 \times \cdots \times G_r = D$ onto G with $s_i \beta = s_i$ for each i. Conversely S satisfies the Coxeter relations in D, so by 28.6 there is a homomorphism γ of G onto D with $s_i \gamma = s_i$. Then $\gamma = \beta^{-1}$, so β is an isomorphism and (1) holds. Similarly, by 29.6.2, $x_j \in x_i^{\perp}$, so, if $j \in \Delta_a$ and $i \in \Delta_b$, then

$$V_a = \langle [V, r_k] : k \in \Delta_a \rangle = \langle x_k : k \in \Delta_a \rangle \leq V_b^{\perp},$$

and hence (2) holds.

Because of 29.10 it does little harm to assume the graph of the Coxeter diagram of (G, S) is connected. In that event (G, S) is said to be an *irreducible Coxeter system*.

(29.11) Assume (G, S) is an irreducible Coxeter system. Then
  (1) G acts absolutely irreducibly on $V/V^\perp$.
  (2) If W is finite then (V, Q) is nondegenerate.

Proof. Let U be a proper $\mathbb{R}G$-submodule of V. For $s_i \in S$, $[V, s_i] = \langle x_i \rangle$ is of dimension 1 so either $x_i \in U$ or $U \leq C_V(s_i) = x_i^\perp$. Thus either there exists $i \in \Delta$ with $x_i \in U$ or $U \leq \bigcap_{i \in \Delta} X_i^\perp = V^\perp$, and I assume the former. Claim $x_j \in U$ for each $j \in \Delta$, so that $V = \langle x_j : j \in \Delta \rangle \leq U$, contradicting U proper. As the graph of Δ is connected it suffices to prove $x_j \in U$ for $m_{ij} > 2$. But, for such j, $\langle s_i, s_j \rangle$ is irreducible on $\langle x_i, x_j \rangle$ by 29.7, so as $x_i \in U$ and U is G-invariant, $x_j \in U$.

   I've shown G is irreducible on $\bar{V} = V/V^\perp$. $F = \text{End}_{\mathbb{R}G}(\bar{V})$ acts on $[\bar{V}, s_i] = \langle x_i \rangle$, so, for $a \in F$, $ax_i = bx_i$ for some $b \in \mathbb{R}^\#$, and hence, as G is irreducible on $\bar{V}$ and centralizes a, a acts as a scalar transformation via b on $\bar{V}$. That is $F = \mathbb{R}$. So, by 25.8, G is absolutely irreducible on $\bar{V}$.

   Suppose W is finite. Then, by Maschke's Theorem, $V = V^\perp \oplus Z$ for some $\mathbb{R}G$-submodule Z of V. $\bar{V}$ is $\mathbb{R}G$-isomorphic to Z and $[\bar{V}, s_i] \neq 1$, so $[Z, s_i] \neq 1$. Hence $\langle x_i \rangle = [Z, s_i] \leq Z$, so $V = \langle x_i : 1 \leq i \leq n \rangle \leq Z$. Thus $V^\perp = 0$ and (2) holds.

(29.12) Assume (G, S) is an irreducible Coxeter system and W is finite. Then (V, Q) is isometric to n-dimensional Euclidean space under the usual inner product.

Proof. Let h be the bilinear form on V which makes X into an orthonormal basis and define $g : V \times V \to \mathbb{R}$ by $g(u, v) = \sum_{w \in W} h(uw, vw)$. It is straightforward to check that g is a symmetric bilinear form on V preserved by G. As the quadratic form of h is positive definite, so is the form P of g, so (V, g) is nondegenerate. But, by 29.11, V is an absolutely irreducible $\mathbb{R}G$-module, so, by Exercise 9.1, $P = aQ$ for some $a \in \mathbb{R}^\#$. As P is positive definite and $Q(x_i) = 1/2 > 0$, $a > 0$, so Q is positive definite. By 19.9 there is a basis $Y = (y_i : 1 \leq i \leq n)$ for V with $(y_i, y_j) = 0$ for $i \neq j$. As Q is positive definite, $Q(y_i) > 0$ so, adjusting by a suitable scalar, we can take $Q(y_i) = 1$, since every positive member of $\mathbb{R}$ is a square in $\mathbb{R}$. Thus Y is an orthonormal basis for (V, Q), so (V, Q) is Euclidean space under the usual inner product.

Let $\Delta = \{1, \ldots, n\}$ and for $J \subseteq \Delta$ let $S_J = \{S_j : j \in J\}$ and $G_J = \langle S_J \rangle$. The subgroups G_J and their conjugates under G are the *parabolic subgroups* of the Coxeter system (G, S).

(29.13) Let $J, K \subseteq \Delta$ and $g \in G_J$. Then
  (1) If $l(g) = m$ and $g = s_{i_1} \ldots s_{i_m}$ with $s_{i_k} \in S$ then $i_k \in J$ for each $1 \le k \le m$.
  (2) (G_J, S_J) is a Coxeter system with Coxeter matrix $M_J = (m_{ij}), i, j \in J$.
  (3) $\langle G_J, G_K \rangle = G_{J \cup K}$ and $G_J \cap G_K = G_{J \cap K}$.
  (4) If $G_J = G_K$ then $J = K$.

Proof. Let $g = s_{\alpha_1} \ldots s_{\alpha_k}$ with $\alpha_k \in J$ and k minimal subject to this constraint. Claim $k = m$. By induction on k, $l(s_{\alpha_1} g) = k - 1$, so by the Exchange Condition either $k = l(g)$ or $s_{\alpha_1} \ldots s_{\alpha_{t-1}} = s_{\alpha_2} \ldots s_{\alpha_t}$ for some t, in which case $g = s_{\alpha_2} \ldots s_{\alpha_{t-1}} s_{\alpha_{t+1}} s_{\alpha_k}$, contrary to minimality of k. Hence (1) follows from 29.5.

Next (1) says the Exchange Condition is satisfied by (G_J, S_J), so (G_J, S_J) is a Coxeter system by 29.4. 29.9 says M_J is the Coxeter matrix of (G_J, S_J).

The first remark in (3) and the inclusion $G_{J \cap K} \le G_J \cap G_K$ are trivial. Part (1) gives the inclusion $G_J \cap G_K \le G_{J \cap K}$.

If $G_J = G_K$ then $G_J = G_{J \cap K}$ by (3), so we may assume $K \subseteq J$. By (2) we may assume $J = \Delta$. Hence $G = G_J = G_K$, so, by (1), $S = S_K$. Hence $K = \Delta = J$ by 29.9.1.

30 Root systems

In this section V is a finite dimensional Euclidean space over a field F equal to the reals or the rationals. That is V is an n-dimensional space over F together with a quadratic form Q such that (V, Q) possesses an orthonormal basis. Hence Q is positive definite. Let $(,)$ be the bilinear form defined by Q. For $v \in V^\#$ there exists a unique reflection with center $\langle v \rangle$ by 22.6.2; denote this reflection by r_v.

A *root system* is a finite subset Σ of $V^\#$ invariant under $W(\Sigma) = \langle r_v : v \in \Sigma \rangle$ and such that $|\langle v \rangle \cap \Sigma| \le 2$ for each $v \in \Sigma$. Observe that if $v \in \Sigma$ then $-v = v r_v \in \Sigma$, so, as $|\langle v \rangle \cap \Sigma| \le 2$, $\langle v \rangle \cap \Sigma = \{v, -v\}$. We call $W(\Sigma)$ the *Weyl group* of Σ.

Here's one way to obtain root systems:

(30.1) Let G be a finite subgroup of $O(V, Q)$ generated by a G-invariant set R of reflections. Let Σ consist of those $v \in V^\#$ with $Q(v) = 1/2$ and $\langle v \rangle$ the center of some member of R. Then Σ is a root system and $G = W(\Sigma)$.

Proof. For $v \in \Sigma$, $\langle v \rangle$ is the center of some $r \in R$, so $r = r_v$. Thus $G = \langle R \rangle = W(\Sigma)$, so it remains to show Σ is a root system. If $u = av \in \Sigma$ then, by definition of Σ, $1/2 = Q(u) = a^2 Q(v) = a^2/2$, so $a = \pm 1$. Thus $|\Sigma \cap \langle v \rangle| \le 2$.

Also if $g \in G$ then, as R is G-invariant, $r^g \in R$ and $\langle vg \rangle$ is the center of r^g with $Q(vg) = Q(v) = 1/2$, so $vg \in \Sigma$. Hence $G = W(\Sigma)$ acts on Σ. Finally $|\Sigma| = 2|R| \leq 2|G| < \infty$. So Σ is a root system.

By 29.12 and 30.1, every finite Coxeter group can be represented as the Weyl group of some root system. We'll see later in this section that on the one hand this representation is faithful and on the other that the Weyl group of each root system is a finite Coxeter group. Thus the finite Coxeter groups are precisely the Weyl groups of root systems.

For the remainder of this section let Σ be a root system and $W = W(\Sigma)$ its Weyl group. The elements of Σ will be called *roots*.

(30.2) (1) $\langle \Sigma \rangle$ is a nondegenerate subspace of V and W centralizes $\Sigma^{\perp}$, so W is faithful on $\langle \Sigma \rangle$.

(2) The permutation representation of W on Σ is faithful.

(3) W is finite.

Proof. Let $U = \langle \Sigma \rangle$. Then $H = C_W(U) = C_W(\Sigma)$ and W/H is faithful on Σ, so W/H is finite. Thus to prove (2) and (3) it suffices to show $H = 1$. As Q is positive definite, U is nondegenerate. Thus $V = U \oplus U^{\perp}$. But, for $v \in \Sigma$, r_v centralizes $v^{\perp} \supseteq U^{\perp}$, so $W = \langle r_v : v \in \Sigma \rangle$ centralizes $U^{\perp}$. Hence H centralizes V, so $H = 1$, completing the proof of the lemma.

An *ordering* of V is a total ordering of V preserved by addition and multiplication by positive scalars; that is if $u, v, w \in V$ with $u \geq v$, and $0 < a \in F$, then $u + w \geq v + w$ and $au \geq av$. I leave the following lemma as an exercise.

(30.3) (1) If $\leq$ is an ordering on V and $V^+ = \{v \in V : v > 0\}$ then

(i) V^+ is closed under addition and multiplication by positive scalars, and
(ii) for each $v \in V^{\#}$, $|\{v, -v\} \cap V^+| = 1$.

(2) If $S \subseteq V^{\#}$ satisfies (i) and (ii) of (1) then there exists a unique ordering of V with $S = V^+$.

(3) If $T \subseteq V^{\#}$ satisfies (1.i) and $|\{v, -v\} \cap T| \leq 1$ for each $v \in V^{\#}$, then $T \subseteq V^+$ for some ordering of V.

A subset P of Σ is a *positive system* if $P = \Sigma^+ = \Sigma \cap V^+$ for some ordering of V. A subset π of Σ is a *simple system* if π is linearly independent and each $v \in \Sigma$ can be written $v = \sum_{x \in \pi} a_x x$ such that either $0 \leq a_x \in F$ for all $x \in \pi$ or $0 \geq a_x \in F$ for all $x \in \pi$.

(30.4) Each simple system is contained in a unique positive system and each positive system contains a unique simple system.

Proof. If π is a simple system let T consist of those elements $\sum_{x \in \pi} a_x x \in V^{\#}$ with $a_x \geq 0$. By 30.3.3, $T \subseteq V^+$ for some ordering of V. By definition of simple systems and 30.3.1, $P = T \cap \Sigma$ is the unique positive system containing π. Let S be the set of those $v \in P$ such that, if $v = \sum_{y \in Y} c_y y$ for some $Y \subseteq P$ and $0 < c_y \in F$, then $Y = \{v\}$. Evidently $S \subseteq \pi$. I'll show $\pi \subseteq S$ which will prove $\pi = S$ is the unique simple system in P.

So assume $v \in \pi$ and $v = \sum_{y \in Y} c_y y$ for some $Y \subseteq P$ and $0 < c_y \in F$. Now, for $y \in Y$, $y = \sum_{x \in \pi} b_{yx} x$ for some $0 \leq b_{yx} \in F$, so $v = \sum_{y,x} c_y b_{yx} x$ and hence, by the linear independence of π, $0 = \sum_y c_y b_{yx}$ for $x \in \pi - \{v\}$. Therefore $b_{yx} = 0$ for $x \neq v$, so $y = b_{yv} v$ for each $y \in Y$. Then, as $P \cap \langle v \rangle = \{v\}$, $Y = \{v\}$, completing the proof of the claim.

It remains to show that if P is a positive system then P contains a simple system. In the process I'll show:

(30.5) If π is a simple system and x and y are distinct members of π, then $(x, y) \leq 0$.

Indeed, returning to the proof of 30.4, let π be a subset of P minimal subject to P being contained in the nonnegative linear span of π. I'll show π satisfies 30.5 and then use this fact to show π is linearly independent and hence a simple system. This will complete the proof of 30.4.

Suppose $(x, y) > 0$. By 22.6.2, $z = yr_x = y - 2(x, y)x/(x, x)$, so $z = y - cx$ with $c > 0$. Now $z = \sum_{t \in \pi} a_t t$ with $a_t \geq 0$ if $z \in P$ and $a_t \leq 0$ if $z \in -P$. If $z \in P$, consider the equation:

$$(1 - a_y)y = \sum_{t \neq x, y} a_t t + (a_x + c)x.$$

The right hand side is in V^+, so, as $y \in V^+$, $1 - a_y > 0$. But then y is in the nonnegative span of $\pi - \{y\}$, contradicting the minimality of π. If $z \in -P$ consider:

$$(a_x + c)x = \sum_{t \neq x, y} (-a_t)t + (1 - a_y)y$$

and apply the same argument for a contradiction.

So π satisfies 30.5 and it remains to show π is linearly independent. Suppose $0 = \sum_{x \in \pi} a_x x$ and let $\alpha = \{x \in \pi : a_x > 0\}$ and $\beta = \{y \in \pi : a_y < 0\}$. Then $\sum_{x \in \alpha} a_x x = z = \sum_{y \in \beta} (-a_y) y$ so $0 \leq (z, z) = (\sum_{x \in \alpha} a_x x, \sum_{y \in \beta} (-a_y) y) \leq 0$ by 30.5. Hence as Q is positive definite $z = 0$. But if $a_x \neq 0$ for some x

then $z \in V^+$, a contradiction. So π is linearly independent and the proof is complete.

For the remainder of this section let π be a simple system and P the positive system containing π.

(30.6) For each $v \in P$ there is $x \in \pi$ with $(v, x) > 0$.

Proof. Write $v = \sum_{x \in \pi} a_x x$ with $a_x \geq 0$ and observe $0 < (v, v) = \sum_x a_x (v, x)$.

(30.7) For each $x \in \pi$, $x r_x = -x$ and r_x acts on $P - \{x\}$.

Proof. Let $r = r_x$. By definition of r, $xr = -x$. Let $v \in P - \{x\}$, so that $v = \sum_{y \in \pi} a_y y$ with $a_y \geq 0$. As $\{x\} = \langle x \rangle \cap P$, $a_z > 0$ for some $z \in \pi - \{x\}$. Now

$$vr = \sum_y a_y yr = \sum_y a_y(y - 2(y, x)x) = \sum_{y \neq x} a_y y + bx$$

for some $b \in F$. As $a_z > 0$ it follows that $vr \in P$ from the definition of simple system.

(30.8) W is transitive on simple systems and on positive systems.

Proof. By 30.3, W permutes positive systems, while it is evident from the definitions that W permutes simple systems. By 30.4 it suffices to show W is transitive on positive systems. Assume not and let P and R be positive systems in different orbits of W, and subject to this constraint with $|P \cap (-R)| = n$ minimal. As $R \neq P$, $n > 0$. Let π be the simple system in P. By 30.4 there is $x \in \pi \cap (-R)$. By 30.7, $|Pr_x \cap (-R)| = n - 1$, so, by minimality of n, $R \in (Pr_x)W = PW$, a contradiction.

For $v \in \Sigma$, $v = \sum_{z \in \pi} a_x x$ for unique $a_x \in F$; define the *height* of v to be $h(v) = \sum_{x \in \pi} a_x$. Evidently the height function h depends on π. Notice for $v \in P$ that $h(v) > 0$ and $h(-v) = -h(v) < 0$. Also $h(x) = 1$ for $x \in \pi$.

(30.9) (1) $h(v) > 1$ for each $v \in P - \pi$.
   (2) $W = \langle r_x : x \in \pi \rangle$.
   (3) Each member of Σ is conjugate to an element of π under W.

Proof. Let $G = \langle r_x : x \in \pi \rangle$ and $v \in P$. Pick $u \in P \cap vG$ such that $h(u)$ is minimal. By 30.6 there is $y \in \pi$ with $(u, y) > 0$. Then $ur_y = u - cy$ with

$c = 2(u, y) > 0$ and if $u \neq y$ then, by 30.7, $ur_y \in P \cap vG$ and $h(ur_y) = h(u) - c < h(u)$, a contradiction. So $u = y \in \pi$. In particular if $v \notin \pi$ then $h(v) > h(y) = 1$, so (1) holds. Further it follows that each member of Σ is conjugate to an element of π under G, so that (3) holds. Finally $v = yg$ for some $g \in G$, so $r_v = r_{yg} = (r_y)^g \in G$ and hence $W = \langle r_v : v \in P \rangle = G$.

For $w \in W$ let $N(w) = |Pw \cap (-P)|$. Then $N(1) = 0$ and, by 30.7, $N(r_x) = 1$ for $x \in \pi$. Let $R = \{r_x : x \in \pi\}$ and $l = l_R$ the length function with respect to R; that is, for $w \in W$, $l(w)$ is the minimal length of a word u in the members of R such that $u = w$. It will develop shortly that the functions l and N agree on W.

(30.10) Let $w \in W$ and $x \in \pi$. Then
  (1) $N(r_x w) = N(w) + 1$ if $xw > 0$, and
  (2) $N(r_x w) = N(w) - 1$ if $xw < 0$.

Proof. $Pr_x w = (\{-x\} \cup (P - \{x\}))w = \{-xw\} \cup (Pw - \{xw\})$ by 30.7, so

$$Pr_x w \cap (-P) = \begin{cases} (Pw \cap (-P)) \cup \{-xw\} & \text{if } xw > 0, \\ (Pw \cap (-P)) - \{xw\} & \text{if } xw < 0. \end{cases}$$

(30.11) Let $r_i \in R, 0 \leq i \leq n$ and $w = r_1 \ldots r_n \in W$ with $N(w) = n$ and $N(r_0 w) \leq n$. Then there exists k, $1 \leq k \leq n$, such that $r_0 r_1 \ldots r_{k-1} = r_1 \ldots r_k$.

Proof. By 30.10, $xw < 0$, where $x \in \pi$ with $r_0 = r_x$. Let k be minimal subject to $xr_1 \ldots r_k < 0$. Then, by 30.7, $y = xr_1 \ldots r_{k-1} \in \pi$ and $r_k = r_y$. Thus $r_0^{r_1 \ldots r_{k-1}} = r_{xr_1 \ldots r_{k-1}} = r_y = r_k$, which can be rewritten as $r_0 r_1 \ldots r_{k-1} = r_1 \ldots r_k$.

(30.12) $l(w) = N(w)$ for each $w \in W$.

Proof. I've already observed that $l(1) = 0 = N(1)$. Then 30.10 and induction on $l(w)$ implies $l(w) \geq N(w)$. Suppose that the lemma is false and pick $u = r_0 r_1 \ldots r_n \in W$ with $l(u) = n + 1 > N(u)$, and subject to this constraint with n minimal. Let $w = r_1 \ldots r_n$. As $l(u) = n + 1$, $l(w) = n$, and, by minimality of $l(u)$, $N(w) = l(w)$. As $N(r_0 w) = N(u) < n + 1$, there exists k with $r_0 r_1 \ldots r_{k-1} = r_1 \ldots r_k$ by 30.11. Thus $u = r_0 \ldots r_{k-1} r_k \ldots r_n = r_1 \ldots r_k^2 r_{k+1} \ldots r_n = r_1 \ldots r_{k-1} r_{k+1} \ldots r_n$ is of length at most $n - 1$, contrary to the choice of u.

(30.13) (W, R) is a Coxeter system.

Proof. This is immediate from 29.4, 30.11, and 30.12.

(30.14) W is regular on the positive systems and on the simple systems.

Proof. By 30.8 and 30.4 it suffices to show $N_W(P) = 1$. But $N_W(P) = \{w \in W : N(w) = 0\}$ so 30.12 completes the proof.

Let D consist of those $v \in V$ such that $(v, x) \geq 0$ for all $x \in \pi$.

(30.15) (1) $vW \cap D$ is nonempty for each $v \in V$.
    (2) $(d, u) \geq 0$ for each $d \in D$ and $u \in P$.

Proof. Pick $z \in vW$ maximal with respect to the ordering defining P. Then, for $x \in \pi, z \geq zr_x = z - 2(z, x)x$, so $(z, x) \geq 0$ as $x > 0$. That is $z \in vW \cap D$. Let $d \in D$ and $u \in P$. Then $u = \sum_{x \in \pi} a_x x$ with $a_x \geq 0$ so $(d, u) = \sum_x a_x(d, x) \geq 0$.

(30.16) Let $d \in D$. Then $C_W(d)$ is the Weyl group of the root systems $\Sigma \cap d^{\perp}$ and has simple system $\pi \cap d^{\perp}$.

Proof. $U = \langle r_y : x \in \pi \cap d^{\perp} \rangle \leq C_W(d)$, so assume $w \in C_W(d) - U$ and subject to this constraint with $n = l(w)$ minimal. $w \neq 1$ so $n > 0$. Then $N(w) = n > 0$ so there is $x \in \pi$ with $xw < 0$. Now, by 30.15.2, $0 \geq (xw, d) = (x, dw^{-1}) = (x, d)$, so $x \in d^{\perp}$ by another application of 30.15.2. Thus $r_x \in U$ and, by 30.10, $l(r_xw) < n$, so, by minimality of n, $r_xw \in U$. But then $w \in U$.
    So $U = C_W(d)$. Evidently $\Sigma \cap d^{\perp}$ is a root system and an easy calculation using 30.15.2 shows $\pi \cap d^{\perp}$ is a simple system.

(30.17) Let $S \subseteq V$. Then $C_W(S)$ is the Weyl group of the root system $\Sigma \cap S^{\perp}$.

Proof. Let $U = \langle r_v : v \in \Sigma \cap S^{\perp} \rangle$. It suffices to show $U = C_W(S)$. $C_W(S) = C_W(\langle S \rangle) = C_W(S_0)$ where S_0 is a basis for $\langle S \rangle$, so without loss S is finite. Replacing S by a suitable conjugate under W and appealing to 30.15, we may take $d \in D \cap S$. Let $G = C_W(d)$ and $\Sigma_0 = \Sigma \cap d^{\perp}$. Then, by 30.16, $G = W(\Sigma_0)$ and $C_W(S) = C_G(S)$, while, by induction on the order of Σ, $C_G(S) = U$.

(30.18) Let (G, S) be a Coxeter system with G finite. Then the representation α of 29.8 is faithful and G is the Weyl group of a root system.

Proof. By 29.12 and 30.1, $G\alpha$ is the Weyl group of a root system. By 29.7 and 30.13, $G\alpha$ is a Coxeter group of type M, so $|G\alpha| = |G|$. Hence α is an isomorphism.

(30.19) Let H be the symmetric group on $\Omega = \{1, \dots, n\}$, $\langle e \rangle \cong \mathbb{Z}_2$, and $G = \langle e \rangle \mathrm{wr}_\Omega H$ the wreath product of $\langle e \rangle$ by H. Choose notation so that $C_H(e)$ is the stabilizer in H of $n \in \Omega$. Let $G_1 = H$, $S_1 = \{(1, 2), (2, 3), \dots, (n - 1, n)\}$, $G_2 = G$, $S_2 = S_1 \cup \{e\}$, $S_3 = S_1 \cup \{(n - 1, n)^e\}$, and $G_3 = \langle S_3 \rangle$. Then (G_i, S_i) is a Coxeter system of type A_{n-1}, C_n, D_n, for $i = 1, 2, 3$, respectively. Further G_3 is of index 2 in G_2 and is the semidirect product of $E_{2^{n-1}}$ by S_n.

Proof. Let V be n-dimensional Euclidean space with orthonormal basic $X = (x_i : 1 \le i \le n)$, e_i the reflection with center $\langle x_i \rangle$, and $E = \langle e_i : 1 \le i \le n \rangle$. Represent H on V via $x_i h = x_{ih}$ for $h \in H$. Then $G = \langle H, E \rangle \le O(V)$ and G is the wreath product $\langle e \rangle \mathrm{wr}_\Omega H$, where $e = e_n$. Hence G is the semidirect product of $E \cong E_{2^n}$, by $H \cong S_n$, and G_3 is the semidirect product of $D = E \cap G_3$ by H where $D = \langle e_i e_j : 2 \le i \le j \le n \rangle \cong E_{2^{n-1}}$. The transposition (i, j) is the reflection with center $\langle x_{ij} \rangle$, where $x_{ij} = x_i - x_j$. Also $G_i = \langle S_i \rangle$ and S_i is a set of reflections, so, by 30.1, $G_i = W(\Sigma_i)$ where Σ_i is the root system consisting of the G_i-conjugates of x_{12} if $i = 1$ or 3, and $\{x_{12}, x_n\}$ if $i = 2$. Thus

$$\Sigma_1 = \{\pm x_{ij} : 1 \le i < j \le n\},$$

$$\Sigma_3 = \Sigma_1 \cup \{\pm(x_i + x_j) : 1 \le i < j \le n\},$$

and

$$\Sigma_2 = \Sigma_3 \cup \{\pm x_i : 2 \le i \le n\}.$$

Next π_i is a simple system for Σ_i, where $\pi_i = \{x_{i,i+1} : 1 \le i < n\}$, $\pi_2 = \pi_1 \cup \{x_n\}$, and $\pi_3 = \pi_1 \cup \{x_{n-1} + x_n\}$. Hence, as $S_i = \{r_x : x \in \pi_i\}$, (G_i, S_i) is a Coxeter system by 30.13. Finally it is evident that the Coxeter diagram of (G_i, S_i) is A_{n-1}, C_n, D_n, for $i = 1, 2, 3$, respectively.

For $J \subseteq \pi$ let $V_J = \langle J \rangle$ and $W_J = \langle r_j : j \in J \rangle$. Recall the subgroups W_J and their conjugates under W are called *parabolic subgroups* of W.

(30.20) Let $\varnothing \ne J \subseteq \pi$. Then
  (1) $\Sigma_J = \Sigma \cap V_J$ is a root system with simple system J and Weyl group W_J.
  (2) $W_J = C_W(V_J^\perp)$.

Proof. $W_J \le C_W(V_J^\perp) = U$ and, by 30.17, U is the Weyl group of the root system Σ_J. π is linearly independent and J spans V_J, so J is a basis of V_J.

Hence each member of Σ_J is a linear combination of the members of J, so, as π is a simple system, J is a simple system for Σ_J. Thus $U = \langle r_j : j \in J \rangle = W_J$.

Remarks. The discussion of Coxeter systems in section 29 follows that of Bourbaki [Bo] and Suzuki [Su]. The presentation of root systems given here draws heavily on the appendix of Steinberg [St].

Coxeter groups and root systems play an important role in branches of mathematics other than finite group theory, most particularly in the study of Lie algebras, Lie groups, and algebraic groups. We will find in chapters 14 and 16 that they are crucial to the study of the finite groups of Lie type.

Exercises for chapter 10

1. Prove $D_{2n} = \mathrm{Grp}\,(x, y : x^2 = y^2 = (xy)^n = 1)$. Prove every group generated by a pair of distinct involutions is a dihedral group.
2. Prove lemma 30.3.
3. Let Σ be a root system, π a simple system for Σ, P the positive system of π, $R = (r_\alpha : \alpha \in \pi)$, and $W = \langle R \rangle$ the Weyl group of Σ. Prove
   (1) There exists a unique $w_0 \in W$ with $Pw_0 = -P$.
   (2) w_0 is the unique element of W of maximal length in the alphabet R. Further $l(w_0) = |P|$ and w_0 is an involution.
   (3) $\pi w_0 = -\pi$ and $R^{w_0} = R$.
4. Let Σ be an irreducible root system, π a simple system for Σ, P the positive system for π, $J \subset \pi$, Σ_j, the subset of Σ, spanned by J, and $\psi = P - \Sigma_j$. Prove $\langle \Sigma \rangle = \langle \psi \rangle$.
5. Let Σ be a root system, P a positive system for Σ, and $w \in W$. Prove for each $\alpha \in P$ that Pw contains exactly one of α and $-\alpha$.
6. Assume the hypothesis of Exercise 10.3 and let w_0 be the element of W of maximal length in the alphabet R. Define a relation $\leq$ on W by $u \leq w$ if $w = xu$ with $l(w) = l(x) + l(u)$. Prove
   (1) $\leq$ is a partial order on W.
   (2) w_0 is the unique maximal element of W. That is $w \leq w_0$ for all $w \in W$.
   (3) For $r = r_\alpha \in R$ and $w \in W$ the following are equivalent:
      (a) $rw \leq w$.
      (b) $l(rw) \leq l(w)$.
      (c) $\alpha w < 0$.
      (d) $w = r_1 \ldots r_n$ with $r = r_1$ and $l(w) = n$.
   (4) If $u \leq w$ and $r \in R$ with $l(ur) \leq l(u)$ and $l(wr) \leq l(w)$ then $ur \leq wr$.
   (Hint: To prove (2) let $w_0 \neq w \in W$ and use Exercise 10.3.2, 30.10, and 30.12 to show there exists $r \in R$ with $l(w) < l(rw)$.)

11

The generalized Fitting subgroup

We've seen that the composition factors of a finite group control the structure of the group in part, but that control is far from complete. Section 31 introduces a tool for studying finite groups via composition factors 'near the bottom' of the group. The generalized Fitting subgroup $F^*(G)$ of a finite group G is a characteristic subgroup of G generated by the small normal subgroups of G and with the property that $C_G(F^*(G)) \leq F^*(G)$. This last property supplies a representation of G as a subgroup of $\mathrm{Aut}(F^*(G))$ with kernel $Z(F^*(G))$. G can be effectively investigated via this representation because $F^*(G)$ is a relatively uncomplicated group whose embedding in G is particularly well behaved.

It turns out that $F^*(G)$ is a central product of the groups $O_p(G)$, $p \in \pi(G)$, with a subgroup $E(G)$ of G. To define $E(G)$ requires some terminology. A *central extension* of a group X is a group Y together with a surjective homomorphism of Y onto X whose kernel is in the center of Y. The group Y will also be said to be a central extension of X. A group L is *quasisimple* if L is perfect and the central extension of a simple group. The *components* of G are its subnormal quasisimple subgroups, and $E(G)$ is the subgroup of G generated by the components of G. It develops that $E(G)$ is a central product of the components of G.

Recall that if p is a prime then a *p-local subgroup* of G is the normalizer in G of a nontrivial p-subgroup of G. The local theory of groups investigates finite groups from the point of view of p-locals. A question of great interest in this theory is the relationship between the generalized Fitting subgroup of G and that of its local subgroups. Section 31 contains various results about such relationships. In the final chapter we'll get some idea of how such results are used to classify the finite simple groups.

If $F^*(G)$ is a p-group, it can be particularly difficult to analyze the structure of G. One tool for dealing with such groups is the Thompson factorization. Lemma 32.5 shows that, if G is solvable and the Thompson factorization fails, then the structure of G is rather restricted. This result will be used in later chapters to prove the Thompson Normal p-Complement Theorem and the Solvable Signalizer Functor Theorem. The Normal p-Complement Theorem will be used in turn to establish the nilpotence of Frobenius kernels.

Finally the importance of components focuses attention on quasisimple groups. In section 33 we find there is a largest perfect central extension

$\pi: \tilde{G} \to G$ of each perfect group G. $\tilde{G}$ is the *universal covering group* of G and $\ker(\pi)$ is the *Schur multiplier* of G. In particular if G is simple then $\tilde{G}$ is the largest quasisimple group with G as a homomorphic image and the Schur multiplier of G is the center of $\tilde{G}$. As an illustration of this theory, the covering groups and Schur multipliers of the finite alternating groups are determined.

The Schur multiplier can be defined for nonperfect groups using an alternate definition requiring homological algebra. The presentation given here is group theoretic and restricted to perfect groups; it follows Steinberg [St].

31 The generalized Fitting subgroup

In this section G is a finite group. A group X is *quasisimple* if $X = X^{(1)}$ and $X/Z(X)$ is simple.

(31.1) Let X be a group such that $X/Z(X)$ is a nonabelian simple group. Then $X = X^{(1)} Z(X)$ and $X^{(1)}$ is quasisimple.

Proof. Let $Y = X^{(1)}$ and $X^* = X/Z(X)$. Now $Y^* \trianglelefteq X^*$ and X^* is simple so $Y^* = 1$ or X^*. In the latter case $X = YZ(X)$ and in the former X^* is abelian, contrary to hypothesis.

So $X = YZ(X)$. Thus $X/Y^{(1)}$ is abelian so $Y = Y^{(1)}$. Further $Y/Z(Y) \cong X^*$ is simple, so Y is quasisimple.

(31.2) Let X be a quasisimple group and $H \trianglelefteq \trianglelefteq X$. Then either $H = X$ or $H \leq Z(X)$.

Proof. If $H \not\leq Z(X)$, $X = HZ(X)$, so that X/H is abelian and hence $X = X^{(1)} \leq H$.

The *components* of a group X are its subnormal quasisimple subgroups. Write $\mathrm{Comp}(X)$ for the set of components of X. Set $E(X) = \langle \mathrm{Comp}(X) \rangle$.

(31.3) If $H \trianglelefteq \trianglelefteq X$ then $\mathrm{Comp}(H) = \mathrm{Comp}(X) \cap H$.

(31.4) Let $L \in \mathrm{Comp}(G)$ and $H \trianglelefteq \trianglelefteq G$. Then either $L \in \mathrm{Comp}(H)$ or $[L, H] = 1$.

Proof. Let G be a minimal counterexample. If $L = G$ the lemma holds by 31.2, so by 7.2 we may take $X = \langle L^G \rangle \neq G$. Similarly if $H = G$ the lemma is trivial, so take $Y = \langle H^G \rangle \neq G$. $X \cap Y \trianglelefteq X$, so, by minimality of G, either $L \in \mathrm{Comp}(X \cap Y)$ or $[L, X \cap Y] = 1$. In the first case $L \in \mathrm{Comp}(Y)$ by 31.3

and then, as $H \leq Y < G$, the lemma holds by minimality of G. In the second $[Y, L, L] \leq [Y \cap X, L] = 1$, so, by 8.9, $[Y, L] = 1$.

(31.5) Distinct components of G commute.

Proof. This is a direct consequence of 31.2 and 31.4.

(31.6) Let $L \in \mathrm{Comp}(G)$ and H an L-invariant subgroup of G. Then
    (1) Either $L \in \mathrm{Comp}(H)$ or $[L, H] = 1$.
    (2) If H is solvable then $[L, H] = 1$.
    (3) If $R \leq G$ then either $L \in \mathrm{Comp}([R, L])$ or $[R, L] = 1$.

Proof. Part (1) follows from 31.4 applied to LH in the role of G. Then (1) implies (2). If $R \leq G$ then $[L, R]$ is L-invariant by 8.5.6, so by (1) either $L \in \mathrm{Comp}([L, R])$ or $[R, L, L] = 1$. In the latter case $[R, L] = 1$ by 8.9.

(31.7) Let $E = E(G)$, $Z = Z(E)$, and $E^* = E/Z$. Then
    (1) $Z = \langle Z(L): L \in \mathrm{Comp}(G)\rangle$.
    (2) E^* is the direct product of the groups $(L^*: L \in \mathrm{Comp}(G))$.
    (3) E is a central product of its components.

The *Fitting subgroup* of G, denoted by $F(G)$, is the largest nilpotent normal subgroup of G. $O_\infty(G)$ denotes the largest solvable normal subgroup of G.

(31.8) Let G be a finite group. Then $F(G)$ is the direct product of the groups $(O_p(G): p \in \pi(G))$.

Proof. See 9.11.

(31.9) $O_\infty(C_G(F(G))) = Z(F(G))$.

Proof. Let $Z = Z(F(G))$. $G^* = G/Z$, and $H = O_\infty(C_G(F(G)))$. Assume $H^* \neq 1$ and let X^* be a minimal normal subgroup of H^*. Then X^* is a p-group for some prime p, so $X = PZ$, where $P \in \mathrm{Syl}_p(X)$. X centralizes Z, so $P \trianglelefteq X$. Thus $P \leq O_p(G) \leq F(G)$, So $P \leq C_{F(G)}(F(G)) = Z$, contradicting $P^* = X^* \neq 1$.

(31.10) If G is solvable then $C_G(F(G)) \leq F(G)$.

Define the *socle* of G to be the subgroup generated by all minimal normal subgroups of G, and write $\mathrm{Soc}(G)$ for the socle of G.

(31.11) Let $Z = Z(F(G))$, $G^* = G/Z$, and $S^* = \mathrm{Soc}(C_G(F(G))^*)$. Then $E(G) = S^{(1)}$ and $S = E(G)Z$.

Proof. Let $H = C_G(F(G))$. By 31.9, $O_\infty(H^*) = 1$. So, by 8.2 and 8.3, each minimal normal subgroup of H^* is the direct product of nonabelian simple groups and by 31.3 these factors are components of H^*. Thus $S^* \leq E(H^*)$. Let K^* be a component of H^*. By 31.1, $K = K^{(1)}Z$ with $K^{(1)}$ quasisimple. By 31.3, $K^{(1)} \in \mathrm{Comp}(G)$, so $S \leq E(G)Z$. By 31.6.2, $E(G) \leq H$. Let $L \in \mathrm{Comp}(G)$, and $M = \langle L^H \rangle$. Then M^* is a minimal normal subgroup of H^* by 31.4, so $M \leq S$. Thus $S = E(G)Z$. By 31.1, $E(G) = S^{(1)}$.

Define the *generalized Fitting subgroup* of G to be $F^*(G) = F(G)E(G)$.

(31.12) $F^*(G)$ is a central product of $F(G)$ with $E(G)$.

Proof. See 31.6.2.

(31.13) $C_G(F^*(G)) \leq F^*(G)$.

Proof. Let $H = C_G(F^*(G))$, $K = C_G(F(G))$, $Z = Z(F(G))$, and $G^* = G/Z$. Then $H^* \trianglelefteq K^*$, so if $H^* \neq 1$ then $1 \neq H^* \cap \mathrm{Soc}(K^*)$, and then, by 31.11, $H \cap E(G) \neq Z$, a contradiction.

Recall $O_{p',E}(G)$ is defined by $O_{p',E}(G)/O_{p'}(G) = E(G/O_{p'}(G))$.

(31.14) Let p be a p-subgroup of G. Then
  (1) $O_{p',E}(N_G(P)) \leq C_G(O_p(G))$, and
  (2) if $P \leq O_p(G)$ then $O^p(F^*(N_G(p))) = O^p(F^*(G))$.

Proof. Let $X = O_{p',E}(N_G(p))$. To prove (1), it suffices to show X centralizes $R = C_{O_p(G)}(P)$ by the $A \times B$ Lemma. But $[R, O_{p'}(X)] \leq R \cap O_{p'}(X) = 1$ and $[R, X] \leq O_{p'}(X)$ by 31.6.2, so $[R, X] = 1$ by coprime action, 18.7.

So take $P \leq O_p(G)$. Then $Y_0 = O^p(F^*(G)) \leq Y = O^p(F^*(N_G(P)))$. If L is a component of $N_G(P)$ then $[L, O_p(G)] = 1$ by (1) and, as $Y_0 \leq Y$, either $[L, Y_0] = 1$ or $L \in \mathrm{Comp}(Y_0)$ by 31.4. In the first case $L \leq C_G(F^*(G)) = Z(F(G))$, a contradiction. So $E(N_G(P)) = E(G)$. Let $q \neq p$ and $Q = O_q(Y)$. We must show $Q \leq O_q(G)$. Passing to $G/O_q(G)$ and appealing to coprime action, 18.7, we may take $O_q(G) = 1$, and it remains to show $Q = 1$. But, by (1) and as $Y_0 \leq Y$, $Q \leq C_G(O^q(F^*(G))) = C_G(F^*(G)) = Z(F(G))$, so indeed $Q = 1$.

(31.15) If G is solvable and P is a p-subgroup of G then $O_{p'}(N_G(P)) \le O_{p'}(G)$.

Proof. Passing to $G/O_{p'}(G)$ and appealing to coprime action, 18.7, we may take $O_{p'}(G) = 1$, and it remains to show $O_{p'}(N_G(P)) = 1$. This follows from 31.10 and 31.14.1.

(31.16) Let $F^*(G) = O_p(G)$ for some prime p. Then $F^*(N_G(P)) = O_p(N_G(P))$ for each p-subgroup P of G.

Proof. This follows from 31.13 and 31.14.1.

The *Schreier Conjecture* says $\mathrm{Out}(L)$ is solvable for each finite simple group L.

(31.17) Let $O_{p'}(G) = 1$ and P a p-subgroup of G. Then
 (1) $O_{p',E}(N_G(P))$ fixes each component of G.
 (2) If each component of G satisfies the Schreier conjecture, then $O_{p',E}(N_G(P))^\infty \le E(G)$.

Proof. Let $H = N_G(P)$, $X = O_{p'}(H)$, $H^* = H/X$, and $Y = O_{p',E}(H)^\infty$. Let $K \le X$ or $K \le H$ with $K^* \in \mathrm{Comp}(H^*)$, and subject to these constraints pick K minimal subject to moving a component of G. Let $P \le P_0 \in \mathrm{Syl}_p(C_G(K))$. As K satisfies the same hypothesis with respect to P_0, we may take $P = P_0$. In particular, by 31.14, $O_P(G) \le P$. Let $R \in \mathrm{Syl}_p(H \cap E(G))$.

Suppose first $K \le X$. Then K is a q-group for some prime q, and by coprime action, 18.7, there exists an R-invariant Sylow q-group Q of $O_{p'}(H)$. Replacing K by a suitable conjugate, we may assume $K \le Q$. By 24.4, $Q = [R, Q]C_Q(R)$. Now $[R, Q] \le [E(G), Q] \le E(G)$, so $[R, Q]$ fixes each component of G. Hence we may take $[K, R] = 1$. Thus, by choice of P, $R \le P$. So $P \cap E(G) \in \mathrm{Syl}_p(E(G))$ by Exercise 3.2. As $O_{p'}(G) = 1$, $p \in \pi(L)$ for each $L \in \mathrm{Comp}(G)$, so $1 \ne P \cap L \in \mathrm{Syl}_p(L)$ by 6.4. Then $P \cap L \not\le Z(L)$, so $L = [E(G), P \cap L]$ is K-invariant.

So $K \not\le X$. By 31.4, either $K^* \le [K^*, R^*]$ or $[K^*, R^*] = 1$. In the latter case by coprime action, 18.7, $K = O_{p'}(K)C_K(R)$ so, by minimality of K, $[K, R] = 1$. But then an argument in the last paragraph supplies a contradiction. In the former, $K \le X[K, R] \le XE(G)$, so K fixes each component of G.

Let L be a component of G. I have shown $Y \le N(L)$. If L satisfies the Schreier conjecture, then $\mathrm{Aut}_Y(L) \le \mathrm{Inn}(L)$ as $Y = Y^\infty$, so $Y \le LC(L)$. Hence, under the hypothesis of (2), $Y \le E(G)C(E(G))$ and then, by 31.14, $Y \le E(G)C(F^*(G)) \le F^*(G)$, so $Y \le F^*(G)^\infty = E(G)$.

G is *balanced* for the prime p if $O_{p'}(C_G(X)) \leq O_{p'}(G)$ for each X of order p in G.

(31.18) Let $O_{p'}(G) = 1$, x of order p in G, $L \in \mathrm{Comp}(G)$, and $Y = O_{p'}(C_G(x))$. Then

(1) If $L \neq [L, x]$ then $[L, Y] = 1$ and either $L \in \mathrm{Comp}(C_G(x))$ or $L \neq L^x$ and $C_{[L,x]}(x)^{(1)} = K \in \mathrm{Comp}(C_G(x))$ with K a homomorphic image of L.

(2) If $L = [L, x]$ and $\mathrm{Aut}_{Y\langle x\rangle L}(L)$ is balanced for the prime p, then $[Y, L] = 1$.

Proof. Assume $L \neq [L, x]$. Then either $[L, x] = 1$ or $L \neq L^x$. In the first case $L \in \mathrm{Comp}(C(x))$, so $[L, Y] = 1$. In the second let $M = [L, x]$ and $M^* = M/Z(M)$. By 8.9, $[M, L] \neq 1$, so, by 31.4, $L \in \mathrm{Comp}(M)$, and we conclude $M = \langle L^{\langle x\rangle}\rangle$ is the central product of the groups $(L^{x^i} : 0 \leq i < p)$ from 31.5. Hence, by Exercise 3.5, $K = C_M(x)^{(1)}$ is a homomorphic image of L. As L is quasisimple, so is its homomorphic image K. Also $M \trianglelefteq \trianglelefteq G$, so $K \trianglelefteq \trianglelefteq C_G(x)$, and hence $K \in \mathrm{Comp}(C(x))$. So $[K, Y] = 1$. Y acts on L^* by 31.17, so, as $[K^*, Y] = 1$, $[L^*, Y] = 1$ by Exercise 3.5. Then $[L, Y] = 1$ by 31.6, completing the proof of (1).

So assume $L = [L, x]$, let $U = Y\langle x\rangle L$, and let $\pi: U \to \mathrm{Aut}_G(L)$ be the conjugation map. Let $P = O_p(U)\langle x\rangle$ so that $P \in \mathrm{Syl}_p(\ker(\pi)\langle x\rangle)$. By 31.14, $[P, Y] = 1$, so, as $C_U(P) \leq C_U(x)$, $Y \leq O_{p'}(C_U(P)) = O_{p'}(N_U(P))$. Then, as $P \in \mathrm{Syl}_p(\ker(\pi)\langle x\rangle)$, $N_{U\pi}(\langle x\pi\rangle) = N_U(P)\pi$ by a Frattini Argument, so $Y\pi \leq O_{p'}(C_{U_\pi}(x\pi))$. Hence, if $\mathrm{Aut}_U(L)$ is balanced for the prime p, then $[Y, L] = 1$, so that (2) holds.

(31.19) Let $O_{p'}(G) = 1$ and assume $\mathrm{Aut}_H(L)$ is balanced for the prime p for each $L \in \mathrm{Comp}(G)$ and each $H \leq G$ with $L \trianglelefteq H$. Then G is balanced for the prime p.

Proof. Let X be a subgroup of G of order p and $Y = O_{p'}(C_G(X))$. I must show $Y = 1$. By 31.18 and the hypothesis on the components of G, $[Y, E(G)] = 1$. By 31.14, $[Y, O_p(G)] = 1$. So $Y \leq C_G(F^*(G)) = Z(F(G))$, so, as $O_{p'}(G) = 1$, $Y = 1$.

Here's technical lemma to be used in chapter 15.

(31.20) Let A be an elementary abelian r-group acting on a solvable r'-group G and let $a \in A$. Let $p \in \pi \subseteq \pi(G)$, $p^\pi = \pi' \cup \{p\}$, and P an A-invariant p-subgroup of G.

(1) Suppose $P \leq O_p(K)$ for some a-invariant π-subgroup K of G such that $C_K(a)$ is a Hall π-subgroup of $C_G(a)$. Then $P \leq O_{p^\pi}(N)$ for each a-invariant subgroup N of G with $N = \langle P, C_N(a) \rangle$.

(2) Assume A is noncyclic and let Δ be the set of hyperplanes of A. Assume for each $B \in \Delta$ that $C_P(B) \leq O_\pi(\langle C_G(a), C_P(B) \rangle)$. Then $P \leq O_\pi(\langle C_G(a), P \rangle)$.

(3) If B is a noncyclic subgroup of $C_A(P)$ and $P \leq O_\pi(\langle C_G(\langle a, b \rangle), P \rangle)$ for each $b \in B^\#$, then $P \leq O_\pi(\langle C_G(a), P \rangle)$.

Proof. Let G be a minimal counterexample to (1), (2), or (3). Without loss, $A = \langle a \rangle$ in (1) and $A = \langle B, a \rangle$ in (3). In each case it suffices to assume $G = \langle P, C_G(a) \rangle$ and prove $P \leq O_\alpha(G)$, where $\alpha = p^\pi$ in (1) and $\alpha = \pi$ in (2) and (3).

Let H be a minimal A-invariant normal subgroup of G and $G^* = G/H$. Then $P^* \leq O_\alpha(G^*) = S^*$ by minimality of G and coprime action 18.7.4, and H is a q-group for some prime q. Hence if $P \leq O_\alpha(HP)$ we are done, so this is not the case, and in particular $q \notin \alpha$.

Let $I = C_G(a)$. Then $C_H(a) \trianglelefteq I$. We show $[P, C_H(a)] = 1$. Then $G = \langle I, P \rangle \leq N(C_H(a))$, so, by minimality of H, $C_H(a) = 1$ or H. In the latter case $[P, H] = 1$, contradicting $P \not\leq O_\alpha(PH)$. Thus $C_H(a) = 1$.

Now to verify that $[P, C_H(a)] = 1$. In (1), as $q \in \pi$ and $C_H(a) \trianglelefteq I$, $C_H(a)$ is contained in each Hall π-group of $C_G(a)$ by Hall's Theorem 18.5, and hence in K. So $[C_H(a), P] \leq O_p(K) \cap O_q(G) = 1$. In (2), $[C_P(B), C_H(a)] \leq O_\pi(G_B) \cap O_q(G_B) = 1$, where $G_B = \langle C_G(a), C_P(B) \rangle$ and $B \in \Delta$. Hence, by Exercise 8.1, $P = \langle C_P(B) : B \in \Delta \rangle \leq C(C_H(a))$. Finally, in (3), $[P, C_H(\langle a, b \rangle)] \leq O_\pi(G_b) \cap O_q(G_b) = 1$, where $G_b = \langle P, C_G(\langle a, b \rangle) \rangle$. So, again by Exercise 8.1, $C_H(a) = \langle C_H(\langle a, b \rangle) : b \in B^\# \rangle \leq C(P)$.

We've shown $C_H(a) = 1$. In (2) and (3) let R be an A-invariant Hall π-subgroup of S containing P. By a Frattini Argument, $G = HN_G(R)$. As $C_H(a) = 1$, $C_G(a) \cong C_{G^*}(a) \cong N_G(R) \cap C(a)$, so $C_G(a) \leqslant N_G(R)$. But then $G = \langle C_G(a), P \rangle \leq N(R)$, so $[P, H] \leq R \cap H = 1$, a contradiction.

This leaves (1). Here we may take $P = O_p(K) \trianglelefteq K$, so, if U is a Hall π'-group of I, then $R = \langle P^I \rangle = \langle P^{(I \cap K)U} \rangle = \langle P^U \rangle$ and $R \trianglelefteq \langle I, P \rangle = G$. Thus we may assume $H \leq R$. Let X be an a-invariant Hall q'-subgroup of US. As $C_H(a) = 1$, $C_{US}(a)$ is a q'-group, so X is the unique a-invariant Hall q'-subgroup of US by Exercise 6.2. Hence $\langle P, U \rangle \leq X$. But then $H \leq R \leq X$, a contradiction.

32 Thompson factorization

In this section p is a prime and G is a finite group. Denote by $\mathscr{A}(G)$ the set of elementary abelian p-subgroups of G of p-rank $m_p(G)$. Set $J(G) = \langle \mathscr{A}(G) \rangle$.

We call $J(G)$ the *Thompson subgroup* of G. Of course $J(G)$ depends on the choice of p.

(32.1) (1) $J(G)$ char(G).

(2) If A is in $\mathscr{A}(G)$ and $A \leq H \leq G$ then $\mathscr{A}(H) \subseteq \mathscr{A}(G)$.

(3) Let $P \in \mathrm{Syl}_p(G)$. Then $J(P) = J(Q)$ for each p-subgroup Q of G containing $J(P)$.

If V is a GF$(p)G$-module define $\mathscr{P}(G, V)$ to consist of the nontrivial elementary abelian p-subgroups A of G such that

$$m(A) + m(C_V(A)) \geq m(B) + m(C_V(B))$$

for each $B \leq A$. Notice that if B is a nontrivial subgroup of $A \in \mathscr{P}(G, V)$ for which this inequality is an equality, then B is in $\mathscr{P}(G, V)$. A subset $\mathscr{P}$ of $\mathscr{P}(G, V)$ is *stable* if G permutes $\mathscr{P}$ via conjugation and, whenever A is in $\mathscr{P}$ and B is a nontrivial. subgroup of A with

$$m(A) + m(C_V(A)) = m(B) + m(C_V(B)),$$

then B is in $\mathscr{P}$. As a final remark note that if A is in $\mathscr{P}(G, V)$ then $m(A) \geq m(V/C_V(A))$; to see this just take $B = 1$ in the inequality defining membership in $\mathscr{P}(G, V)$.

(32.2) Let V be a normal elementary abelian p-subgroup of G, $G^* = G/C_G(V)$, and $\mathscr{P} = \{A^*: A \in \mathscr{A}(G) \text{ and } A^* \neq 1\}$. Then $\mathscr{P}$ is a stable subset of $\mathscr{P}(G^*, V)$.

Proof. Let A be in $\mathscr{A}(G)$ and $C_A(V) \leq B \leq A$. $A_0 = AC_V(A)$ is an elementary abelian p-group, so, as $A \in \mathscr{A}(G)$, $m(A_0) \leq m(A)$. Hence $A = A_0$ since $A \leq A_0$. Thus $C_V(A) = A \cap V$. Similarly $m(BC_V(B)) \leq m(A)$. As $C_B(V) = C_A(V)$, we have

$$m(BC_V(B)) = m(B^*) + m(C_A(V)) + m(C_V(B)) - m(A \cap V)$$

while

$$m(BC_V(B)) \leq m(A) = m(A^*) + m(C_A(V))$$

so as $C_V(A) = A \cap V$ it follows that

$$m(A^*) + m(C_V(A)) \geq m(B^*) + m(C_V(B))$$

with equality only if $m(BC_V(B)) = m(A)$. Thus if $A^* \neq 1$ then $A^* \in \mathscr{P}(G, V)$ while if $m(B^*) + m(C_V(B)) = m(A^*) + m(C_V(A))$ then $m(A) = m(BC_V(B))$, so $B_0 = BC_V(B)$ is in $\mathscr{A}(G)$ and hence $B^* = B_0^*$ is in $\mathscr{P}(G^*, V)$.

(32.3) Let G be a solvable group, V a faithful $GF(p)G$-module, and P a Sylow p-group of G. Assume $O_p(G) = 1$, $\mathscr{P}$ is a stable subset of $\mathscr{P}(G, V)$ and $H = \langle \mathscr{P} \rangle \neq 1$. Then $p \leq 3$, $H = H_1 \times \cdots \times H_n$, $V = [V, H] \oplus C_V(H)$ with $[V, H] = [V, H_1] \oplus \cdots \oplus [V, H_n]$, G permutes $\{H_i : 1 \leq i \leq n\}$, $H_i \cong SL_2(p)$, $m([V, H_i]) = 2$, and $\langle \mathscr{P} \cap P \rangle$ is a Sylow p-group of H.

Proof. Let Γ consist of the minimal members of $\mathscr{P}$, pick $A \in \Gamma$, and set $U = C_V(A)$. If B is a hyperplane of A then $m(A) + m(U) \geq m(B) + m(C_V(B))$ with $B \in \mathscr{P}$ or $B = 1$ in case of equality. So by minimality of A either $|A| = p$ or the inequality is strict, and in that event, as $m(A) = m(B) + 1$, it follows that $U = C_V(B)$.

Suppose $|A| > p$ and let $K = [F(G), A]$. As $O_p(G) = 1$, $F(G)$ is a p'-group and A is faithful on $F(G)$ by 31.10, so $K \neq 1$. By Maschke's Theorem, $V = [V, K] \oplus C_V(K)$, and, as $K \neq 1$, $0 \neq [V, K]$. A acts on K and hence on $[V, K]$, so $C_{[V,K]}(A) \neq 0$. Thus $U \not\leq C_V(K)$. On the other hand, by Exercise 8.1, $K = \langle C_K(B) : |A : B| = p \rangle$ and by the last paragraph, $U = C_V(B)$, so $C_K(B) \leq N(U)$. Thus $K \leq N(U)$, so, by Exercise 3.6, $K = [K, A] \leq C_G(U)$, contradicting $U \not\leq C_V(K)$.

So $|A| = p$ for each $A \in \Gamma$. As $A \in \mathscr{P}(G, V)$, $m(V/C_V(A)) \leq m(A) = 1$, so A is generated by a transvection. Let Ω consist of the subgroups $L = \langle A_1, A_2 \rangle$ with $A_i \in \Gamma$ and $1 \neq O^p(L) \leq F(G)$. For each $A_0 \in \Gamma$ there is $L_0 \in \Omega$ with $A_0 \leq L_0$ as $[A_0, F(G)] \neq 1$. Let $L = \langle A, B \rangle \in \Omega$ and $W = [V, L]$. $W = [V, A] + [V, B]$ and $C_V(L) = C_V(A) \cap C_V(B)$, so, as $1 = m([V, A]) = m(V/C_V(A))$, $m(W) \leq 2 \geq m(V/C_V(A))$. If $0 \neq C_W(L)$ then A and B centralize $C_W(L)$, $W/C_W(L)$, and V/W, so L centralizes $C_W(L)$, $W/C_W(L)$, and V/W, which is impossible by Exercise 3.1 since L is not a p-group. Therefore $W = [V, A] \oplus [V, B]$ is of rank 2 and $V = W \oplus C_V(L)$. In particular L is faithful on W so $L \leq GL(W) \cong GL_2(p)$. Now A fixes only $[V, A]$ amongst the set θ of $p + 1$ points of W and hence is transitive on the remaining p points. So as B moves $[V, A]$, L is transitive on θ. Thus as L contains the group A of transvections with center $[V, A]$, L contains all transvections in $GL(W)$, so, by 13.7, $L = SL(W) \cong SL_2(p)$. In particular, as G is solvable, $p \leq 3$ by 13.8.

Next let $Y = O^p(L)$ and $M = N_G(Y)$. Notice M acts on W and $C_V(L)$ as $W = [V, Y]$ and $C_V(L) = C_V(Y)$. Suppose $L \neq L_0 \in \Omega$ with $Y_0 = O^p(L_0) \leq M$. Then either $W = [V, Y_0]$ or $[V, Y_0] \leq C_V(Y)$, since Y_0 is irreducible on $[V, Y_0]$. If $W = [V, Y_0]$ then $C_V(Y) = C_V(Y_0)$, so $L = SL(W) = L_0$, contrary to the choice of L_0. Thus $[V, Y_0] \leq C_V(Y)$, so $W \leq C_V(L_0)$ and, by Exercise 3.6, $[L, L_0] \leq C_G(W) \cap C_G(C_V(L)) = C_G(V) = 1$.

Let Δ be a maximal set of commuting members of Ω and $D = \langle \Delta \rangle$. By the last paragraph, D is the direct product of $(I : I \in \Delta)$ and $V = [V, D] \oplus C_V(D)$ with $[V, D] = \bigoplus_{I \in \Delta}[V, I]$. So it remains to show $\Omega = \Delta$ and $J = \langle \mathscr{P} \cap P \rangle \leq D$.

If $\Omega \neq \Delta$ then, by Exercise 11.4, there is $\bar{L} \in \Omega - \Delta$ with $\bar{Y} = O^p(\bar{L}) \leq N_{F(G)}(O^p(D))$. Thus $\bar{Y}$ permutes $\Sigma = \{[V, I]: I \in \Delta\}$ and, as $m([V, \bar{Y}]) = 2$, it follows that $\bar{Y}$ acts on each member of Σ, and hence also on each member of Δ. But now by the next-to-last paragraph and maximality of Δ, $\bar{L} \in \Delta$, contrary to the choice of $\bar{L}$.

So $\Omega = \Delta$. Finally assume $J \not\leq D$ and let $E \in \mathscr{P}$ with E minimal subject to $E \not\leq D$. Let $Z = \langle W^E \rangle$. As E is abelian, $C_E(W) = C_E(Z)$ and $E/N_E(L)$ is regular on W^E. Let S be a set of coset representatives for $N_E(L)$ in E. Then $Z = \bigoplus_{s \in S} Ws$ so $C_Z(E) = \{\sum_{s \in S} ws: w \in C_W(N_E(L))\}$ is of rank $\varepsilon = m(C_W/N_E(L))$. Therefore $m(Z/C_Z(E)) = m(Z) - \varepsilon = 2p^a - \varepsilon$, where $a = m(E/N_E(L))$. Also $m(N_E(W)/C_E(W)) = \delta \leq 1$ with $\varepsilon = 1$ in case of equality. Further $\varepsilon \leq m(W) = 2$. Thus if $a > 0$ then $2p^a - \varepsilon > a + \delta$. Finally, if $E \neq N_E(L)$, we conclude

$$m(C_V(C_E(Z))) - m(C_V(E)) \geq m(Z) - m(C_Z(E))$$

$$= 2p^a - \varepsilon > a + \delta = m(E) - m(C_E(Z)),$$

contradicting $E \in \mathscr{P}(G, V)$.

Therefore E acts on each member of Δ. Hence, as $L = O^{p'}(\text{GL}(W))$, $E \leq DC_G([D, V]) = X$. From the action of D on $[D, V]$ it follows that $m(E/C_E([D, V]) \leq m([D, V]/C_{[D,V]}(E))$, so as $\mathscr{P}$ is stable either $C_E([D, V]) = 1$ or $C_E([D, V])$ is in $\mathscr{P}$. In either case, by minimality of E, $C_E([D, V]) \leq C_D([D, V]) = 1$. Now $m(E) \leq m([D, V]/C_{[D,V]}(E))$, so as $E \in \mathscr{P}(G, V)$, $E \leq C_X(C_V(D)) = D$.

This completes the proof of the lemma. Notice that, as $D = H \trianglelefteq G$, Ω is the set of all subgroups of G generated by a pair of noncommuting members of Γ.

(32.4) Let $F^*(G) = O_p(G)$, $P \in \text{Syl}_p(G)$, $Z = \Omega_1(Z(P))$, $V = \langle Z^G \rangle$ and $G^* = G/C_G(V)$. Then V is elementary abelian and $O_p(G^*) = 1$.

Proof. $F^*(G) = O_p(G) \leq P \leq C(Z)$, so, by 31.13, $Z \leq \Omega_1(Z(O_p(G))) = V_0$. So, as $V_0 \trianglelefteq G$, $V = \langle Z^G \rangle \leq V_0$ and hence, as V_0 is elementary abelian, so is V. Let $K^* = O_p(G^*)$. By 6.4, $Q = P \cap K \in \text{Syl}_p(K)$, so, as K^* is a p-group, $K = C_G(V)Q \leq C_G(Z)$. Hence, as $K \trianglelefteq G$, $V = \langle Z^G \rangle \leq Z(K)$, so $K \leq C_G(V)$. That is $K^* = 1$.

Lemma 32.4 supplies a tool for analyzing groups G with $F^*(G) = O_p(G)$. Namely, as V is elementary abelian, we can regard V as a vector space over $\text{GF}(p)$ and the representation of G on V by conjugation makes V into a faithful $\text{GF}(p)G^*$-module. These observations are used in conjunction with lemmas 32.2 and 32.3 in the proof of the next two lemmas. Both are versions of Thompson Factorization.

(32.5) (Thompson Factorization) Let G be a solvable group with $F(G) = O_p(G)$, let $P \in \mathrm{Syl}_p(G)$, $Z = \Omega_1(Z(P))$, $V = \langle Z^G \rangle$, and $G^* = G/C_G(V)$. Then either

    (1) $G = N_G(J(P))C_G(Z)$, or

    (2) $p \le 3$, $J(G)^*$ is the direct product of copies of $\mathrm{SL}_2(p)$ permuted by G, and $J(P)^* \in \mathrm{Syl}_p(J(G)^*)$.

Proof. If $J(P)^* \neq 1$ then, by 32.2 and 32.4, G^*, V, $\mathscr{A}(G)^*$ satisfies the hypothesis of 32.3, and hence (2) holds by 32.3. So assume $J(P) \le D = C_G(V)$. By 32.1 and a Frattini Argument, $G = N_G(J(P))D \le N_G(J(P))C_G(Z)$.

(32.6) (Thompson Factorization) Let G be a solvable group with $F(G) = O_p(G)$. Let p be odd and if $p = 3$ assume G has abelian Sylow 2-subgroups. Then $G = N_G(J(P))C_G(\Omega_1(Z(P)))$ for $P \in \mathrm{Syl}_p(G)$.

Proof. This follows from 32.5 and the observation that $\mathrm{SL}_2(3)$ has nonabelian Sylow 2-groups.

33 Central extensions

A *central extension* of a group G is a pair (H, π) where H is a group and $\pi \colon H \to G$ is a surjective homomorphism with $\ker(\pi) \le Z(H)$. H is also said to be a central extension of G. Notice that the quasisimple groups are precisely the perfect central extensions of the simple groups.

    A morphism $\alpha \colon (G_1, \pi_1) \to (G_2, \pi_2)$ of central extensions of G is a group homomorphism $\alpha \colon G_1 \to G_2$ with $\pi_1 = \alpha \pi_2$. A central extension $(\tilde{G}, \pi)$ of G is *universal* if for each central extension (H, σ) of G there exists a unique morphism $\alpha \colon (\tilde{G}, \pi) \to (H, \sigma)$ of central extensions.

(33.1) Up to isomorphism there is at most one universal central extension of a group G.

Proof. If (G_i, π_i), $i = 1, 2$, are universal central extensions of G then there exist morphisms of central extensions $\alpha_i \colon (G_i, \pi_i) \to (G_{3-i}, \pi_{3-i})$. As $\alpha_i \alpha_{3-i}$ and 1 are morphisms of (G_i, π_i) to (G_i, π_i), the uniqueness of such a morphism says $\alpha_1 \alpha_2 = 1 = \alpha_2 \alpha_1$. Thus α_i is an isomorphism.

(33.2) If $(\tilde{G}, \pi)$ is a universal central extension of G then both $\tilde{G}$ and G are perfect.

Proof. Let $H = \tilde{G} \times (\tilde{G}/\tilde{G}^{(1)})$ and define $\alpha\colon H \to G$ by $(x, y)\alpha = x\pi$. Then (H, α) is a central extension of G and $\alpha_i\colon (\tilde{G}, \pi) \to (H, \alpha)$, $i = 1, 2$, are morphisms, where $x\alpha_1 = (x, 1)$ and $x\alpha_2 = (x, x\tilde{G}^{(1)})$. So, by the uniqueness of such a morphism, $\alpha_1 = \alpha_2$ and hence $\tilde{G} = \tilde{G}^{(1)}$. Thus $\tilde{G}$ is perfect, so, by 8.8.2, $G = \tilde{G}\pi$ is perfect.

(33.3) Let G be perfect and (H, π) a central extension of G. Then $H = \ker(\pi)H^{(1)}$ with $H^{(1)}$ perfect.

Proof. By 8.8, $H^{(1)}\pi = (H\pi)^{(1)} = G^{(1)}$, so, as G is perfect, $H^{(1)}\pi = G$. Hence $H = \ker(\pi)H^{(1)}$. As (H, π) is a central extension, $\ker(\pi) \leq Z(H)$, so $H/H^{(2)} = Z(H/H^{(2)})(H^{(1)}/H^{(2)})$ is abelian, and hence $H^{(1)} = H^{(2)}$ by 8.8.4. Thus $H^{(1)}$ is perfect.

(33.4) G possesses a universal central extension if and only if G is perfect.

Proof. By 33.2, if G possesses a universal central extension then G is perfect.

Conversely assume G is perfect. Let $g \mapsto \bar{g}$ be a bijection of G with a set $\bar{G}$ and let F be the free group on $\bar{G}$. Let Γ be the set of words $\bar{x}\bar{y}(\overline{xy})^{-1}$, $x, y \in G$, and let M be the normal subgroup of F generated by Γ. Next let Δ be the set of words $[w, \bar{z}]$, $w \in \Gamma$, $z \in G$, and let N be the normal subgroup of F generated by Δ. As $M \trianglelefteq F$, $N = [M, F] \trianglelefteq F$ and $M/N \leq Z(F/N)$ by 8.5.2. Then by 28.6 and 28.7 there is a unique homomorphism $\pi\colon F/N \to G$, with $(\bar{x}N)\pi = x$ for all $x \in G$, and indeed $M/N = \ker(\pi)$. Therefore $(F/N, \pi)$ is a central extension of G.

Let (H, σ) be a central extension of G. For $x \in G$, let $h(x) \in H$ with $h(x)\sigma = x$. Then, for $x, y, z \in G$, $w = h(x)h(y)h(xy)^{-1} \in \ker(\sigma) \leq C_H(h(z))$, so $[w, h(z)] = 1$. Hence by 28.6 there exists a unique homomorphism $\alpha\colon F/N \to H$ with $(\bar{x}N)\alpha = h(x)$ for each $x \in G$. Notice $\alpha\colon (F/N, \pi) \to (H, \sigma)$ is a morphism.

Now let $\tilde{G} = (F/N)^{(1)}$. By 33.3, $F/N = \ker(\pi)\tilde{G}$ and $\tilde{G}$ is perfect. Hence $(\tilde{G}, \pi)$ is also a central extension of G, and $\alpha\colon (\tilde{G}, \pi) \to (H, \sigma)$ a morphism. Suppose $\beta\colon (\tilde{G}, \pi) \to (H, \sigma)$ is a second morphism, and define $\gamma\colon \tilde{G} \to H$ by $u\gamma = u\alpha(u\beta)^{-1}$, for $u \in \tilde{G}$. Then $\alpha\sigma = \pi = \beta\sigma$, so $\tilde{G}\gamma \subseteq Z(H)$. Thus $(uv)\gamma = (uv)\alpha((uv)\beta)^{-1} = u\alpha v\alpha(v\beta)^{-1}(u\beta)^{-1} = u\alpha v\gamma(u\beta)^{-1} = u\alpha(u\beta)^{-1}v\gamma = u\gamma v\gamma$, so γ is a homomorphism. Moreover $\tilde{G}\gamma$ is abelian, so, as $\tilde{G}$ is perfect, γ is trivial by 8.8.4. Thus $\alpha = \beta$.

$(\tilde{G}, \pi)$ has been shown to be a universal central extension, so the proof is complete.

If G is a perfect group and $(\tilde{G}, \pi)$ its universal central extension, then $\tilde{G}$ is called the *universal covering group* of G and ker(π) the *Schur multiplier* of G. Notice that, by 33.2, $\tilde{G}$ is perfect.

A *perfect central extension* or *covering* of a perfect group G is a central extension (H, α) of G with H perfect.

(33.5) Let (H, α) be a central extension of a group G, and (K, β) a perfect central extension of H. Then $(K, \beta\alpha)$ is a perfect central extension of G.

Proof. $\beta\alpha: K \to G$ is the composition of surjective homomorphisms and hence a surjective homomorphism. Let $x \in$ ker$(\beta\alpha)$ and $y \in K$. $x\beta \in$ ker$(\alpha) \le Z(H)$, so $[x, y]\beta = [x\beta, y\beta] = 1$. Thus $[x, y] \in$ ker$(\beta) \le Z(K)$. Thus [ker$(\beta\alpha)$, $K, K] = 1$, so, by 8.9, ker$(\beta\alpha) \le Z(K)$.

(33.6) Let (H, α) and (K, β) be central extensions of a group G with K perfect, and $\gamma: (H, \alpha) \to (K, \beta)$ a morphism of central extensions. Then (H, γ) is a central extension of K.

Proof. $\gamma: H \to K$ is a homomorphism with $\alpha = \gamma\beta$. The latter fact implies ker$(\gamma) \le Z(H)$, so it remains to show γ is a surjection. As $\alpha = \gamma\beta$ is a surjection, $K = (H\gamma)$ker(β), so, as ker$(\beta) \le Z(K)$, $H\gamma \trianglelefteq K$ and $K/H\gamma$ is abelian. Hence $K = H\gamma$ as K is perfect.

(33.7) Let $\tilde{G}$ be the covering group of a perfect group G and let (H, α) be a perfect central extension of $\tilde{G}$. Then α is an isomorphism.

Proof. Let $\pi: \tilde{G} \to G$ be the universal covering. By 33.5, $(H, \alpha\pi)$ is a perfect central extension of G, so by the universal property there is a morphism $\beta: (\tilde{G}, \pi) \to (H, \alpha\pi)$. Then $\beta\alpha\pi = \pi$ so by the uniqueness property of universal extensions, $\beta\alpha = 1$. Hence $\beta: \tilde{G} \to H$ is an injection, while β is a surjection by 33.6. Thus β is an isomorphism and as $\beta\alpha = 1$, $\alpha = \beta^{-1}$ is too.

(33.8) Let G be perfect, $(\tilde{G}, \pi)$ the universal central extension of G, and (H, σ) a perfect central extension of G. Then
  (1) There exists a covering $\alpha: \tilde{G} \to H$ with $\pi = \alpha\sigma$.
  (2) $(\tilde{G}, \alpha)$ is the universal central extension of H.
  (3) The Schur multiplier of H is a subgroup of the Schur multiplier of G.

(4) If $Z(G) = 1$ then $Z(\tilde{G})$ is the Schur multiplier of G, and $Z(H) = \ker(\sigma) \cong \ker(\pi)/\ker(\alpha)$ is the quotient of the Schur multiplier of G by the Schur multiplier of H.

Proof. By the universal property there exists a morphism $\alpha: (\tilde{G}, \pi) \to (H, \sigma)$. Then $\pi = \alpha\sigma$ and, by 33.6, α is a covering. Let $(\tilde{H}, \beta)$ be the universal covering of H. By the universal property there is a morphism $\gamma: (\tilde{H}, \beta) \to (\tilde{G}, \alpha)$. By 33.6 and 33.7, γ is an isomorphism so (2) holds. Now (3) and (4) are straightforward.

(33.9) Let G be a group with $G/Z(G)$ finite. Then $G^{(1)}$ is finite.

Proof. Let $n = |G/Z(G)|$. For $z \in Z(G)$ and $g, h \in G$, $[g, hz] = [g, h] = [gz, h]$, so the set Δ of commutators is of order at most n^2. I claim each $g \in G^{(1)}$ can be expressed as a word $g = x_1 \ldots x_m$ in the members of Δ of length $m \leq n^3$. The claim together with the finiteness of Δ show $G^{(1)}$ is finite.

It remains to establish the claim. Pick an expression for g of minimal length m. If $m > n^3$ then, as $|\Delta| \leq n^2$, there is some $d \in \Delta$ with $\Gamma = \{i: x_i = d\}$ of order $k > n$. As $x_i x_{i+1} = x_{i+1} x_i^{x_{i+1}}$ with $x_i^{x_{i+1}} \in \Delta$ we can assume $\Gamma = \{1, \ldots, k\}$. Hence it remains to show that d^{n+1} can be written as a product of n commutators, since then the minimality of m will be contradicted. Let $d = [x, y]$. As $|G/Z(G)| = n$, $d^n \in Z(G)$, so $d^{n+1} = (d^n)^x d = (d^{n-1})^x d^x d = (d^x)^{n-1}[x^2, y]$ by 8.5.4. In particular d^{n+1} is a product of n commutators.

(33.10) Let G be a perfect finite group. Then the universal covering group of G and the Schur multiplier of G are finite.

Proof. This is a direct consequence of 33.9.

(33.11) Let (H, σ) be a perfect central extension of a finite group G, p a prime, and $P \in \mathrm{Syl}_p(H)$. Then $P \cap \ker(\sigma) \leq \Phi(P)$.

Proof. Passing to $H/(\Phi(P) \cap \ker(\sigma))$ we may assume $\Phi(P) \cap \ker(\sigma) = 1$ and it remains to show $X = P \cap \ker(\sigma) = 1$. But as $P/\Phi(P) = P^*$ is elementary abelian there is a complement Y^* to X^* in P^*. Then $P = X \times Y$ so P splits over X. Hence by Gaschütz' Theorem, 10.4, H splits over X. Hence, as H is perfect and $X \leq Z(H)$, $X = 1$.

(33.12) Let G be a perfect finite group and M the Schur multiplier of G. Then $\pi(M) \subseteq \pi(G)$.

Proof. This is a consequence of 33.11.

(33.13) Let (H, σ) be a perfect central extension of a finite group G with $\ker(\sigma)$ a p-group, let G_0 be a perfect subgroup of G containing a Sylow p-group of G, let $H_0 = \sigma^{-1}(G_0)$, and $\sigma_0 = \sigma_{|H_0}: H_0 \to G_0$. Then (H_0, σ_0) is a perfect central extension of G_0 with $\ker(\sigma_0) = \ker(\sigma)$. Hence a Sylow p-group of the Schur multiplier of G is a homomorphic image of a Sylow p-group of the Schur multiplier of G_0.

Proof. Evidently (H_0, σ_0) is a central extension of G_0 with $\ker(\sigma) = \ker(\sigma_0)$, so, by 33.3, $H_0 = \ker(\sigma) H_0^{(1)}$ with $H_0^{(1)}$ perfect. As G_0 contains a Sylow p-group of G, H_0 contains a Sylow p-group P of H and, as $\ker(\sigma)$ is a p-group, $\ker(\sigma) \leq P$. Then, by 33.11, $\ker(\sigma) \leq \Phi(P)$, so $H_0 = \Phi(P) H_0^{(1)}$. Thus $P = P \cap H_0 = \Phi(P)(P \cap H_0^{(1)})$ by the modular property, 1.14, so $P = P \cap H_0^{(1)} \leq H_0^{(1)}$ by 23.1. Thus $H_0 = P H_0^{(1)} = H_0^{(1)}$, so H_0 is perfect, completing the proof.

(33.14) If G is a perfect finite group with cyclic Sylow p-groups then the Schur multiplier of G is a p'-group.

Proof. Let (H, σ) be a perfect central extension of G with $\ker(\sigma)$ a p-group; I must show $\ker(\sigma) = 1$. Let $P \in \mathrm{Syl}_p(H)$, so that $Z = \ker(\sigma) \leq P$ and P/Z is cyclic. By 33.11, $Z \leq \Phi(P)$, so $P/\Phi(P)$ is cyclic and hence, by 23.1, P is cyclic. At this point I appeal to 39.1, which says some $h \in H$ induces a nontrivial p'-automorphism on P. But then, by 23.3, $[\Omega_1(P), h] \neq 1$, so, as $Z \leq Z(H)$, $Z \cap \Omega_1(P) = 1$. As P is cyclic and $Z \leq P$ this says $Z = 1$.

The section on the generalized Fitting group focused attention on quasisimple groups. Observe that the finite quasisimple groups are precisely the perfect central extensions of the finite simple groups. Hence, for each finite simple group G, the universal covering group of G is the largest quasisimple group with G as its simple factor, and the center of any such quasisimple group is a homomorphic image of the Schur multiplier of G. Thus it is of particular interest to determine the covering groups and Schur multipliers of the finite simple groups. This section closes with a description of the covering groups and Schur multipliers of the alternating groups.

(33.15) Let $\bar{G} = A_n$, $n \geq 5$, $\tilde{G}$ the universal covering group of $\bar{G}$, and $\tilde{Z} = Z(\tilde{G})$ the Schur multiplier of $\bar{G}$. Represent $\tilde{G}$ on $X = \{1, \dots, n\}$. Then
   (1) $\tilde{Z} \cong \mathbb{Z}_6$ if $n = 6$ or 7, while $\tilde{Z} \cong \mathbb{Z}_2$ otherwise.

(2) Let $\tilde{t}$ be a 2-element in $\tilde{G}$ such that the image $\bar{t}$ of $\tilde{t}$ in $\bar{G}$ is an involution. Then $\bar{t}$ has $2k$ cycles of length 2, $\tilde{t}$ is an involution if k is even, and $\tilde{t}$ is of order 4 if k is odd.

(3) If $n = 6$ or 7 then 3^{1+2} is a Sylow 3-group of $\tilde{G}$.

This result is usually proved using homological algebra, but I'll take a group theoretical approach here.

There are two parts to the proof. First show the order of the Schur multiplier of A_n is at least 2 (or 6 if $n = 6$ or 7). Second show the multiplier has order at most 2 (or 6) and establish 33.15.2 and 33.15.3. Exercise 11.5 handles part one (unless $n = 6$ or 7 where the proof that 3 divides the order of the multiplier is omitted). The second part is more difficult and appears below.

Assume for the remainder of the section that (G, π) is a perfect central extension of $\bar{G} = A_n$ and write $\bar{S}$ for the image $S\pi$ of $S \subseteq G$. Let $Z = Z(G)$ and assume Z is a nontrivial p-group for some prime p. Let $P \in \mathrm{Syl}_p(G)$. It will suffice to show $p \leq 3$, $|Z| = p$, and 33.15.2 and 33.15.3 hold. Assume otherwise and choose G to be a counter example to one of these statements with n minimal. The idea of the proof is simple: exhibit a perfect subgroup $\bar{H}$ of $\bar{G}$ containing $\bar{P}$, use the induction assumption to show the multiplier of $\bar{H}$ is a p'-group, and hence obtain a contradiction from 33.13. When $p = 2$, and sometimes when $p = 3$, the situation is more complicated but the same general idea works.

I begin a series of reductions.

(33.16) $n \geq 2p$.

Proof. By 33.14, $\bar{P}$ is not cyclic.

(33.17) $n > 5$.

Proof. Assume $n = 5$. By 33.16, $p = 2$. Assume $|Z| = 2.\bar{P} \leq \bar{H} \leq \bar{G}$ with $\bar{H} \cong A_4$, so $\bar{P} \cong E_4$ and $\bar{H}$ is transitive on $\bar{P}^{\#}$. By 33.11, $\Phi(P) = Z$, so by 1.13 there is an element of order 4 in P. Hence, as $\bar{H}$ is transitive on $\bar{P}^{\#}$, every element in $P - Z$ is of order 4. So, by Exercise 8.4, P is quaternion of order 8. Notice this gives 33.15.2 in this case, so it remains to show $|Z| \leq 2$. Now $\bar{P} = \langle \bar{g}, \bar{h} \rangle$ so $P^{(1)} = \langle z \rangle$ where $z = [g, h]$. If U is of index 2 in Z we've seen that $P/U \cong Q_8$, so $z \notin U$. Thus $Z/\Phi(Z)$ is cyclic so Z is cyclic. So, as Z/U is the unique subgroup of P/U of order 2, P also has a unique involution. Hence, by Exercise 8.4, P is quaternion. So, as $|P : Z(P)| = 4$, $P \cong Q_8$, That is $|Z| = 2$.

For $S \subseteq G$ or $\bar{G}$ write $\mathrm{Fix}(S)$ and $M(S)$ for the set of points in X fixed by S and moved by S, respectively.

(33.18) Let $\mathbb{Z}_3 \cong \bar{Y} \leq \bar{G}$ with $|M(\bar{Y})| = 3$. Then $Y = \langle y \rangle \times Z$ for some y of order 3, and if $8 \leq n \neq 9$ or 10 and $p = 3$ then $O^{3'}(C_G(\bar{Y})) = Y \times L$ with $L \cong A_{n-3}$.

Proof. $\bar{Y} \leq \bar{H} \cong A_5$ with $|M(\bar{H})| = 5$. Without loss $p = 3$, so, by 33.17, $H = Z \times H^{(1)}$. Thus $Y = \langle y \rangle \times Z$ where $y \in H^{(1)}$. Also, if $8 \leq n \neq 9$ or 10, then $O^{3'}(C_{\bar{G}}(\bar{Y})) = \langle \bar{y} \rangle \times \bar{K}$, $\bar{K} \cong A_{n-3}$, and, by minimality of n, $K = Z \times K^{(1)}$.

(33.19) If $p = 3$ then $n > 6$.

Proof. Assume $p = 3$ and $n = 6$. Then $E_9 \cong \bar{P} = \langle \bar{y}, \bar{g} \rangle$ with y and g moving 3 points. By 33.18 we may take y and g of order 3. Let $R = \langle y, g \rangle$. Then R is of class at most 2 so, by 23.11, R is of exponent 3. By 33.11, $R = P$ and $Z \leq \Phi(P)$. $P^{(1)} = \langle [g, y] \rangle$ so, as P is of exponent 3, $Z = P^{(1)} = Z(P)$. Thus P is extraspecial of order 27 and exponent 3, so, by 23.13, $P \cong 3^{1+2}$. Hence all parts of 33.15 hold in this case.

I now consider two cases: Case I: n is not a power of p; Case II: n is a power of p. In Case I P is not transitive on X so there is a partition $\mathscr{P} = \{X_1, X_2\}$ of X with P acting on X_1 and X_2. Let H be the subgroup acting on X_1. Then $\bar{H} = \langle \bar{a} \rangle \bar{H}_0$ with $\bar{H}_0 = \bar{H}_1 \times \bar{H}_2$, where $\bar{H}_i = \bar{G}_{X_{3-i}} \cong A_{n_i}$ is the subgroup of $\bar{G}$ fixing each point of X_{3-i}, $n_i = |X_i|$, and $\langle \bar{a} \rangle \bar{H}_i \cong S_{n_i}$ (unless n_1 or n_2 is 1). In case II, P is transitive on X and there is a partition $\mathscr{P}$ of X into p subset X_i, $1 \leq i \leq p$, of order n/p, with P transitive on $\mathscr{P}$. The subgroup of G preserving $\mathscr{P}$ contains a subgroup H with $P \leq H$ and $\bar{H} \cong (A_{n/p}) \mathrm{wr}\, A_p$ if $p \neq 2$, while $\bar{H}$ is of index 2 in $S_{n/2}\, \mathrm{wr}\, Z_2$ if $p = 2$. Notice that, by 33.16, $n/p \geq p$.

Observe next that:

(33.20) One of the following holds:

(1) $\bar{K} = O^{p'}(\bar{H})$ is perfect.
(2) $p = 2$ and, in Case I, $n_1 > 1 < n_2$.
(3) $p = 3$ and n_1 or n_2 is 3 or 4 in Case I.

Moreover, by minimality of n, Exercises 11.2 and 11.3, and 33.13, if $\bar{K}$ is perfect then $p \leq 3$, and if $p = 3$ and Case I holds, then n_1 or n_2 is 6 or 7. In particular:

(33.21) $p \leq 3$ and if $p = 3$ and Case I holds then n_1 or n_2 is 3, 4, 6, or 7.

(33.22) $p = 2$.

Proof. If not $p = 3$. Suppose Case I holds. Then n_1 or n_2 is 3, 4, 6, or 7; say n_1. Then P acts on a subset X_1' of X_1 of order 3, so replacing X_1 by X_1' we may take $n_1 = 3$. But now 33.18 implies $K = \langle y \rangle \times Z \times L$ if $8 \leq n \neq 9$ or 10, contradicting 33.11. Hence $n = 7$, 9, or 10. As n is not a power of 3, $n \neq 9$, while if $n = 10$ we could have chosen $n_1 = 1$, $n_2 = 9$. So $n = 7$, where we choose $n_1 = 1$, so that $\bar{K} \cong A_6$ and, by 33.13 and minimality of n, $|Z| = 3$ and $P \cong 3^{1+2}$.

So Case II holds. Here $\bar{K} = \langle \bar{a} \rangle \bar{K}_0$ where $\bar{K}_0$ is the direct product of 3 copies of $A_{n/3}$, and $\bar{a}$ is of order 3 with $|M(\bar{a})| = n$. Then $\bar{a} = \bar{b}^g$ for some $b \in K_0$, and, by minimality of n and Exercise 11.2, $K_0 = Z \times K_0^{(1)}$ if $n > 9$. If $n = 9$ then, by 33.18, $K_0 = \langle Z, g_i : 1 \leq i \leq 3 \rangle$ with $|M(g_i)| = 3$. Further $g_i \in Z(O^{3'}(C_G(C_G(\bar{g}_i))))$ so K_0 is abelian. Finally there is a 2-group $\bar{T} = [\bar{T}, \bar{a}] \leq \bar{G}$ with $\bar{K}_0 = [\bar{K}_0, \bar{T}]$, so, as K_0 is abelian, 24.6 implies $K_0 T = Z \times O^3(K_0 T)$ and a of course acts on $O^3(K_0 T)$. Let $L = K_0^{(1)}$ if $n > 9$ and $L = O^3(K_0 T)$ if $n = 9$. We can choose $b \in L$, so that a is of order 3. But now $P = Z \times ((P \cap L) \langle a \rangle)$ contradicting 33.11.

This leaves $p = 2$. The proof here is similar to the case $p = 3$. If n is odd then Case I holds with $n_1 = 1$, so $\bar{H} \cong A_{n-1}$ is perfect and 33.13 and minimality of n complete the proof. Thus n is even and $\bar{H} = \bar{H}_0 \bar{U}$ with $\bar{H}_0 = \bar{H}_0 = \bar{H}_1 \times \bar{H}_2$, where, in Case I, $\bar{H}_i \cong A_{n_i}$, n_i is even, and $\bar{U} = \langle \bar{u} \rangle \cong Z_2$, while, in Case II, n is a power of 2. $\bar{H}_i \cong A_{n/2}$ and $\bar{U} = \langle \bar{u}, \bar{v} \rangle \cong E_4$. We can take $U^g \leq H_0$ for some $g \in G$. Let $n_i = n/2$ in Case II. By minimality of n, $H_i = Z * H_i^{(1)}$ with $Z_i = H_i^{(1)} \cap Z$ of order at most 2, unless $n_i \leq 4$. If $n_i = 4$ then $\bar{H}_i \leq \bar{K} \cong A_5$ with $|M(K)| = 5$, and $K = Z * K^{(1)}$, so $H_i = Z * O^2(H_i)$ with $Z_i = O^2(H_i) \cap Z$ of order at most 2. Let $L_i = O^2(H_i)$ in either case. If $n_i > 4$ set $L_i = H_i^{(1)}$.

If $n_1 \geq 4 \leq n_2$ then there is $y_i \in L_i$ with $|M(y_i)| = 4$ and $y_1^a = y_2$ for some $a \in G$. Notice $Z_i = \langle y_i^2 \rangle$ so, as $Z \leq Z(G)$, $Z_2 = Z_1^a = Z_1$. Thus $Z_0 = Z \cap L_0 = Z_1$ where $L_0 = L_1 L_2$. This of course also holds if $n_i < 4$ for some i, since then $n_i = 2$ and $H_i = 1$. L_0 char H_0 so U acts on L_0, and we may take $U^g \leq L_0$, so $Z \cap U^g \leq Z_1$. Thus $H = Z * U L_0$ with $Z \cap U L_0 = Z_1$, so, by 33.11, $Z = Z_1$ is of order 2. Thus 33.15.1 is established and it remains to establish 33.15.2.

Let t be a 2-element in G with $\bar{t}$ an involution in $\bar{G}$. As $\bar{G} = A_n$ consists of even permutations on X, t has $2k$ cycles of length 2 on X for some positive integer k. Then there is a partition $X = Y_0 \cup Y_1 \cup \cdots \cup Y_k$ of X with $Y_0 = \text{Fix}(t)$, $|Y_i| = 4$ for $1 \leq i \leq k$, and t acting on Y_i. We've seen that for $i > 0$ there is K_i fixing $X - Y_i$ pointwise with $K_i \cong SL_2(3)$ and K_i centralizing G_{Y_i}. Thus $\langle K_i : 1 \le i \le k \rangle = K_1 * \cdots * K_k$ is a central product with identified centers and $t = t_1 \ldots t_k$ with $t_i \in K_i$. As $K_i \cong SL_2(3)$, t_i is of order 4 with $\langle t_i^2 \rangle = Z$, so t is of order 4 if and only if k is odd. Hence the proof of 33.15 is at last complete.

Remarks. I'm not sure who shares credit for the notion of 'components'. Certainly Bender [Be 1], Gorenstein and Walter [GW], and Wielandt [Wi 1] should be included. I use Bender's notation of $E(G)$ for the subgroup generated by the components of a group G. The Gorenstein and Walter notation of $L(G)$ is also used in the literature to denote this subgroup. I believe Bender [Be 1] was the first to formally define the generalized Fitting subgroup.

Thompson factorization was introduced by Thompson in [Th 1] and [Th 2], although 32.5 was presumably first proved by Glauberman [Gl 2]. There are analogues of Thompson factorization for arbitrary finite groups G with $F^*(G) = O_p(G)$. Such results require deep knowledge of the GF(p)-representations of nearly simple groups. In particular one needs to know the pairs (G, V) with G a nearly simple finite group and V an irreducible GF(p)G-module such that $\mathscr{P}(G, V)$ is nonempty. Generalized Thompson factorization plays an important role in the classification; see in particular the concluding remarks in section 48. Thompson factorization for solvable groups is used in this book to establish the Thompson Normal p-Complement Theorem, 39.5, and the nilpotence of Frobenius kernels (cf. 35.24 and 40.8).

Exercises for chapter 11

1. Let r be a prime, G a solvable r'-group, and A an r-group acting on G. Prove
   (1) If $\pi \subseteq \pi(G)$ and H is a Hall π-subgroup of G, then $O_p(H) \le O_{p^\pi}(G)$ for each $p \in \pi$, where $p^\pi = \pi' \cup \{p\}$.
   (2) If $[A, F(G)] = 1$ then $[A, G] = 1$.
   (3) If A centralizes a Sylow p-group of G then $[A, G] \le O_{p'}(G)$.
2. Let $(G_i : i \in I)$ be perfect groups and $\tilde{G}_i$ the universal covering group of G_i. Prove the universal covering group of the direct product D of the groups G_i, $i \in I$, is the direct product of the covering groups $\bar{G}_i$, $i \in I$, and hence the Schur multiplier of G is the direct product of the Schur multipliers of the groups G_i, $i \in I$.
3. Let H and G be perfect groups with universal covering groups $\tilde{H}$ and $\tilde{G}$, let H act transitively on a set I, and let $W = G \operatorname{wr}_I H$. Prove the universal covering group of W is the semidirect product of a central product of $|I|$ copies of $\tilde{G}$ with $\tilde{H}$, and the Schur multiplier of W is the direct product of the multipliers of G and H.
4. Let G be a finite group, p a prime, Ω a G-invariant collection of p-subgroups of G, P a p-subgroup of G, and $\Delta \subseteq P \cap \Omega$ with $\Delta \subseteq N_P(\Delta)$.

Prove that either $\Delta = P \cap \Omega$ or there exists $X \in (P \cap \Omega) - \Delta$ with $X \leq N_P(\Delta)$.

5. Let U be a 2-dimensional vector space over $\mathbb{C}$ with basis $X = (x, y)$ and define α, β, and γ to be the elements of $GL(V)$ such that

$$M_X(\alpha) = \begin{pmatrix} 0 & 1 \\ 1 & 0 \end{pmatrix}, \; M_X(\beta) = \begin{pmatrix} 0 & i \\ -i & 0 \end{pmatrix}, \; M_X(\gamma) = \begin{pmatrix} 1 & 0 \\ 0 & -1 \end{pmatrix},$$

where i is in $\mathbb{C}$ with $i^2 = -1$. Let $V = U_1 \otimes \cdots \otimes U_n$ be the tensor product of n copies U_i of U and let s_k in $\text{End}_{\mathbb{C}}(V)$ be defined (cf. 25.3.5) as follows:

$$s_{2j} = (1 \otimes \cdots \otimes 1 \otimes ((1 \otimes \beta) + (\alpha \otimes \gamma)) \otimes \gamma \otimes \cdots \otimes \gamma)/\sqrt{2},$$

$$1 \leq j \leq n,$$

$$s_{2j-1} = (1 \otimes \cdots \otimes 1 \otimes (\alpha + \beta) \otimes \gamma \otimes \cdots \otimes \gamma)/\sqrt{2}, \quad 1 \leq j \leq n,$$

where $(1 \otimes \beta) + (\alpha \otimes \gamma)$ is in the $(j, j+1)$-position of s_{2j} and $\alpha + \beta$ is in the j-position of s_{2j-1}. Prove
   (1) $s_k^2 = I$ and $(s_k s_{k+1})^3 = -I$ for $1 \leq k < 2n$.
   (2) $(s_k s_j)^2 = -I$, for k, j with $|k - j| > 1$.
   (3) Let G be the subgroup of $GL(V)$ generated by $(s_k : 1 \leq k < 2n)$. Prove $G/\langle -I \rangle \cong S_{2n}$ with $-I \in G^{(1)}$, and $G^{(1)}$ is quasisimple if $n > 2$.
6. Let G be a nonabelian simple group and $\tilde{G}$ the universal covering group of G. Prove $\text{Aut}(G) \cong \text{Aut}(\tilde{G})$.
7. Let G be a solvable group with $F(G) = O_p(G) = T$, let R be a p-subgroup of G containing T, let $\langle N_G(C) : 1 \neq C \text{ char } R \rangle \leq H \leq G$, let $X = J(RO_{p,F}(G))$, $V = \langle \Omega_1(Z(R))^X \rangle$, and $XR^* = XR/C_{XR}(V)$. Prove either
   (1) $O_p(H) = T$, or
   (2) $p \leq 3$, $X^* = X_1^* \times \cdots \times X_n^*$ with $X_i^* \cong SL_2(p)$, $R \cap X \in \text{Syl}_p(X)$, and $N_G(R)$ permutes $\{X_i : 1 \leq i \leq n\}$.
8. Let G be a finite group, r a prime, π a set of primes with $r \notin \pi$, and $E_{r^3} \cong A \leq G$. For $E_{r^2} \cong B \leq G$ define

$$\alpha(B) = \alpha_G(B) = \langle [O_\pi(C_G(b)), B] : b \in B^\# \rangle \bigcap_{b \in B^\#} O_\pi(C_G(b)).$$

Assume for each component L of G that $A \leq N_G(L)$ and for each $H \leq G$ with $L \trianglelefteq H$ that $\sigma_{\text{Aut}_H(L)}(D) = 1$ for each $E_{r^2} \cong D \leq \text{Aut}_H(L)$. Prove $\alpha(B) = 1$ for each $E_{r^2} \cong B \leq A$ if $O_\pi(G) = 1$.
9. Let G be a finite group, p a prime, and $H \trianglelefteq \trianglelefteq G$. Prove $O_p(G) \leq N_G(O^p(H))$.
10. Let Γ be a geometry of rank 2 and G a flag transitive group of automorphism of Γ (cf. section 3). For $x \in \Gamma$ let Z_x be the pointwise stabilizer in G of $\bigcup_{y \in \Gamma_x} \Gamma_y$. Assume for each $x \in \Gamma$ and $y \in \Gamma_x$ that G_x is finite, $Z_x \neq 1$, $G_x \neq G_y$, and $N_{G_y}(H) \leq G_x$ for each nontrivial normal subgroup

H of G_x contained in G_y. Prove there exists a prime p such that for each $x \in \Gamma$ and $y \in \Gamma_x$, $F^*(G_x)$ and $F^*(G_{xy})$ are p-groups and either Z_x or Z_y is a p-group.

(Hint: Let $x \in \Gamma$, $y \in \Gamma_x$, $z \in \Gamma_y$, and q a prime such that Z_x is not a q-group. For $H \leq G$ set $\theta(H) = O_q(H)E(H)$ and let $Q_x = G_{x,\Gamma_x}$. Prove $Z_x \trianglelefteq\trianglelefteq G_y$ and G_{yz}. Then use 31.4 and Exercise 11.9 to show $\theta(G_y)$, $\theta(G_{yz}) \leq G_x$. Conclude:

$$(*) \qquad \theta(G_y) \leq \theta(G_{xy}) \leq \theta(Q_y) \leq \theta(G_y)$$

Then (interchanging the roles of x and y if necessary) conclude Z_x is a p-group for some prime p. Using (*) show

$$\theta(Z_y) \leq \theta(Q_x) \leq \theta(G_{x,y}) = \theta(Q_y)$$

and hence conclude $F^*(Z_y)$ is a p-group. Finally use (*) to show $\theta(G_x) = \theta(G_{xy}) = \theta(G_y)$ for each prime $q \neq p$. This proof is due to P. Fan.)

11. (Thompson [Th 3]) Prove there exists a function f from the positive integers into the positive integers such that, for each finite set X, each primitive permutation group G on X, each $x \in X$, and each nontrivial orbit Y of G_x on X, either
   (a) $|G_x| \leq f(|Y|)$, or
   (b) $F^*(G_x)$ is a p-group for some prime p.
   Remark: Let $y \in Y$, $\mathscr{F} = (G_x, G_y)$, and apply Exercise 11.10 to the action of G on the geometry $\Gamma(G, \mathscr{F})$. The *Sims Conjecture* says that a function f exists with $|G_x| \leq f(|Y|)$ even when $F^*(G_x)$ is a p-group. The Sims Conjecture is established in [CPSS] using the classification.

12

Linear representations of finite groups

Chapter 12 considers FG-representations where G is a finite group, F is a splitting field for G, and the characteristic of F does not divide the order of G. Under these hypotheses, FG-representation theory goes particularly smoothly. For example Maschke's Theorem says each FG-representation is the sum of irreducibles, while, as F is a splitting field for G, each irreducible FG-representation is absolutely irreducible.

Section 34 begins the analysis of the characters of such representations. We find that, if m is the number of conjugacy classes of G, then G has exactly m irreducible characters ($\chi_i: 1 \leq i \leq m$), and that these characters form a basis for the space of class functions from G into F.

A result of Brauer (which is beyond the scope of this book) shows that, under the hypothesis of the first paragraph, the representation theory of G over F is equivalent to the theory of G over $\mathbb{C}$, so section 35 specializes to the case $F = \mathbb{C}$. The character table of G over $\mathbb{C}$ is defined; this is the m by m complex matrix $(\chi_i(g_j))$, where $(g_j: 1 \leq j \leq m)$ is a set of representatives for the conjugacy classes of G. Various numerical relations on the character table are established; among these the orthogonality relations of lemma 35.5 are most fundamental. The concepts of induced representations and induced characters are also discussed. These concepts relate the representations and characters of subgroups of G to those of G. Induced characters and relations among characters on the one hand facilitate the calculation of the character table, and on the other make possible the proof of deep group theoretical results. Specifically chapter 35 contains a proof of Burnside's $p^a q^b$-Theorem, which says a group whose order is divisible by just two primes is solvable, and of Frobenius' Theorem on the existence of Frobenius kernels in Frobenius groups.

Section 36 applies some of the results in section 35 and previous chapters to analyze certain minimal groups and their representations. This section is in the spirit of the fundamental paper of Hall and Higman [HH], which showed how many group theoretic questions could be reduced to questions about the FG-representations of certain minimal groups, particularly extensions of elementary abelian or extraspecial p-groups by groups of prime order.

34 Characters in coprime characteristic

In this section G is a finite group and F is a splitting field for G with characteristic not dividing the order of G. Let $R = F[G]$ be the group ring of G over F.

(34.1) R is a semisimple ring and

(1) R is the direct product of ideals $(R_i : 1 \le i \le m)$ which are simple as rings. In particular $R_i R_j = 0$ for $i \ne j$ and $1 = \sum_{i=1}^{m} e_i$, where $e_i = 1_{R_i}$.

(2) R_i is isomorphic to the ring $F^{n_i \times n_i}$ of all n_i by n_i matrices over F.

(3) $(R_i : 1 \le i \le m)$ is the set of homogeneous components of R, regarded as a right module over itself.

(4) Let S_i consist of the matrices in R_i with 0 everywhere except in the first row. Then $(S_i : 1 \le i \le m)$ is a set of representatives for the equivalence classes of simple right R-modules.

(5) $F = \text{End}_R(S_i)$ and $\dim_F(S_i) = n_i$.

Proof. As $\text{char}(F)$ does not divide the order of G, R is semisimple by Maschke's Theorem. Hence we can appeal to the standard theorems on semisimple rings (e.g. Lang [La], chapter 17) which yield 34.1, except that R_i is the ring $D_i^{n_i \times n_i}$ of matrices over the division ring $D_i = \text{End}_R(T_i)$ where T_i is a simple submodule of the ith homogeneous component R_i. But by hypothesis F is a splitting field for G, so $F = \text{End}_R(T)$ for each simple R-module T by 25.8, completing the proof.

Throughout this section m, R_i, S_i, n_i, and e_i will be as in 34.1. Notice that $|G| = \dim_F(R) = \sum_{i=1}^{m} \dim_F(R_i) = \sum_{i=1}^{m} n_i^2$, which I record as:

(34.2) $|G| = \sum_{i=1}^{m} n_i^2$.

I have already implicitly used the equivalence between FG-modules and R-modules discussed in section 12, and will continue to do so without further comment.

Let $(C_i : 1 \le i \le m')$ be the conjugacy classes of G, $g_i \in C_i$, and define $z_i \in R$ by $z_i = \sum_{g \in C_i} g$. It develops in the next lemma that $m = m'$.

(34.3) (1) The number m of equivalence classes of irreducible FG-representations is equal to the number of conjugacy classes of G.

(2) $(z_i : 1 \le i \le m)$ and $(e_i : 1 \le i \le m)$ are bases for $Z(R)$ over F.

Proof. Recall $e_i = 1_{R_i}$. As R is the direct product of the ideals $(R_i : 1 \le i \le m)$, $Z(R) = \bigoplus_{i=1}^{m} Z(R_i) = \bigoplus_{i=1}^{m} F e_i$, with the last equality following from the isomorphism $R_i \cong F^{n_i \times n_i}$ and 13.4.1. Thus if I show $(z_i : 1 \le i \le m') = B$ is also a basis for $Z(R)$, the proof will be complete. As G is a basis for R, B is linearly independent, so it remains to show B spans $Z(R)$ over F.

Let $z = \sum_{g \in G} a_g g \in R$, $a_g \in F$. Then $z \in Z(R)$ precisely when $z^h = z$ for each $h \in G$, which holds in turn precisely when $a_g = a_{(g^h)}$ for all $g, h \in G$. Thus $z \in Z(R)$ if and only if $a_g = a_i$ is independent of the choice of $g \in C_i$, or equivalently, when $z = \sum a_i z_i$. So indeed B spans $Z(R)$ over F.

A *class function* on G (over F) is a function from G into F which is constant on conjugacy classes. Denote by $\mathrm{cl}(G) = \mathrm{cl}(G, F)$ the set of all class functions on G and make $\mathrm{cl}(G)$ into an F-algebra by defining:

$$g(\alpha + \beta) = g\alpha + g\beta \quad \alpha, \beta \in \mathrm{cl}(G),$$
$$g(a\alpha) = a(g\alpha) \quad a \in F, g \in G,$$
$$g(\alpha\beta) = g\alpha \cdot g\beta.$$

Evidently $\dim_F(\mathrm{cl}(G)) = m$. By 14.8, if π is an FG-representation and χ its character, then χ is a class function. Let π_i be the representation of G on S_i by right multiplication and χ_i its character. Then $(\pi_i : 1 \leq i \leq m)$ is a set of representatives for the equivalence classes of irreducible FG-representations. By 14.8, equivalent representations have the same character, so $\{\chi_i : 1 \leq i \leq m\}$ is the *set of irreducible characters* of G over F; that is the set of all characters of irreducible FG-representations.

The *degree* of a representation π is just the dimension of its representation module. As this is also the trace of the d by d identity matrix which is in turn $\chi(1)$, it follows that:

(34.4) $\chi_i(1) = n_i$ is the degree of π_i.

Each group G possesses a so-called *principal representation* of degree 1 in which each element of G acts as the identity on the representation module V. As $\dim(V) = 1$, the principal representation is certainly irreducible. By convention π_1 is taken to be the principal representation. Hence:

(34.5) Subject to the convention that π_1 is the principal representation, $\chi_1(g) = n_1 = 1$ for all $g \in G$.

Observe that there is a faithful representation of the F-algebra $\mathrm{cl}(G)$ on R defined by $(\sum a_g g)\alpha = \sum a_g(g\alpha)$, for $\alpha \in \mathrm{cl}(G)$ and $\sum a_g g \in R$. Recall also that π_i is just the restriction to G of the representation of R on S_i and $\chi_i(r) = \mathrm{Tr}(r\pi_i)$ for $r \in R$. Finally notice that $\chi_i(r)$ is also the value of χ_i at r obtained from the representation of $\mathrm{cl}(G)$ on R.

(34.6) (1) $R_j \pi_i = 0 = \chi_i(R_j)$ for $i \neq j$.

(2) $\chi_i(R_i) = F$ for each i.

(3) $\chi_i(e_i r) = \chi_i(r)$ for each i and each $r \in R$. In particular $\chi_i(e_i) = \chi_i(1) = n_i$.

Proof. By 34.1 $R_i R_j = 0$ for $i \neq j$, so (1) holds. As $e_i = 1_{R_i}$, (1) implies (3). Part (2) is an easy exercise given the description of S_i in 34.1.4.

(34.7) The irreducible characters form a basis for $\mathrm{cl}(G)$ over F.

Proof. As $\dim_F(\mathrm{cl}(G)) = m =$ number of irreducible characters, it suffices to show $(\chi_i : 1 \leq i \leq m)$ is a linearly independent subset of $\mathrm{cl}(G)$. But this is immediate from 34.6.1 and 34.6.3.

The sum of FG-representations was defined in section 12. An immediate consequence of that definition is that $\chi_\alpha + \chi_\beta = \chi_{\alpha+\beta}$ for FG-representations α and β and their sum $\alpha + \beta$, where χ_γ denotes the character of the representation γ. Denote by $\mathrm{char}(G)$ the Z-submodule of $\mathrm{cl}(G)$ spanned by the irreducible characters of G. $\mathrm{char}(G)$ is the set of *generalized characters* of G. As each FG-representation is the sum of the irreducible representations π_i, $1 \leq i \leq m$, (cf. 12.10), it follows from the preceding remarks that each character of G is a nonnegative Z-linear combination of the irreducible characters. Thus characters are generalized characters, although by 34.7 not all generalized characters are characters.

Further if $(m_i : 1 \leq i \leq m)$ are nonnegative integers not all zero then $\sum_i m_i \chi_i$ is the character of the representation $\sum_i m_i \pi_i$. Finally, by Exercise 9.3, $\chi_{\alpha \otimes \beta} = \chi_\alpha \chi_\beta$, so the product of characters is a character. Hence $\mathrm{char}(G)$ is a Z-subalgebra of $\mathrm{cl}(G)$. These remarks are summarized in:

(34.8) The Z-submodule $\mathrm{char}(G)$ of $\mathrm{cl}(G)$ spanned by the irreducible characters is a Z-subalgebra of G. The members of $\mathrm{char}(G)$ are called generalized characters. The characters are precisely the nonnegative Z-linear span of the irreducible characters, and hence a subset of the generalized characters.

Representations and characters of degree 1 are said to be *linear.*

(34.9) Let G be an extraspecial p-group of order p^{1+2n} and $Z = Z(G) = \langle z \rangle$. Then

(1) G has p^{2n} linear representations.

(2) G has $p - 1$ faithful irreducible representations $\phi_1, \ldots, \phi_{p-1}$. Notation can be chosen so that $z\phi_i$ acts via the scalar w^i on the representation module

V_i of ϕ_i, where w is some fixed primitive pth root of 1 in F. The ϕ_i are quasiequivalent for $1 \le i \le p - 1$.

(3) ϕ_i is of degree p^n.

(4) G has exactly $p^{2n} + p - 1$ irreducible representations: those described in (1) and (2).

(5) Let E be the enveloping algebra of ϕ_i and Y a set of coset representatives for Z in G. Then $E \cong F^{p^n \times p^n}$ and Y is a basis for E over F.

Proof. By Exercise 12.1, G has exactly $|G/G^{(1)}|$ linear representations and each is irreducible. As G is extraspecial of order p^{1+2n}, $G^{(1)} = Z$ and $|G/Z| = p^{2n}$, so (1) holds.

For $x \in G - Z$, $x^G = xZ$, so G has $m = p^{2n} - 1 + p$ conjugacy classes. As G has p^{2n} irreducible linear representations, this leaves exactly $p - 1$ nonlinear irreducible representations, by 34.3. Let ϕ be such a representation. By Exercise 12.1, $Z = G^{(1)} \not\le \ker(\phi)$. But Z is the unique minimal normal subgroup of G, so $\ker(\phi) = 1$. That is ϕ is faithful. By 27.16, $z\phi = a(\phi)I$ for some primitive pth root of unity $a(\phi)$, say $a(\phi) = \omega$. By Exercise 8.5, there is an automorphism α of G of order $p-1$ regular on $Z^{\#}$. Let $\phi_i = \alpha^{i-1}\phi$, $1 \le i < p$. Then

$$z\phi_i = z^{j^{(i-1)}}\phi = \omega^{j^{(i-1)}}I$$

where $z\alpha = z^j$ and $\{j^{(i-1)}: 1 \le i < p\} = \{i: 1 \le i < p\} \bmod p$. So, renumbering, we may take $z\phi_i = \omega^i I$, $1 \le i < p$. Hence, for $i \ne j$, ϕ_i is not equivalent to ϕ_j, so we have found our remaining $p - 1$ irreducibles, and established (2) and (4).

Let d be the degree of ϕ. As each ϕ_i is quasiequivalent to ϕ, it too has degree d. We now appeal to 34.2 to conclude $p^{1+2n} = |G| = p^{2n} + (p-1)d^2$, keeping in mind that our first p^{2n} irreducibles are linear. Of course (3) follows from this equality.

Finally let E be the enveloping algebra of ϕ. As $p^n = \deg(\phi)$, $E \cong F^{p^n \times p^n}$ by 12.16. In particular $\dim_F(E) = p^{2n}$. Let Y be a set of coset representatives for Z in G. Then $|Y| = |G/Z| = p^{2n} = \dim_F(E)$, so it suffices to show Y spans E over F. But for $e \in E$, $e = \sum_{g \in G} a_g(g\phi)$, for some $a_g \in F$. Further each $g \in G$ is of the form $z^i y$ for some $y \in Y$ and $(z^i y)\phi = \omega^i y\phi$ with $\omega^i \in F$, so $e = \sum_g a_g(g\phi) = \sum_y b_y y$, for some $b_y \in F$, as desired.

35 Characters in characteristic 0

The hypothesis and notation of the last section are continued in this section. In addition assume $F = \mathbb{C}$.

(35.1) FG-representations π and ϕ are equivalent if and only if they have the same character.

Proof. By 14.8, equivalent representations have the same character. Conversely assume π and ϕ have the same character χ. Now $\pi = \sum m_i \pi_i$ and $\phi = \sum k_i \pi_i$ for some nonnegative integers m_i, k_i. It suffices to show $m_i = k_i$ for each i. But $\sum m_i \chi_i = \chi = \sum k_i \chi_i$, so, by 34.7, $m_i = k_i$ for each i.

The *regular representation* of G is the representation of G by right multiplication on R.

(35.2) Let $e_i = \sum (a_{i,g})g$, $a_{i,g} \in F$. Then

$$a_{i,g} = \chi(e_i g^{-1})/|G| = n_i \chi_i(g^{-1})/|G|,$$

where χ is the character of the regular representation of G.

Proof. Observe $\chi(e_i g^{-1}) = \chi(\sum_h (a_{i,h})hg^{-1}) = \sum_h a_{i,h}\chi(hg^{-1})$. But, by Exercise 12.2, $\chi(x) = 0$ if $x \neq 1$ and $\chi(1) = |G|$, so $\chi(e_i g^{-1}) = |G|(a_{i,g})$, yielding the first equality in the lemma. Next, by Exercise 12.2, $\chi = \sum_{i=1}^{m} n_i \chi_i$, so

$$\chi(e_i g^{-1}) = \sum_{j=1}^{m} n_j \chi_j(e_i g^{-1}) = n_i \chi_i(e_i g^{-1}) = n_i \chi_i(g^{-1})$$

with the last two equalities holding by 34.6.1 and 34.6.3, respectively.

We now define a hermitian symmetric sesquilinear form $(\ ,\ )$ on $\mathrm{cl}(G)$ with respect to the complex conjugation map on $\mathbb{C}$. (Recall $\bar{c}$ denotes the complex conjugate of c in $\mathbb{C}$.) Namely for $\chi, \theta \in \mathrm{cl}(G)$ define

$$(\chi, \theta) = \left(\sum_{g \in G} \chi(g)\bar{\theta}(g) \right) \Big/ |G|.$$

It is straightforward to check that $(\ ,\ )$ is hermitian symmetric and sesquilinear. Indeed the next lemma shows the form is nondegenerate.

(35.3) The irreducible characters form an orthonormal basis for the unitary space $(\mathrm{cl}(G), (\ ,\ ))$. That is $(\chi_i, \chi_j) = \delta_{ij}$.

Proof. By Exercise 9.4, $\bar{\chi}(g) = \chi(g^{-1})$ for each character χ and each $g \in G$. By 35.2:

$$\chi_i(e_j)/n_j = \left(\sum_{g \in G} \chi_i(g)\chi_j(g^{-1}) \right) \Big/ |G| = (\chi_i, \chi_j).$$

Hence $(\chi_i, \chi_j) = \delta_{ij}$ by 34.6.1 and 34.6.3.

Recall $(C_i: 1 \le i \le m)$ are the conjugacy classes of G, $g_i \in C_i$, and by convention $g_1 = 1$. The *character table* of G (over $\mathbb{C}$) is the m by m matrix $(\chi_i(g_j))$. Thus the rows of the character table are indexed by the irreducible characters of G and the columns by the conjugacy classes of G. In particular the character table is defined only up to a permutation of the rows and columns, except that by convention χ_1 is always the principal character and $g_1 = 1$. Subject to this convention, the character table has each entry in the first row equal to 1, while the entries in the first column are the degrees of the irreducible $\mathbb{C}G$-representations.

(35.4) Let A be the character table of G and B the matrix $(|C_i|\bar{\chi}_j(g_i)/|G|)$. Then $B = A^{-1}$.

Proof. $(AB)_{ij} = \sum_k \chi_i(g_k)|C_k|\bar{\chi}_j(g_k)/|G| = (\sum_{g \in G} \chi_i(g)\bar{\chi}_j(g))/|G| = (\chi_i, \chi_j) = \delta_{ij}$. Therefore $AB = I$ is the identity matrix. So A is nonsingular and $B = A^{-1}$.

Observe that, by 5.12, $|C_i| = |G:C_G(g_i)|$, so $|C_i|/|G| = |C_G(g_i)|^{-1}$ in lemma 35.4.

(35.5) (Orthogonality Relations) Let $h_k = |C_k|$ be the order of the kth conjugacy class of G. Then the character table of G satisfies the following orthogonality relations:

(1) $\displaystyle\sum_{k=1}^{n} h_k \chi_i(g_k)\bar{\chi}_j(g_k) = \begin{cases} 0 & \text{if } i \ne j, \\ |G| & \text{if } i = j, \end{cases}$

(2) $\displaystyle\sum_{i=1}^{n} \chi_i(g_k)\bar{\chi}_i(g_l) = \begin{cases} 0 & \text{if } k \ne l, \\ |C_G(g_k)| = |G|/h_k & \text{if } k = l, \end{cases}$

(3) $\displaystyle |G| = \sum_{i=1}^{n} n_i^2,$

(4) $\displaystyle\sum_{i=1}^{n} n_i \chi_i(g_k) = 0 \quad \text{if } k > 1,$

(5) $\displaystyle\sum_{k=1}^{n} h_k \chi_i(g_k) = 0 \quad \text{if } i > 1,$

(6) $\displaystyle\sum_{k=1}^{n} h_k |\chi_i(g_k)|^2 = |G|.$

Proof. Part (1) is a restatement of 35.3, while part (2) is a restatement of 35.4. Parts (3) and (4) are the special cases of (2) with $k=l=1$ and $l=1$, respectively (using $\chi_i(1)=n_i$). Similarly (5) and (6) are just (1) with $j=1$ and $j=i$, respectively.

The orthogonality relations may be interpreted as follows. Part (1) says the inner product of the rows of the character table (weighted by the factor h_k) is 0 or $|G|$, while part (2) says the inner product of the columns (twisted by complex conjugation) is 0 or $|G|/h_k$.

(35.6) Let $h, g \in G$. Then $h \in g^G$ if and only if $\chi_i(g)=\chi_i(h)$ for each i, $1 \le i \le n$.

Proof. If $h \in g^G$ then $\chi_i(g)=\chi_i(h)$ since characters are class functions. Conversely assume $\chi_i(g)=\chi_i(h)$ for each i, but $h \notin g^G$. By Exercise 9.7, $\bar{\chi}_i$ is also an irreducible character, so $\bar{\chi}_i(g)=\bar{\chi}_i(h)$. Hence, by 35.5.2:

$$0 = \sum_{i=1}^{n} \chi_i(g)\bar{\chi}_i(h) = \sum_{i=1}^{n} \chi_i(g)\bar{\chi}_i(g) = |C_G(g)|$$

which is of course a contradiction.

At this point we'll need a few facts about algebraic integers. Recall an *algebraic integer* is an element of $\mathbb{C}$ which is a root of a monic polynomial in $\mathbb{Z}[x]$. The following facts are well known and can be found for example in chapter 9 of Lang [La].

(35.7) (1) The algebraic integers form a subring of $\mathbb{C}$.
    (2) $\mathbb{Z}$ is the intersection of the algebraic integers with $\mathbb{Q}$.
    (3) $|\text{Norm}(z)| \ge 1$ for each algebraic integer $z \ne 0$.
    (4) An element $c \in \mathbb{C}$ is an algebraic integer if and only if there exists a faithful $\mathbb{Z}[c]$-module which is finitely generated as a $\mathbb{Z}$-module.

(35.8) n_i divides n for each i.

Proof. Let $a=n/n_i$. By 35.2, $e_i = \sum_{g \in G}(a_{i,g})g$ with $a_{i,g}=\bar{\chi}_i(g)/a$. Hence $ae_i = \sum_{g \in G} \bar{\chi}_i(g)g$. As $e_i^2=e_i$, also $ae_i = \sum_{g \in G} \bar{\chi}_i(g)ge_i$. By 27.13.1, $\chi(g)$ is the sum of $|G|$-th roots of unity for each character χ and each $g \in G$. Let M be the $\mathbb{Z}$-submodule of R generated by the elements

$$(\zeta ge_i: g \in G, \zeta \text{ is a } |G|\text{-th root of } 1).$$

Then M is a finitely generated $\mathbb{Z}$-module and $a\zeta g e_i = \zeta g a e_i = \sum_{h \in G} \bar{\chi}_i(h)$ $\zeta g h e_i \in M$, as $\bar{\chi}_i(h)$ is the sum of $|G|$-th roots of unity. Hence M is a $\mathbb{Z}[a]$-submodule of R, and certainly M is faithful as $\mathbb{Z}[a] \le \mathbb{C}$ and R is a vector space over $\mathbb{C}$. Therefore a is an algebraic integer by 35.7.4, and then a is an integer by 35.7.2.

Define the *kernel* of χ to be

$$\ker(\chi) = \{g \in G: \chi(g) = \chi(1)\}.$$

(35.9) Let χ be the character of a $\mathbb{C}G$-representation α and let $g \in G$. Then
  (1) $|\chi(g)/\chi(1)| \le 1 \ge |\mathrm{Norm}(\chi(g)/\chi(1))|$ with equality if and only if $g\alpha = \omega I$ for some root of unity ω.
  (2) $\ker(\chi) = \ker(\alpha) \trianglelefteq G$.

Proof. By 27.13.1, $\chi(g) = \sum_{i=1}^n \omega_i$, where $n = \deg(\alpha) = \chi(1)$ and ω_i is a root of unity. Thus $|\chi(g)| \le \sum_{i=1}^n |\omega_i| \le \chi(1)$ with equality if and only if $\omega_i = \omega$ is independent of i. Similarly, if $\mathrm{Norm}(\chi)$ is the norm of χ then

$$\mathrm{Norm}(\chi) = \prod_{\sigma \in \Sigma} \chi^\sigma,$$

where Σ is the set of embeddings of $Q(\chi)$ into $\bar{Q}$. But $\chi^\sigma = \sum_i \omega_i^\sigma$, so $|\chi^\sigma| \le \chi(1)$ and hence

$$|\mathrm{Norm}(\chi)|/n \le \prod_\sigma |\chi^\sigma|/n \le 1$$

with equality if and only if $\omega_i = \omega$ for all i. Recall in this last case from the proof of 27.13 that as $\mathbb{C}$ is algebraically closed, $g\alpha = \omega I$. So (1) holds. Also if $\chi(g) = n$ then $\omega = 1$ so $g\alpha = I$. Hence (2) holds.

(35.10) Let $z_i = \sum_{g \in C_i} g$, $a_{ij} = h_i \chi_j(g_i)/n_j$, and $b_{ijk} = |\{(g, h): g \in C_i, h \in C_j, gh = g_k\}|$. Then
  (1) $z_i = \sum_{j=1}^n a_{ij} e_j$,
  (2) $a_{ij} = \chi_j(z_i)/n_j$,
  (3) $a_{il} a_{jl} = \sum_k b_{ijk} a_{kl}$,
  (4) a_{ij} is an algebraic integer.

Proof. As $R = \bigoplus_{j=1}^m R_j$, $z_i = \sum_{j=1}^m z_{ij}$ for suitable $z_{ij} \in R_j$. By 34.3, $z_i \in Z(R)$, so $z_{ij} \in Z(R_j)$. Now $R_j = \mathbb{C}^{n_j \times n_j}$ so $Z(R_j) = \mathbb{C}e_j$ is 1-dimensional and consists of the scalar matrices in $\mathbb{C}^{n_j \times n_j}$. Thus $z_{ij} = c_{ij} e_j$ for some c_{ij} in $\mathbb{C}$, and $\chi_j(z_i) = \chi_j(\sum_k c_{ik} e_k) = \sum_k c_{ik} \chi_j(e_k) = c_{ij} n_j$ by 34.6. Thus to prove (1) and (2) it remains to show $a_{ij} = c_{ij}$. But $c_{ij}n_j = \chi_j(z_i) = \chi_j(\sum_{g \in C_i} g) = \sum_{g \in C_i} \chi_j(g) = h_i \chi_j(g_i)$, so the proof of (1) and (2) is complete.

Next $z_i z_j \in Z(R)$, so, as $(z_i : 1 \le i \le m)$ is basis of $Z(R)$ by 34.3.2, $z_i z_j = \sum_k c_{ijk} z_k$. Also

$$z_i z_j = \left(\sum_{g \in C_i} g \right) \left(\sum_{g \in C_j} h \right) = \sum_{\substack{g \in C_i \\ h \in C_j}} gh$$

and, in the sum on the right, the coefficient of g_k is b_{ijk}. As $z_i z_j$ is a linear combination of the z_k, and as G is a basis for R, the coefficient of $x \in C_k$ in the sum on the right is equal to that of g_k and that coefficient is c_{ijk}. That is $b_{ijk} = c_{ijk}$. Next

$$z_i z_j = \left(\sum_k z_{ik} \right) \left(\sum_i z_{jl} \right) = \sum_{k,l} z_{ik} z_{jl} = \sum_i z_{il} z_{jl}$$

as $R_k R_l = 0$ for $k \ne l$. Also $z_{il} z_{jl} = a_{il} e_l a_{jl} e_l = a_{il} a_{jl} e_l$ so $z_i z_j = \sum_i a_{il} a_{jl} e_l$. On the other hand $z_i z_j = \sum_k b_{ijk} z_k = \sum_{k,l} b_{ijk} a_{kl} e_l$ by the last paragraph and (1), so $\sum_k b_{ijk} a_{kl} = a_{il} a_{jl}$, as $(e_i : 1 \le i \le m)$ is a basis for $Z(R)$. Hence (3) holds.

Fix (i, l) and set $a = a_{i,l}$. Then

(*) $$a a_{j,l} = \sum_k b_{ijk} a_{k,l} \quad l \le j \le n.$$

Consider the following system of n equations in n unknowns $x = (x_1, \ldots, x_n)$:

(**) $$0 = \left(\sum_{k \ne j} b_{ijk} x_k \right) + (b_{ijj} - a) x_j \quad 1 \le j \le n.$$

Observe (**) has the solution $a = (a_{1,l}, \ldots, a_{m,l})$ by (*). $\chi_l(g_1) = n_l \ne 0$, so $a_{1,l} = h_1 \chi_l(g_1)/n_l \ne 0$. So, as $(0, \ldots, 0)$ is also a solution to (**), the matrix M of (**) is of determinant 0. Consider the matrix N with entries in $\mathbb{Z}[x]$ obtained by replacing a by x in M. (Observe that by definition the b_{ijk}s are nonnegative integers.) Let $f(x) = \det(N) \in \mathbb{Z}[x]$. Then $f(a) = 0$ and f is a monic polynomial, so a is an algebraic integer. Thus (4) holds.

(35.11) Let $(n_i, h_j) = 1$. Then either

(1) $\chi_i(g_j) = 0$, or
(2) $|\chi_i(g_j)| = n_i$, so $g_j \ker(\chi_i) \in Z(G/\ker(\chi_i))$.

Proof. By 27.13.1 and 35.7.1, $a = \chi_i(g_j)$ is an algebraic integer, as is $b = a_{ij} = h_j a/n_i$ by 35.10.4. Assume $a \ne 0$ and let f be the minimal polynomial of a over $\mathbb{Q}$. Say $f(x) = \sum_{k=0}^d a_k x^k$, $a_k \in \mathbb{Q}$. Now the field extensions $\mathbb{Q}(a)$ and $\mathbb{Q}(b)$ are equal as $r = h_j/n_i$ is in $\mathbb{Q}$. So the minimal polynomial $f(x)$ of b over $\mathbb{Q}$ is also of degree d. As b is a root of $g(x) = \sum_{k=1}^d a_k r^{d-k} x^k$, it follows that $g = f$. As b is an algebraic integer, $g \in \mathbb{Z}[x]$, so $a_k r^{d-k}$ is an integer. As $r = h_j/n_i$ with $(h_j, n_i) = 1$, it follows that n_i^{d-k} divides a_k for each k. Hence

$$h(x) = \sum_{k=1}^{d} \left(a_k/n_i^{d-k}\right)x^k \in \mathbb{Z}[x].$$

But a/n_i is a root of $h(x)$ and $h(x)$ is monic, so a/n_i is an algebraic integer. On the other hand $|\text{Norm}\,(a/n_i)| \leq 1$ by 35.9.1. Thus $|\text{Norm}\,(a)| = n_i$ by 35.7.3, so, by 35.9.1, $g_j\pi_i = \omega I$. Therefore (2) holds, completing the proof.

(35.12) Assume $h_j = p^e$ is a prime power for some $j > 1$ and G is simple. Then G is of prime order.

Proof. We may assume G is not of prime order. Then, as G is simple, $Z(G) = 1$ and $\ker(\chi_i) = 1$ for each $i > 1$. So, by 35.11, for each $i, j > 1$ either $\chi_i(g_j) = 0$ or p divides n_i. Next, by the orthogonality relations:

$$0 = \sum_{i=1}^{n} n_i \chi_i(g_j) = 1 + pc$$

for some algebraic integer c. Hence $c = -p^{-1}$, impossible as $-p^{-1}$ is not an algebraic integer, by 35.7.2.

(35.13) (Burnside's $p^a q^b$-Theorem). Let $|G| = p^a q^b$, with p, q prime. Then G is solvable.

Proof. Let G be a minimal counterexample. If $1 \neq H \trianglelefteq G$ then G/H and H are of order less then G and both are $\{p, q\}$-groups, so, by minimality of G, each is solvable. But now 9.3.2 contradicts the choice of G. So G is simple.

Let $Q \in \text{Syl}_q(G)$. If $Q = 1$ then G is a p-group and hence G is solvable. So $Q \neq 1$, so in particular there is $g_j \in Z(Q)^{\#}$. Then $Q \leq C_G(g_j)$, so $h_j = |G : C_G(g_j)|$ divides $|G : Q| = p^a$. So, by 35.12, G is of prime order, and hence G is solvable.

(35.14) Let $\psi \in \text{cl}(G)$. Then
  (1) $\psi = \sum_{i=1}^{m} (\psi, \chi_i)\chi_i$, and
  (2) $(\psi, \psi) = \sum_{i=1}^{m} (\psi, \chi_i)^2$.

Proof. As the irreducible characters are a basis for $\text{cl}(G)$, $\psi = \sum_{i=1}^{m} a_i \chi_i$ for some complex numbers a_i. By 35.3, $(\psi, \chi_j) = (\sum_i a_i \chi_i, \chi_j) = \sum_i a_i (\chi_i, \chi_j) = a_j$, so (1) holds. Similarly $(\psi, \psi) = \sum_i a_i a_j (\chi_i, \chi_j) = \sum_i a_i^2$, so (2) holds.

We next consider induced representations. Let $H \leq G$, F a field, and α an FG-representation. It will be convenient to regard the image $h\alpha$ of $h \in H$ under

α as a matrix rather than a linear transformation. Extend α to G by defining $g\alpha = 0$ for $g \in G - H$. Let $X = (x_i : 1 \le i \le n)$ be a set of coset representatives for H in G; hence $n = |G : H|$. For $g \in G$ define $g\alpha^G = ((x_i g x_j^{-1})\alpha)$ to be the n by n matrix whose (i, j)-th entry is the $\deg(\alpha)$ by $\deg(\alpha)$ matrix $(x_i g x_j^{-1})\alpha$. We can also regard $g\alpha^G$ as a square matrix of size $\deg(\alpha)n$ over F. If we take this point of view then:

(35.15) α^G is an FG-representation of degree $\deg(\alpha)|G : H|$.

Proof. Let $u, v \in G$, $A = u\alpha^G$, $B = v\alpha^G$, and $C = (uv)\alpha^G$. We must show $AB = C$. Regarding A and B as n by n matrices with entries from $F^{d \times d}$ (where $d = \deg(\alpha)$) we have:

$$(AB)_{ij} = \sum_k A_{ik} B_{kj} = \sum_k (x_i u x_k^{-1})\alpha (x_k v x_j^{-1})\alpha.$$

Most of the terms in the sum on the right are 0, since $(x_i u x_k^{-1})\alpha = 0$ unless $x_i u x_k^{-1} \in H$. But $x_i u x_k^{-1} \in H$ precisely when $Hx_i u = Hx_k$, so $(AB)_{ij} = 0$ unless there exists some k with $Hx_i u = Hx_k$ and $Hx_k v = Hx_j$. This holds precisely when $Hx_i uv = Hx_j$, and in that event there is a unique k with the property; namely k is defined by $Hx_k = Hx_i u$. Moreover in this case $(x_i u x_k^{-1})\alpha \cdot (x_k v x_j^{-1})\alpha = (x_i u v x_j^{-1})\alpha = C_{ij}$. Thus $(AB)_{ij} = C_{ij}$ if $Hx_i uv = Hx_j$ and $(AB)_{ij} = 0$ otherwise. Finally if $Hx_i uv \ne Hx_j$ then $x_i uv x_j^{-1} \notin H$, so $C_{ij} = (x_i uv x_j^{-1})\alpha = 0 = (AB)_{ij}$. Hence $AB = C$, so α^G is a homomorphism, and the proof is complete.

α^G is called the *induced representation* of α to G.

(35.16) (1) Up to equivalence, α^G is independent of the choice of coset representatives for H in G.

(2) If α and β are equivalent FH-representations then α^G and β^G are equivalent FG-representations.

Proof. Let $Y = (y_i : 1 \le i \le n)$ be a second set of coset representatives for H in G. Then $y_i = h_i x_i$ for some $h_i \in H$. Now, if θ is the induced representation of α to G defined with respect to Y, then $(g\theta)_{ij} = (h_i x_i g x_j^{-1} h_j^{-1})\alpha = (h_i \alpha)(x_i g x_j^{-1})\alpha(h_j \alpha)^{-1}$. Let B be the n by n diagonal matrix over $F^{d \times d}$ with $B_{ii} = h_i \alpha$. Then $(g\theta) = (g\alpha^G)^B$, so, by a remark at the beginning of section 13, θ is equivalent to α^G as an FG-representation.

Similarly if α is equivalent to β there is $D \in F^{d \times d}$ with $(h\alpha)^D = h\theta$ for each $h \in H$. Let E be the n by n diagonal matrix over $F^{d \times d}$ with $E_{ii} = D$ for each i. Then $(g\alpha^G)^E = g\beta^G$ for each $g \in G$, so (2) holds.

(35.17) Let χ be the character of α and extend χ to G by defining $\chi(g)=0$ for $g \in G - H$. Then
  (1) The character χ^G of the induced representation α^G is defined by

$$\chi^G(g) = \left(\sum_{v\in G} \chi(g^v)\right) \Big/ |H|.$$

  (2) $\chi^G(g)=0$ if $g \notin H^v$ for some $v \in G$.

Proof. $\chi^G(g) = \mathrm{Tr}(\alpha^G) = \sum_{i=1}^n \mathrm{Tr}((g^{x_i^{-1}})\alpha) = \sum_{i=1}^n \chi(g^{x_i^{-1}})$. Moreover $\chi(g^{x_i^{-1}})=0$ unless $g \in H^{x_i}$, so (2) holds. Finally $G = \bigcup_{i=1}^n Hx_i$ and, as χ is a class function, $\chi(g^{x_i^{-1}h}) = \chi(g^{x_i^{-1}})$ for $h \in H$, so $\sum_{y\in Hx_i} \chi(g^{y^{-1}}) = |H|\chi(g^{x_i^{-1}})$. Hence (1) holds.

χ^G is called the *induced character* of χ to G.

(35.18) Let $H \le K \le G$ and χ a character of H. Then $\chi^G = (\chi^K)^G$.

Proof. Let $\theta = \chi^K$. Then, for $g \in G$,

$$\theta^G(g) = \left(\sum_{v\in G}\theta(g^v)\right)\Big/|K| = \left(\sum_{\substack{g\in G\\u\in K}} \chi(g^{vu})\right)\Big/|K||H|.$$

Further the map $(v, u) \mapsto vu$ is a surjection of $G \times K$ onto G whose fibres are of order $|K|$, so $\theta^G(g) = (\sum_{w\in G} \chi(g^w))/|H| = \chi^G(g)$.

For $\psi \in \mathrm{cl}(H)$ define $\psi^G: G \to F$ by $\psi^G(g) = (\sum_{v\in G} \psi(g^v))/|H|$, where as usual $\psi(x)=0$ for $x \in G - H$. If ψ is a character of H then, by 35.17 , ψ^G is a character of G, and evidently the map $\psi \mapsto \psi^G$ preserves addition and scalar multiplication, so as the irreducible characters form a basis for $\mathrm{cl}(H)$ we conclude:

(35.19) The map $\psi \mapsto \psi^G$ is a linear transformation of $\mathrm{cl}(G)$ into $\mathrm{cl}(H)$ which maps characters to characters and generalized characters to generalized characters.

The map $\psi \mapsto \psi^G$ is called the *induction map* of $\mathrm{cl}(G)$ into $\mathrm{cl}(H)$. Notice that 35.17 and the definition of ψ^G show:

(35.20) Let $\psi \in \mathrm{cl}(H)$. Then
  (1) $\psi^G(g)=0$ for $g \in G - (\bigcup_{v\in G} H^v)$, and
  (2) $\psi^G(1) = |G : H|\psi(1)$.

Recall that there are hermitian symmetric forms $(\,,\,)_H$ and $(\,,\,)_G$ defined on $\mathrm{cl}(H)$ and $\mathrm{cl}(G)$, respectively.

(35.21) (Frobenius Reciprocity Theorem) Let $\psi \in \mathrm{cl}(H)$ and $\chi \in \mathrm{cl}(G)$. Then $(\psi, \chi|_H)_H = (\psi^G, \chi)_G$.

Proof.
$$(\psi^G, \chi)_G = \left(\sum_{g\in G} \psi^G(g)\bar{\chi}(g) \right) \Big/ |G|$$
$$= \left(\sum_{v,g\in G} \psi(g^v)\bar{\chi}(g) \right) \Big/ |G||H|$$
$$= \left(\sum_{v,g\in G} \psi(g^v)\bar{\chi}(g^v) \right) \Big/ |G||H|$$

since $\bar{\chi}$ is a class function. As the map $g \mapsto g^v$ is a permutation of G for each $v \in G$, it follows that $(\psi^G, \chi)_G = (\sum_{g\in G} \psi(g)\bar{\chi}(g))/|H|$. As $\psi(g) = 0$ for $g \in G - H$, this sum reduces to $(\sum_{h\in H} \psi(h)\bar{\chi}(h))/|H| = (\psi, \chi|_H)_H$.

A subset T of G is said to be a *TI-set* in G if $T \cap T^g \subseteq \{1\}$ for each $g \in G - N_G(T)$.

(35.22) Let T be a *TI*-set in G, $H = N_G(T)$, and $\psi, \theta \in \mathrm{cl}(H)$ with ψ and θ equal to 0 on $H - T$. Then
   (1) $\psi^G(t) = \psi(t)$ for each $t \in T^\#$.
   (2) If $\psi(1) = 0$ then $(\psi, \theta)_H = (\psi^G, \theta^G)_G$.

Proof. Let $g \in G^\#$. Then $\psi^G(g) = (\sum_{v\in G} \psi(g^v))/|H|$ with $\psi(g^v) = 0$ unless $g^v \in T$. Thus $\psi^G(g) = 0$ unless $g \in T^u$ for some $u \in G$, in which case Exercise 12.3 says T^u is the unique conjugate of T containing g while $g^v \in T$ if and only if $v \in u^{-1}H$. In particular if $g \in T$ then $\psi^G(g) = (\sum_{h\in H} \psi(g^h))/|H| = \psi(g)$. Thus (1) holds. Also, as ψ^G is a class function, $\psi^G(g) = \psi(g^v)$ if $g^v \in T$ for some $v \in G$, and $\psi^G(g) = 0$ otherwise.

Assume $\psi(1) = 0$. Then $\psi^G(1) = 0$ by 35.20.2, so

$$(\psi^G, \theta^G)_G = \left(\sum_{g\in G} \psi^G(g)\bar{\theta}^G(g) \right) \Big/ |G|$$
$$= \left(\sum_{g\in \Delta} \psi(g^{v(g)})\bar{\theta}^G(g^{v(g)}) \right) \Big/ |G|$$

where Δ consists of those $g \in G^\#$ with $g^{v(g)} \in T$ for some $v(g) \in G$. Then, by Exercise 12.3,

$$(\psi^G, \theta^G)_G = \left(\sum_{t \in T^*} \psi(t)\bar{\theta}(t) \right) \Big/ |H| = (\psi, \theta)_H$$

as ψ is 0 on $H - T^\#$.

A *Frobenius group* is a transitive permutation group G on a finite set X such that no member of $G^\#$ fixes more than one point of X and some member of $G^\#$ fixes at least one point of X. The following lemma is left as Exercise 12.4.

(35.23) (1) If G is a Frobenius group on a set X, $x \in X$, and $H = G_x$, then H is a proper nontrivial subgroup of T, H is a *TI*-set in G, and $H = N_G(H)$. Let K be the subset of G consisting of 1 together with the elements of G fixing no points of X. Then $|K| = |X|$.

(2) Assume H is a proper nontrivial subgroup of a finite group G such that H is a *TI*-set in G and $H = N_G(H)$. Then G is faithfully represented as a Frobenius group by right multiplication on the coset space G/H.

Notice that, under the hypothesis of Lemma 35.23.1, the representation of G on X is equivalent to the representation of G by right multiplication on the coset space G/H by 5.8. Hence the permutation group theoretic hypotheses of 35.23.1 are equivalent to the group theoretic hypothesis of 35.23.2 by that lemma. I'll refer to a group G satisfying either hypothesis as a *Frobenius group* and call the subgroup H the *Frobenius complement* of G. Exercise 12.4 says H is determined up to conjugacy in G. The subset K of 35.23.1 can be described group theoretically by

$$K = G - \left(\bigcup_{g \in G} (H^g)^\# \right).$$

K will be called the *Frobenius kernel* of G. The following important theorem of Frobenius shows K is a normal subgroup of G.

(35.24) (Frobenius' Theorem) The Frobenius kernel of a Frobenius group G is a normal subgroup of G.

Proof. Let H and K be the Frobenius complement and kernel of G, respectively. The idea is to produce a character χ of G with $K = \ker(\chi)$; then $K \trianglelefteq G$ by 35.9.2. This will be achieved by applying 35.22 to the *TI*-set H of G.

Let θ_i, $1 \le i \le k$, be the irreducible characters of H, $d_i = \theta_i(1)$, and, for $1 < i \le k$, let $\psi_i = d_i\theta_1 - \theta_i$. As $H \ne 1$, $k > 1$. $\psi_i(1) = d_i\theta_1(1) - \theta_i(1) = d_i - d_i = 0$. Hence, by 35.22.2 and the orthogonality relations:

$$\left(\psi_i^G, \psi_j^G\right)_G = (\psi_i, \psi_j)_H = (d_i\theta_1 - \theta_i, d_j\theta_1 - \theta_j) = d_i d_j + \delta_{ij}.$$

Next, by 35.19, $\psi_i^G = \sum_{j=1}^m c_{ij}\chi_j$ for suitable integers c_{ij}. By 35.14 and Frobenius reciprocity, $c_{i1} = (\psi_i^G, \chi_1)_G = (\psi_i, \chi_1|_H)_H = (\psi_i, \theta_1) = d_i$. So $c_{i1} = d_i$. Also $d_i^2 + 1 = (\psi_i^G, \psi_i^G) = \sum_{j=1}^m (c_{ij})^2 = d_i^2 + \sum_{j=2}^m (c_{ij})^2$, so $\psi_i^G = d_i\chi_1 + \varepsilon_i \chi_{t(i)}$ for some irreducible character $\chi_{t(i)}$ with $t(i) > 1$, and some $\varepsilon_i = \pm 1$. Next $d_i d_j + \delta_{ij} = (\psi_i^G, \psi_j^G)_G = d_i d_j + \varepsilon_i \varepsilon_j(\chi_{t(i)}, \chi_{t(j)})$, so the map $i \mapsto t(i)$ is an injection, and without loss we may take $t(i) = i$. Finally, by 35.20.2, $\psi_i^G(1) = \psi_i(1)|G : H| = 0$, so, as $\psi_i^G(1) = d_i + \varepsilon_i\chi_i(1)$ and $\chi_i(1) = \deg(\pi_i)$ is a positive integer, it follows that $\varepsilon_i = -1$ and $\chi_i(1) = d_i$.

Define $\chi = \sum_{i=1}^k d_i\chi_i$. Then $\chi(1) = \sum_{i=1}^k d_i^2 = |H|$ by 35.5.3. By 35.20.1, $\psi_i^G(g) = 0$ for $g \in K^{\#}$, so $\chi_i(g) = d_i - \psi_i^G(g) = d_i$. Thus $\chi(g) = \sum_{i=1}^n d_i^2 = \chi(1)$ for all $g \in K$, so $K \subseteq \ker(\chi)$. On the other hand each member of $G - K$ is conjugate to an element of $H^{\#}$, so to show $K = \ker(\chi)$ (and complete the proof) it remains to show $\chi(h) = 0$ for $h \in H^{\#}$. By 35.22, $\psi_i^G(h) = \psi_i(h) = d_i - \theta_i(h)$, so $\chi_i(h) = d_i - \psi_i^G(h) = \theta_i(h)$. Therefore $\chi(h) = \sum_i d_i\theta_i(h) = 0$ by 35.5.4.

(35.25) Let G be a Frobenius group with Frobenius complement H and kernel K. Then

(1) $K \trianglelefteq G$ and H is a complement to K in G. Thus G is a semidirect product of K by H.

(2) H acts semiregularly on K; that is $C_H(x) = 1$ for each $x \in K^{\#}$, or equivalently $C_K(y) = 1$ for each $y \in H^{\#}$.

(3) K is a regular normal subgroup of G in its representation as a Frobenius group.

Proof. By Frobenius' Theorem, $K \trianglelefteq G$. By definition of H and K, $H \cap K = 1$. By 35.23.1, $|K| = |G : H|$, so, by 1.7, $|KH| = |K||H| = |G|$. Thus $G = HK$. Hence (1) holds. Notice (1) and 15.10 imply (3), while (3) and 15.11 imply (2).

36 Some special actions

In this section representation theory developed in previous sections is used to derive various results on the representations of certain minimal groups, and these results are used in turn to prove a number of group theoretic lemmas. Some of these lemmas will be used in chapter 15 and one is used in 40.7 to prove the nilpotence of Frobenius kernels.

(36.1) Let p, q, and r be distinct primes, X a group of order r acting on an extra-special q-group Q with $C_Q(X) = Z(Q)$, and V a faithful $\mathrm{GF}(p)XQ$-module such that $C_V(X) = 0$. Then $r = 2^n + 1$ is a Fermat prime, $q = 2$, and Q is of width n.

Proof. Replacing V by an irreducible XQ-submodule of $[V, Z(Q)]$, we may assume V is an irreducible XQ-module. Let F be a splitting field for XQ over $\mathrm{GF}(p)$; by 27.13 we may take F to be finite. Pass to $V^F = V \otimes_{\mathrm{GF}(p)} V$. By 26.2, F is the direct sum of c Galois conjugates of an irreducible FXQ-module W. By 27.12, $C_W(X) = 0$. In particular $\dim_F(W) \neq r \dim_F(C_W(X))$, so by 27.17, W is an irreducible FQ-module. Therefore by 34.9, $\dim_F(W) = q^e$, where e is the width of Q, and for any set Y of coset representatives for $Z = Z(Q)$ in Q, Y is a basis for $E = \mathrm{End}_F(W)$ over F. As $C_Q(X) = Z$, we may pick Y to be invariant under X via conjugation and we may pick $1 \in Y$.

Next, XQ is a subgroup of E and hence acts on E via conjugation, and as Y is a basis for E over F, E is the permutation module for the permutation representation of X on Y. As $C_Q(X) = Z$, X is semiregular on $Y - \{1\}$, so, by Exercise 4.6.1, $C_E(X)$ is of dimension

$$d = 1 + \frac{|Y| - 1}{r} = 1 + \frac{q^{2e} - 1}{r}.$$

On the other hand we may assume F contains a primitive rth root of unity, so a generator x of X can be diagonalized on W. Let a_i, $1 \leq i \leq r$, be the rth roots of 1, and m_i the multiplicity of a_i as an eigenvalue of x. As x is diagonalizable, $C_E(x)$ is isomorphic as an F-algebra to the direct product of algebras $F^{m_i \times m_i}$, $1 \leq i \leq r$, so $d = \sum_{i=1}^{r} m_i^2$. Also $q^e = \dim_F(W) = \sum_i m_i$. Thus

$$q^{2e} = \left(\sum_{i=1}^{r} m_i \right)^2 = \sum_{i=1}^{r} m_i^2 + \sum_{i>j} 2 m_i m_j.$$

Therefore

$$\sum_{i>j} (m_i - m_j)^2 = (r-1) \left(\sum_i m_i^2 \right) - \sum_{i>j} 2 m_i m_j$$

$$= r \left(\sum_i m_i^2 \right) - \left(\sum_i m_i \right)^2 = rd - q^{2e} = r - 1.$$

Pick l in the range $1 \leq l \leq r$, and let $s = s(l) = |\{i : m_i = m_l\}|$. If $s = r$ then $q^e = \dim(W) = rm_l$, impossible as r and q are distinct primes. So there are at least two multiplicities. Now since there are $s(r - s)$ differences of the form $m_l - m_j$ with $m_l \neq m_j$,

(*) $$r - 1 = \sum_{i>j} (m_i - m_j)^2 \geq s(r - s)$$

with equality only if there are exactly two multiplicities m_l and m_k and $|m_l - m_k| = 1$. Then

(**) $$r(s - 1) \leq s^2 - 1 = (s - 1)(s + 1),$$

so either $s = 1$ or $r \leq s + 1$. Then as $s < r$, it follows that $s = 1$ or $r - 1$ and (**) is an equality. Thus (*) is also an equality, so by the remark after (*), there are exactly two multiplicities and $|m_l - m_k| = 1$. Hence we may take $s(l) = 1$, $m = m_l$, $s(k) = r - 1$, and $m_k = m + \epsilon$, where $\epsilon = \pm 1$. Therefore

$$\dim(W) = \sum_i m_i = m_l + (r - 1)m_k = m + (r - 1)(m + \epsilon).$$

Let $a_1 = 1$. As $C_V(X) = 0$, $m_1 = 0$. As $Z = C_Q(X)$, X does not act by scalar multiplication on W, so $s(1) \neq r - 1$. Thus $s(1) = 1$ and $m_k = 1$ for $k > 1$. Therefore $q^e = \sum_i m_i = r - 1$, completing the proof of the lemma.

(36.2) Let p, q, and r be distinct primes, X a group of order r acting faithfully on a q-group Q, and V a faithful $GF(p)XQ$-module. If $q = 2$ and r is a Fermat prime assume Q is abelian. Then $C_V(X) \neq 0$.

Proof. Assume otherwise and choose a counterexample with $m(V)$ and $|XQ|$ minimal. Then XQ is irreducible on V by minimality of $m(V)$, and, by minimality of $|XQ|$, X centralizes every proper subgroup of Q. The latter remark and Exercise 8.10 imply Q is elementary abelian or Q is extraspecial and $Z(Q) = C_Q(X)$. Moreover in either case X is irreducible on $Q/\Phi(Q)$. Now by 36.1, Q is elementary abelian, and then Exercise 4.4 and 27.18 supply a contradiction.

(36.3) Let a be an involution acting on a solvable group G of odd order, let $p \in \pi \subseteq \pi(G)$, and $p^\pi = \{p\} \cup \pi'$. Assume K is an a-invariant subgroup of G such that $C_K(a)$ contains a Hall π-subgroup of $C_G(a)$ and $X = [X, a]$ is a p-subgroup of $O_{p^\pi}(K)$. Then $X \leq O_{p^\pi}(G)$.

Proof. Take G to be a minimal counterexample. Then $O_{p^\pi}(G) = 1$ and it remains to show $X = 1$. Let V be a minimal normal subgroup of $G\langle a \rangle$ contained in G and $G^* = G/V$. By minimality of G, $X^* \leq O_{p^\pi}(G^*)$, so $C_X(V) \leq O_{p^\pi}(G) = 1$. Also V is a q-group for some prime q, and as $O_{p^\pi}(G) = 1$, $q \in \pi - \{p\}$.

By coprime action, Exercise 6.2, X is contained in an a-invariant Hall π-subgroup H of K, and as $C_K(a)$ contains a Hall π-subgroup of $C_G(a)$, $C_H(a)$ is a Hall π-subgroup of $C_G(a)$. As $X \leq O_{p^\pi}(K)$, $X \leq O_p(H)$. Thus setting

$K_0 = \langle C_{XV}(a), X \rangle$, $X \le O_{p^\pi}(K_0)$ by 31.20.1. Therefore (K_0, XV) satisfies the hypotheses of (K, G), so $G = XV$ by minimality of $|G|$. In particular X is irreducible on V and G is a π-group. As G is a π-group, $C_G(a) \le K$ and $X \le O_p(K)$, so $[C_V(a), X] \le V \cap O_p(K) = 1$. Therefore, as X is faithful and irreducible on V, $C_V(a) = 1$. But now 36.2 supplies a contradiction, completing the proof of the lemma.

Remarks. The representation theory in sections 34 and 35 is basic and belongs in any introductory course on finite groups. The results in section 36 are more technical. They are in the spirit of the fundamental paper of Hall and Higman [HH]. In particular lemma 36.1 is due to Shult [Sh] although the proof given here is from Collins [Co] and uses techniques of Hall and Higman [HH]. Lemma 36.3 will be used in the proof of the Solvable 2-Signalizer Functor Theorem in chapter 15. In the first edition of this text, section 36 contained stronger results used in the proof of the Solvable Signalizer Functor Theorem given in the first edition. These results have been omitted, since they are unnecessary for 2-signalizers.

Exercises for chapter 12

1. Let G be a finite group and F a splitting field for G whose characteristic does not divide the order of G. Prove
   (1) A character χ of G is linear if and only if χ is irreducible and $G^{(1)} \le \ker(\chi)$.
   (2) If $G = \langle g \rangle$ is cyclic of order n then F contains a primitive nth root of unity ω and the irreducible characters of G are χ_i, $1 \le i \le n$, with $\chi_i(g^j) = \omega^{ij}$.
   (3) G has exactly $|G/G^{(1)}|$ linear characters.

2. Let π be the regular representation of a finite group G over a splitting field F whose characteristic does not divide the order of G and let χ be the character of π. Prove
   (1) π is the representation induced by the regular permutation representation.
   (2) $\pi = \sum_{i=1}^m n_i \pi_i$, where $(\pi_i : 1 \le i \le m)$ are the irreducible FG-representations and $n_i = \deg(\pi_i)$.
   (3) $\chi(g) = 0$ for $g \in G^\#$ and $\chi(1) = |G|$.

3. Let G be a finite group, $T \subseteq G$, $\Delta = \{t^g : t \in T^\#, g \in G\}$, and $H = N_G(T)$. Prove
   (1) T is a *TI*-set in G if and only if, for each $t \in T^\#$, $t^G \cap T = t^H$ and $C_G(t) \le H$.

(2) If T is a *TI*-set in G then the set of conjugates of $T^{\#}$ under G partitions Δ, and, for each $t \in T^{\#}$, $|t^G| = |G:H||t^H|$.

4. (1) Prove lemma 35.23.

(2) Prove a finite group has at most one faithful permutation representation as a Frobenius group, and conclude a Frobenius group has at most one class of Frobenius complements. (You may use the fact (proved in 40.8) that Frobenius kernels are solvable.)

5. Let $G \cong S_5$ be the symmetric group on $X = \{1, \ldots, 5\}$.

(1) Use 15.3 to determine the conjugacy classes of G.

(2) As in Exercise 5.1.2, show G has 2-transitive permutation representations of degree 5 and 6, and find their permutation characters (cf. Exercise 4.5).

(3) Find all linear characters of G.

(4) Determine the character table of G.

(Hint: Use (2) and Exercise 12.6 to determine two irreducible characters of G. Then use the nonprincipal linear character from (3) and Exercises 9.3 and 9.10 to produce two more irreducible characters. Finally, given these characters and the linear characters, use the orthogonality relations to complete the table.)

6. Let π be a permutation representation of the finite group G on a set X, α the $\mathbb{C}G$-representation induced by π, and χ the character of α (cf. Exercise 4.5). We say χ is the *permutation character* of π. Prove

(1) (χ, χ_1) is the number of orbits of G on X.

(2) If π is transitive then (χ, χ) is the permutation rank of G on X.

(3) G is doubly transitive on X if and only if $G = \chi_1 + \chi_i$ for some $i > 1$. G is of permutation rank 3 if and only if $G = \chi_1 + \chi_i + \chi_j$ for some $i > j > 1$. (Hint: See Exercise 4.5.)

7. Let a be an element of prime order r acting on an r'-group G. Let p be a prime, with $p = 2$ if r is a Fermat prime, let $P = O_p(G)$, and assume $C_P(a) = 1$. Prove $[a, O^p(G)] \leq C_G(P)$.

8. Let a be an element of prime order r acting on an r'-group G. Let $1 \neq X = [X, a]$ be a q-subgroup of G with X abelian if $q = 2$, and let p be a prime distinct from q. Prove $[O_p(G), X] = \langle C_{[O_p(G),X]}(a)^X \rangle$.

13

Transfer and fusion

If G is a finite group, $H \leq G$, and $\alpha\colon H \to A$ is a homomorphism of H into an abelian group A, then it is possible to construct a homomorphism $V\colon G \to A$ from α in a canonical way. V is called the *transfer* of G into A via α. If we can show there exists $g \in G - \ker(V)$, then, as $G/\ker(V)$ is abelian, $g \notin G^{(1)}$ the commutator group of G. In particular G is not nonabelian simple.

It is however in general difficult to calculate gV explicitly and decide whether $g \in \ker(V)$. To do so we need information about the *fusion* of g in H; that is information about $g^G \cap H$. Hence chapter 13 investigates both the transfer map and techniques for determining the fusion of elements in subgroups of G.

Section 38 contains a proof of Alperin's Fusion Theorem, which says that p-local subgroups control the fusion of p-elements. To be somewhat more precise, if P is a Sylow p-subgroup of G then we can determine when subsets of P are *fused* in G (i.e. conjugate in G) by inspecting the p-locals H of G with $P \cap H$ Sylow in H.

Section 39 investigates normal p-complements. A *normal p-complement* for a finite group G is a normal Hall p'-subgroup of G. Various criteria for the existence of such objects are generated, The most powerful is the Thompson Normal p-Complement Theorem, which is used in the next section to establish the nilpotence of Frobenius kernels. Section 39 also contains a proof of the Baer–Suzuki Theorem which says a p-subgroup X of a finite group G is contained in $O_p(G)$ if and only if $\langle X, X^g \rangle$ is a p-group for each g in G.

A group A is said to act *semiregularly* on a group G if A is a group of automorphisms of G with $C_G(a) = 1$ for each $a \in A^\#$. Such actions are investigated in section 40.

37 Transfer

Let G be a finite group, $H \leq G$, and $\alpha\colon H \to A$ a homomorphism of H into an abelian group A. Given a set X of coset representatives for H in G, define $V\colon G \to A$ by

$$gV = \prod_{x \in X} ((xg)(x\bar{g})^{-1})\alpha$$

where $x\bar{g}$ denotes the unique member of X in the coset Hxg. We say V is the *transfer* of G into A via α.

(37.1) The transfer map V is independent of the choice of the set X of coset representatives.

Proof. Let Y be a second set of coset representatives for H in G. Then there is a bijection $x \mapsto y(x)$ of X with Y such that $y(x) = h(x)x$ for some $h(x) \in H$. For $y \in Y$ and $g \in G$, write $y\tilde{g}$ for the member of Y in Hyg. Observe that $\{y(x)\tilde{g}\} = Hy(x)g \cap Y = Hxg \cap Y = Hx\bar{g} \cap Y = \{y(x\bar{g})\}$; that is $y(x)\tilde{g} = y(x\bar{g})$. Also

$$y(x)g = h(x)xg = h(x)xg(x\bar{g})^{-1}(x\bar{g})$$
$$= h(x)xg(x\bar{g})^{-1}h(x\bar{g})^{-1}y(x\bar{g})$$

so that

$$y(x)g(y(x)\tilde{g})^{-1} = h(x)xg(x\bar{g})^{-1}h(x\bar{g})^{-1}.$$

Therefore

$$\prod_{y \in Y}(yg(y\tilde{g})^{-1})\alpha = \prod_{x \in X}(y(x)g(y(x)\tilde{g})^{-1})\alpha$$
$$= \prod_{x \in X} h(x)\alpha(xg(x\bar{g})^{-1})\alpha h(x\bar{g})^{-1}\alpha = \prod_{x \in X}(xg(x\bar{g})^{-1})\alpha$$

with the last equality holding as A is abelian and the map $x \mapsto x\bar{g}$ is a permutation of X. So the lemma holds.

(37.2) The transfer V is a group homomorphism of G into A.

Proof. Let $s, t \in G$. Observe $x(\overline{st}) = (x\bar{s})\bar{t}$. Hence

$$(st)V = \prod_{x \in X}((xst)(x\overline{st})^{-1})\alpha = \prod_{x \in X}(xs(x\bar{s})^{-1}(x\bar{s})t(x\bar{s}\bar{t})^{-1})\alpha$$
$$= \prod_{x \in X}(xs(x\bar{s})^{-1})\alpha((x\bar{s})\bar{t}(x\bar{s}\bar{t})^{-1})\alpha$$
$$= \left(\prod_{x \in X}(xs(x\bar{s})^{-1})\alpha\right)\left(\prod_{x \in X}(xt(x\bar{t})^{-1})\alpha\right) = sVtV,$$

using the fact that A is abelian and the map $x \mapsto x\bar{s}$ is a permutation on X.

(37.3) Let $(Hx_i g^j : 0 \leq j < n_i)$, $1 \leq i \leq r$, be the cycles of $g \in G$ on the coset space G/H. Pick $X = \{x_i g^j : 1 \leq i \leq r, 0 \leq j \leq n_i\}$. Then
  (1) $(g^{n_i})^{x_i^{-1}} \in H$ for $1 \leq i \leq r$.
  (2) $\sum_{i=1}^{r} n_i = |G : H|$.
  (3) $gV = \prod_{i=1}^{r}((g^{n_i})^{x_i^{-1}})\alpha$.

Proof. Parts (1) and (2) are immediate from the definitions. By 37.1 we may calculate V with respect to this particular choice of coset representatives. By definition of X, $(x_i g^j)g = x_i g^{j+1} = (x_i g^j)\bar{g}$ for $j < n_i - 1$. Also $(x_i g^{n_i - 1})g = x_i g^{n_i}$ and $(x_i g^{n_i - 1})\bar{g} = x_i$. So (3) holds.

(37.4) Let G be a finite group, p a prime, $H \leq G$ with $(p, |G : H|) = 1$, $K \trianglelefteq H$ with H/K abelian, and g a p-element in $H - K$ such that $g^{ma} \in g^m K$ for all integers m, and all $a \in G$, such that $g^{ma} \in H$. Then $g \notin G^{(1)}$.

Proof. Let $A = H/K$ and $\alpha : H \to A$ the natural surjection. I'll show $gV \neq 1$. Hence $g \notin \ker(V)$, and of course, as $GV \leq A$, GV is abelian, so $G^{(1)} \leq \ker(V)$.

Choose a set X of coset representatives for H in G as in 37.3. By 37.3.1, $(g^{n_i})^{x_i^{-1}} \in H$, so by hypothesis $g^{n_i x_i^{-1}} \in g^{n_i} K$. Hence $(g^{n_i x_i^{-1}})\alpha = g^{n_i}\alpha = (g\alpha)^{n_i}$. Hence, by 37.3.3, $gV = (g\alpha)^n$, where $n = \sum_{i=1}^r n_i$. Finally, by 37.3.2, $n = |G : H|$, so by hypothesis $(p, n) = 1$. Hence, as g is a p-element and $g \notin K$, also $g^n \notin K$. Thus $1 \neq g^n \alpha = gV$, as desired.

(37.5) Let G be a finite group, p a prime, $H \leq G$ with $(p, |G : H|) = 1$, and assume $g^G \cap H = g^H$ for all p-elements g in H. Then $(O^p(G)G^{(1)}) \cap H = O^p(H)H^{(1)}$.

Proof. Let $G_0 = O^p(G)G^{(1)}$ and $K = O^p(H)H^{(1)}$. Then certainly $K \leq G_0 \cap H$, and each p'-element in $G_0 \cap H$ is in $O^p(H) \leq K$, so it remains to show each p-element g in $G_0 \cap H$ is in K. But if m is an integer and $a \in G$ with $g^{ma} \in H$ then by hypothesis $g^{ma} = g^{mh}$, for some $h \in H$, so, as H/K is abelian, $g^{ma} = g^{mh} \in g^m K$. It follows from 37.4 that if $g \notin K$ then $g \notin G^{(1)}$. Then, as all p-elements in G_0 are in $G^{(1)}$, also $g \notin G_0$, contrary to the choice of g. So the lemma holds.

Let $H \leq G$ and S an H-invariant subset of G. Then H is said to *control fusion* in S if $s^G \cap S = s^H$ for each $s \in S$. For example one of the hypotheses in the last lemma says that H controls fusion of its p-elements.

Let $X \subseteq H \leq G$. Then X is said to be *weakly closed* in H with respect to G if $X^G \cap H = \{X\}$.

(37.6) Let p be a prime, $T \in \mathrm{Syl}_p(G)$, $W \leq T$ with W weakly closed in T with respect to G, and $D = C_G(W)$. Then $N_G(W)$ controls fusion in D.

Proof. Let $d \in D$ and $g \in G$, with $d^g \in D$. Then $W, W^{g^{-1}} \leq C(d)$. By Sylow's Theorem there is $x \in C_G(d)$ with $U = \langle W, W^{g^{-1}x}\rangle$ a p-group. Let

$U \leq S \in \mathrm{Syl}_p(G)$. As W is weakly closed in T, $\{W\} = W^G \cap S = \{W^{g^{-1}x}\}$. Thus $h = x^{-1}g \in N_G(W)$ and $d^g = d^h$.

(37.7) Assume T is an abelian Sylow p-group of G. Then $T \cap O^p(G) = [T, N_G(T)]$.

Proof. Let $H = N_G(T)$. By the Schur–Zassenhaus Theorem there is a complement X to T in H. Then $K = X[T, X] \trianglelefteq XT = H$ and $H/K \cong T/[T, X]$ is an abelian p-group, so $K = O^p(H)H^{(1)}$. Next $G = TO^p(G)$, so $G/O^p(G) \cong T/(T \cap O^p(G))$ is abelian and hence $O^p(G) = O^p(G)G^{(1)}$. Certainly T is weakly closed in itself, so, by 37.6, H controls fusion in $C_G(T)$, and hence, as $T \leq C_G(T)$, H controls fusion in T. But T is the set of p-elements in H so H controls fusion of its p-elements and hence, by 37.5, $O^p(G) \cap H = K$. As $T \cap O^p(G)$ and $[T, H]$ are Sylow in $O^p(G)$ and K, respectively, it follows that $T \cap O^p(G) = [T, N_G(T)]$.

38 Alperin's Fusion Theorem

In this section G is a finite group, p is a prime, and P is some Sylow p-group of G. A p-subgroup X of G is said to be a *tame intersection* of Sylow p-groups Q and R of G if $X = Q \cap R$ and $N_Q(X)$ and $N_R(X)$ are Sylow p-groups of $N_G(X)$.

The main result of this section is:

(38.1) (Alperin's Fusion Theorem) Let $P \in \mathrm{Syl}_p(G)$, $g \in G$, and $A, A^g \subseteq P$. Then there exists $Q_i \in \mathrm{Syl}_p(G)$, $1 \leq i \leq n$, and $x_i \in N_G(P \cap Q_i)$ such that:

(1) $g = x_1 \ldots x_n$.
(2) $P \cap Q_i$ is a tame intersection of P and Q_i for each i, $1 \leq i \leq n$.
(3) $A \subseteq P \cap Q_1$ and $A^{x_1 \cdots x_i} \subseteq P \cap Q_{i+1}$ for $1 \leq i < n$.

Alperin's Theorem will follow from Theorem 38.2 and an easy argument. But first some definitions.

For $R, Q \in \mathrm{Syl}_p(G)$ write $R \to Q$ if there exist Sylow p-groups $(Q_i : 1 \leq i \leq n)$ of G and elements $x_i \in N_G(P \cap Q_i)$ such that:

(1) $P \cap Q_i$ is a tame intersection of P and Q_i for each i with $1 \leq i \leq n$.
(2) $P \cap R \leq P \cap Q_1$ and $(P \cap R)^{x_1 \cdots x_i} \leq P \cap Q_{i+1}$ for each i with $1 \leq i < n$.
(3) $R^x = Q$, where $x = x_1 \ldots x_n$.

I'll also write $R \xrightarrow{x} Q$ when it's necessary to emphasize the role of the element x in (3), and say that $(Q_i, x_i : 1 \leq i \leq n)$ *accomplish* $R \to Q$.

Theorem 38.2. $Q \to P$ for each $Q \in \text{Syl}_p(G)$.

The proof of 38.2 involves several reductions. Observe first that:

(38.3) $P \to P$.

Indeed $P \overset{x}{\to} P$ is accomplished by $P, 1$.

(38.4) $\to$ is a transitive relation.

Proof. Let $(R_i, y_i : 1 \leq i \leq m)$ and $(Q_i, x_i : 1 \leq i \leq n)$ accomplish $S \to R$ and $R \to Q$, respectively. Then $R_1, \ldots, R_m, Q_1, \ldots, Q_n$, and $y_1, \ldots, y_m$, $x_1, \ldots, x_n$ accomplish $S \to Q$.

(38.5) If $S \overset{x}{\to} P$, $Q^x \to P$, and $P \cap Q \leq P \cap S$, then $Q \to P$.

Proof. By 38.4 it suffices to show $Q \to Q^x$. Let $(S_i, x_i : 1 \leq i \leq n)$ accomplish $S \to P$. Then $(S_i, x_i : 1 \leq i \leq n)$ also accomplish $Q \to Q^x$.

(38.6) Assume $R, Q \in \text{Syl}_p(G)$ with $R \to P$ and $P \cap Q < R \cap Q$. Assume further, for all $S \in \text{Syl}_p(G)$ with $|S \cap P| > |Q \cap P|$, that $S \to P$. Then $Q \to P$.

Proof. By hypothesis there is $x \in G$ with $R \overset{x}{\to} P$. Now $P \cap Q^x = R^x \cap Q^x = (R \cap Q)^x$, so $|P \cap Q^x| = |R \cap Q| > |P \cap Q|$. Hence $Q^x \to P$ by hypothesis. Now apply 38.5 to complete the proof.

(38.7) Assume $P \cap Q$ is a tame intersection of P and Q such that $S \to P$ for all $S \in \text{Syl}_p(G)$ with $|S \cap P| > |Q \cap P|$. Then $Q \to P$.

Proof. By 38.3 we may assume $Q \neq P$. Thus $P \cap Q < P_0 = N_P(P \cap Q)$. By hypothesis P_0 and $Q_0 = N_Q(P \cap Q)$ are Sylow in $M = N_G(P \cap Q)$, so there is $x \in M$ with $Q_0^x = P_0$. Notice $Q \to Q^x$ is accomplished by (Q, x). Further $P \cap Q < P_0 \leq P \cap Q^x$, so by hypothesis $Q^x \to P$. Therefore, by 38.4, $Q \to P$.

We are now in a position to prove 38.2. Pick a counterexample Q with $P \cap Q$ of maximal order. By 38.3, $P \neq Q$, so $P \cap Q \neq P$, and hence $P \cap Q < N_P(P \cap Q)$. Let $S \in \text{Syl}_p(G)$ with $N_P(P \cap Q) \leq N_S(P \cap Q) \in \text{Syl}_p(N_G(P \cap Q))$. As $P \cap Q < N_P(P \cap Q) \leq P \cap S$, it follows that $S \to P$ by maximality of $P \cap Q$. Thus there is $x \in G$ with $S \overset{x}{\to} P$.

Evidently $(P \cap Q)^x \leq Q^x$. Also $P \cap Q \leq S$ and $S^x = P$, so $(P \cap Q)^x \leq P$. Thus $(P \cap Q)^x \leq P \cap Q^x$. If $(P \cap Q)^x \neq P \cap Q^x$ then $|P \cap Q| < |P \cap Q^x|$, so, by maximality of $|P \cap Q|$, $Q^x \to P$. But then $Q \to P$ by 38.5, contradicting the choice of Q.

So $(P \cap Q)^x = P \cap Q^x$. Next let $T \in \mathrm{Syl}_p(G)$ with $N_{Q^x}(P \cap Q^x) \leq N_T(P \cap Q^x) \in \mathrm{Syl}_p(N_G(P \cap Q^x))$. Again $P \cap Q^x < N_{Q^x}(P \cap Q^x) \leq T$, so $P \cap Q^x < T \cap Q^x$. Hence if $T \to P$ then, by 38.6, $Q^x \to P$, which we have already observed to be false. Thus we do not have $T \to P$, so, by maximality of $|P \cap Q|$, $P \cap Q^x = P \cap T$.

By choice of T, and as $P \cap Q^x = P \cap T$, we have $N_T(P \cap T) \in \mathrm{Syl}_p(N_G(P \cap T))$. By choice of S, $N_S(P \cap Q) \in \mathrm{Syl}_p(N_G(P \cap Q))$ so, as $(P \cap Q)^x = P \cap Q^x = P \cap T$ and $S^x = P$, we have $N_P(P \cap T) \in \mathrm{Syl}_p(N_G(P \cap T))$. Thus $P \cap T$ is a tame intersection of P and T. But now, by 38.7, $T \to P$, contrary to the last paragraph.

This completes the proof of 38.2.

Now for the proof of Alperin's Fusion Theorem. Assume the hypothesis of Alperin's Theorem. By 38.2, $P^{g^{-1}} \to P$. Let $(Q_i, x_i : 1 \leq i \leq n-1)$ accomplish $P^{g^{-1}} \to P$. As A, $A^g \subseteq P$, $A \subseteq P \cap P^{g^{-1}}$, so, by definition of $\to$, $A \subseteq P \cap P^{g^{-1}} \leq P \cap Q_1$ and $A^{x_1 \cdots x_i} \subseteq (P \cap P^{g^{-1}})^{x_1 \cdots x_i} \leq P \cap Q_{i+1}$ for $1 \leq i < n-1$. Also, setting $x = x_1 \ldots x_{n-1}$, $P^{g^{-1}x} = P$, so $x_n = x^{-1}g \in N_G(P)$ and $g = xx_n = x_1 \ldots x_n$. Finally let $Q_n = P$ and observe $A^{x_1 \cdots x_{n-1}} = A^{gx_n^{-1}} \leq P^{x_n^{-1}} = P = P \cap Q_n$, so the theorem holds.

39 Normal *p*-complements

In this section p is a prime and G is a finite group. A *normal p-complement* for G is a normal Hall p'-subgroup of G; that is a normal p-complement is a normal complement to a Sylow p-subgroup of G.

(39.1) (Burnside Normal *p*-Complement Theorem) If a Sylow *p*-subgroup of G is in the center of its normalizer then G possesses a normal p-complement.

Proof. This is immediate from 37.7.

(39.2) If p is the smallest prime divisor of the order of G and G has cyclic Sylow *p*-groups, then G has a normal p-complement.

Proof. Let $P \in \mathrm{Syl}_p(G)$. By hypothesis P is cyclic. As P is abelian and Sylow in G, $\mathrm{Aut}_G(P)$ is a p'-group, so, by 23.3, $|\mathrm{Aut}_G(P)|$ divides $p - 1$. Hence, as

p is the smallest prime divisor of $|G|$, $\mathrm{Aut}_G(P) = 1$. Equivalently, P is in the center of its normalizer, so 39.1 completes the proof.

G is *metacyclic* if there exists a normal subgroup H of G such that H and G/H are cyclic.

(39.3) Assume each Sylow group of G is cyclic. Then G is metacyclic.

Proof. Let p be the smallest prime divisor of $|G|$. By 39.2, G has a normal p-complement H. By induction on the order of G, H is metacyclic. In particular H is solvable, so, as G/H is cyclic, G is solvable.

Let $K = F(G)$. K is nilpotent with cyclic Sylow groups, so K is cyclic. Thus $\mathrm{Aut}(K)$ is abelian. However, by 31.10, $K = C_G(K)$, so $\mathrm{Aut}_G(K) = G/K$ is abelian. As G/K has cyclic Sylow groups, G/K is cyclic, so G is metacyclic.

A subgroup H of G is a *p-local subgroup* if $H = N_G(P)$ for some nontrivial p-subgroup P of G.

(39.4) (Frobenius Normal p-Complement Theorem) The following are equivalent:

(1) G has a normal p-complement.
(2) Each p-local subgroup of G has a normal p-complement.
(3) $\mathrm{Aut}_G(P)$ is a p-group for each p-subgroup P of G.

Proof. The implications (1) implies (2) and (2) implies (3) are easy and left as exercises. Assume G satisfies (3) but not (1), and, subject to this constraint, choose G minimal.

Observe first that $G = O^p(G)$. For if not, by minimality of G, $O^p(G)$ has a normal p-complement, which is also a normal p-complement for G.

Next let's see that a Sylow p-subgroup P of G controls fusion in P. For let $g \in G$, $a \in P$, and $a^g \in P$. Apply Alperin's Theorem to obtain $Q_i \in \mathrm{Syl}_p(G)$ and $x_i \in N_G(P \cap Q_i)$ satisfying the conditions of that theorem with $A = \{a\}$. In particular $g = x_1 \ldots x_n$. It will suffice to show $a^{x_1 \cdots x_i} \in a^P$ for each $1 \le i \le n$. Assume otherwise and let i be a minimal counterexample. Then $b = a^{x_1 \cdots x_{i-1}} \in P \cap Q_i = U$, $x_i \in H = N_G(U)$, $P \cap H \in \mathrm{Syl}_p(H)$, and $a^{x_1 \cdots x_i} = b^{x_i}$. By hypothesis $\mathrm{Aut}_G(U)$ is a p-group, so $H = C_G(U)(P \cap H)$, as $P \cap H \in \mathrm{Syl}_p(H)$. Thus, as $b \in U$, $b^{x_i} = b^t$ for some $t \in P \cap H$, so, as $b \in a^P$, $a^{x_1 \cdots x_i} = b^{x_i} \in a^P$.

So P controls fusion in P. Hence by 37.5 $(O^p(G)G^{(1)}) \cap P = O^p(P)P^{(1)}$. But $O^p(P) = 1$ and, as $G = O^p(G)$, $(O^p(G)G^{(1)}) \cap P = P$. Thus $P = P^{(1)}$, so, as p-groups are solvable, $P = 1$. But then G is its own normal p-complement.

(39.5) (Thompson Normal p-Complement Theorem) Let p be odd and $P \in \mathrm{Syl}_p(G)$. Assume $N_G(J(P))$ and $C_G(\Omega_1(Z(P)))$ have normal p-complements. Then G has a normal p-complement.

Proof. Assume otherwise and let G be a minimal counterexample. By the Frobenius p-Complement Theorem there is a p-local subgroup H of G which possesses no normal p-complement. By Sylow's Theorem we take $Q = P \cap H \in \mathrm{Syl}_p(H)$. Choose H with Q maximal subject to these constraints.

Claim $P = Q$. If not $Q < N_P(Q)$, so $|N_G(C)|_p > |Q|$ for each C char Q. Hence, by maximality of Q, $N_G(C)$ has a normal p-complement. So $N_H(C)$ also has a normal p-complement. In particular this holds for $C = J(Q)$ and $\Omega_1(Z(Q))$, so the hypotheses of the theorem hold in H. Hence, by minimality of G, H has a normal p-complement, contrary to the choice of H.

So $P \in \mathrm{Syl}_p(H)$. Hence, as H has no p-complement, $H = G$ by minimality of G. Moreover if $O_{p'}(G) \neq 1$ then, by 18.7 and minimality of G, $G/O_{p'}(G)$ has a p-complement, and then G does too. So $O_{p'}(G) = 1$.

As $G = H$ is a p-local, $O_p(G) \neq 1$. Let $G^* = G/O_p(G)$. Then $O_p(G^*) = 1$ so $N_G(J(P^*))$ and $C_G(\Omega_1(Z(P^*)))$ are proper subgroups of G containing P, and hence by minimality of G have p-complements. So, by minimality of G, G^* has a p-complement. Hence $G = O_{p,p',p}(G)$. Thus $E(G) = O_{p,p',p}(E(G))$. But $[E(G), F(G)] = 1$ so $O_p(E(G)) \leq Z(E(G))$, and thus $O_{p,p'}(E(G)) = O_p(E(G)) \times O_{p'}(E(G))$. Now $O_{p'}(E(G)) \leq O_{p'}(G) = 1$, so $E(G)$ is a p-group. Hence, as $E(G)$ is perfect, $E(G) = 1$. So $F^*(G) = F(G)E(G) = O_p(G)$.

Let $p \neq r \in \pi(G)$ and $R \in \mathrm{Syl}_r(G)$. $R \leq O_{p,p'}(G)$, so, by a Frattini Argument, $O_p(G)N_G(Z(R))$ contains a Sylow p-group of G which we may take to be P. In particular P acts on $O_p(G)Z(R)$ so $PZ(R) = K$ is a subgroup of G. Now if $K \neq G$ then, by minimality of G, $Z(R) = O_p(K) \trianglelefteq K$. But then $Z(R) \leq C_G(O_p(G)) \leq O_p(G)$ by 31.10, and hence $Z(R) = 1$, a contradiction.

So $G = PZ(R)$. In particular G is solvable and $R = Z(R)$ is abelian. But now by Thompson Factorization, 32.6, $G = G_1 G_2$ where $G_1 = N_G(J(P))$ and $G_2 = C_G(\Omega_1(Z(P)))$. By hypothesis $G_i = O_{p'}(G_i)P$. As $O_{p'}(G_i) \leq C_G(O_p(G)) \leq O_p(G)$, $O_{p'}(G_i) = 1$, so $G_i = P$. Thus $G = G_1 G_2 = P$, contradicting the choice of G as a counterexample.

The next result does not involve normal p-complements, but it has the same flavor.

(39.6) (Baer-Suzuki Theorem) Let X be a p-subgroup of G. Then either $X \leq O_p(G)$ or there exists $g \in G$ with $\langle X, X^g \rangle$ not a p-group.

Proof. Assume the theorem is false and let $P \in \mathrm{Syl}_p(G)$ and $\Delta = X^G \cap P$. $\langle X^G \rangle \trianglelefteq G$, so, as $X \not\leq O_p(G)$, $\langle X^G \rangle$ is not a p-group. Thus $\Delta \neq X^G$. For $Y \in X^G - \Delta$, $\langle Y, \Delta \rangle$ is not a p-group, by Sylow's Theorem. Let Γ be of maximal order subject to $\Gamma \subseteq \Delta$ and $\langle \Gamma, Y \rangle$ a p-group for some $Y \in X^G - \Delta$. Then $\Gamma \neq \Delta$, but on the other hand as G is a counter example to the theorem, at least Γ is nonempty. Let $Q = \langle \Gamma, Y \rangle$. By maximality of Γ, $\Gamma = \Delta \cap Q$. Hence by Exercise 11.4 we may take $Y \leq N(\Gamma)$ and there is $Z \in N_\Delta(\Gamma) - \Gamma$. As $X \not\leq O_p(G)$, $N_G(\Gamma) \neq G$, so, by induction on the order of G, $\langle X^G \cap N_G(\Gamma) \rangle$ is a p-group. But now $\Gamma \subset \Gamma \cup \{Z\} = \Gamma'$ and $\langle \Gamma', Y \rangle$ is a p-group, contradicting the maximality of Γ.

40 Semiregular action

In this section G is a nontrivial finite group and A is a nontrivial group of automorphisms of G which acts semiregularly on G. Recall this means $C_G(a) = 1$ for each $a \in A^{\#}$.

(40.1) $|G| \equiv 1 \bmod |A|$. In particular $(|A|, |G|) = 1$.

Proof. As A is semiregular, each orbit of A on $G^{\#}$ is of order $|A|$.

We will wish to apply coprime action, 18.6, to the representation of A on G. Thus, until 40.7, assume either G or A is solvable. In 40.7 we prove G is nilpotent, at which point we see the assumption was unnecessary.

(40.2) A is semiregular on each A-invariant subgroup and factor group of G.

Proof. The first remark is trivial and the second follows from coprime action, 18.7.

(40.3) For each $p \in \pi(G)$, there is a unique A-invariant Sylow p-subgroup of G.

Proof. By coprime action, 18.7, the set Δ of A-invariant Sylow p-groups is nonempty and $C_G(A)$ is transitive on Δ. So as $C_G(A) = 1$, the lemma follows.

(40.4) For each $a \in A$, the map $g \mapsto [g, a]$ is a permutation of G.

Proof. $[g, a] = g^{-1}g^a$, So $[g, a] = [h, a]$ if and only if $hg^{-1} \in C_G(a)$. So, as $C_G(a) = 1$, the commutator map is an injection, and hence also a bijection as G is finite.

(40.5) If A is of even order there is a unique involution t in A, $g^t = g^{-1}$ for $g \in G$, and G is abelian.

Proof. As $|A|$ is even there is an involution $t \in A$. Let $g \in G$. By 40.4, $g = [h, t]$ for some $h \in G$. Thus $g^t = (h^{-1}h^t)^t = h^{-t}h = g^{-1}$. Therefore, for each $x \in G$, $x^t = x^{-1}$, so $xg = (xg)^{-t} = (g^{-1}x^{-1})^t = g^{-t}x^{-t} = gx$. So G is abelian. Finally, if s is any involution in A, then I've shown s inverts G, so $st \in C_A(G) = 1$. Thus t is unique.

(40.6) Let $p, q \in \pi(A)$. Then
  (1) If p is odd then Sylow p-groups of A are cyclic.
  (2) Sylow 2-groups of A are cyclic or quaternion.
  (3) Subgroups of A of order pq are cyclic.
  (4) Subgroups of A of odd order are metacyclic.

Proof. By 40.2 and 40.3 we may assume G is an r-group for some prime r, and indeed replacing G by $G/\Phi(G)$ we may assume G is elementary abelian. Observe that, for $B \leq A$, $\langle C_G(b): b \in B^{\#} \rangle \neq G$ as $C_G(b) = 1$ for each $b \in B^{\#}$. So, by Exercise 8.1, $m_p(A) = 1$. Hence Exercise 8.4 implies (1) and (2). Part (3) follows from 27.18, while (1) and 39.3 imply (4).

(40.7) G is nilpotent.

Proof. Let G be a minimal counterexample and a an element of A of prime order. By induction on the order of A we may take $A = \langle a \rangle$. In particular A is solvable, so all lemmas in this section apply.

Suppose $q \in \pi(G)$ and $1 \neq Q$ is an A-invariant normal elementary abelian q-subgroup of G. By 40.2 and minimality of G, G/Q is nilpotent so $RQ \trianglelefteq G$ for each Sylow r-group R of G. In particular if $r = q$ then $R \trianglelefteq G$. But as G is a counter example there exists $r \in \pi(G)$ with Sylow r-groups of G not normal. By symmetry between r and q, $O_r(G) = 1$. As $RQ \trianglelefteq G$, $C_R(Q) \leq O_r(G) = 1$, so $Z(R)$ is faithful on Q. By 40.3 we may choose R to be a-invariant; then a is faithful on $Z(R)$. But now, as $C_Q(a) = 1$, 36.2 supplies a contradiction.

So $F(G) = 1$. Now if H is a proper A-invariant normal subgroup of G then, by minimality of G, H is nilpotent, so, as $F(G) = 1$, $H = 1$.

As G is not nilpotent, G is not a 2-group so there is an odd prime $p \in \pi(G)$. Let P be an A-invariant Sylow p-group of G, $G_1 = N_G(J(P))$ and $G_2 = C_G(\Omega_1(Z(P)))$. As $F(G) = 1$, G_i is a proper A-invariant subgroup of G, so G_i

is nilpotent by minimality of G. In particular G_i has a normal p-complement, so, by the Thompson Normal p-Complement Theorem, G has a normal p-complement. This is impossible as G possesses no nontrivial proper A-invariant normal subgroups.

Recall from section 35 that a Frobenius group is a finite group H which is the semidirect product of a nontrivial group K by a nontrivial group B with B semiregular on K. K and B are the Frobenius kernel and Frobenius complement of H, respectively. Notice that, by 40.7:

(**40.8**) Frobenius kernels are nilpotent.

Notice that the lemmas in this section also give lots of information about Frobenius complements.

Remarks. Transfer and fusion are basic tools in the study of finite groups. One can begin to see the power of these tools in some of the lemmas and exercises in this chapter, but just barely.

The proof of the Baer–Suzuki Theorem essentially comes from Alperin and Lyons [AL]. Alperin's Fusion Theorem is (surprise) due to Alperin [Al]. Thompson was the first to prove the nilpotence of Frobenius kernels in his thesis.

Exercises for chapter 13

1. Let G be a finite group, p a prime, $H \leq G$ with $(|G : H|, p) = 1$, and g a p-element in H. g is *extremal* in H if $|C_H(g)|_p = |C_G(g)|_p$.
   (1) Represent G on G/H by right multiplication and let Hx be a fixed point of g on G/H. Prove the orbit $HxC_G(g)$ of $C_G(g)$ on G/H is of order prime to p if and only if $g^{x^{-1}}$ is extremal in H.
   (2) Assume g is of order p, $K \trianglelefteq H$ with H/K abelian, $g \in H - K$, g is extremal in H, and each H-conjugate of g extremal in H is contained in gK. Prove $g \notin G^{(1)}$.
2. Let G be a finite group with $G = O^2(G)$ and let $T \in \mathrm{Syl}_2(G)$. Assume T is dihedral, semidihedral, or $\mathbb{Z}_{2^n} \mathrm{wr}\, \mathbb{Z}_2$, and prove G has one conjugacy class of involutions.
3. Let G be a finite group with $G = O^2(G)$. Prove that if $m_2(G) > 2$ then $m_2(C_G(x)) > 2$ for each involution x in G. (Hint: Let $T \in \mathrm{Syl}_2(G)$. Prove there is $E_4 \cong U \trianglelefteq T$ and $x^G \cap C_T(U)$ is nonempty for each involution x of G.)

4. Let G be a finite group, p a prime, and assume a Sylow p-subgroup of G is the modular group M_{p^n} of order $p^n \neq 8$. Prove
   (1) $O^p(G)$ has cyclic Sylow p-subgroups.
   (2) If $p = 2$ then G has a normal 2-complement.

5. Let G be a finite group with quaternion Sylow 2-subgroups. Prove that either G has a normal 2-complement or there exists $K \trianglelefteq H \leq G$ with $H/K \cong \mathrm{SL}_2(3)$.

6. Let G be a group of order 60. Prove either G is solvable or G is isomorphic to the alternating group A_5 of degree 5. (Hint: Let $T \in \mathrm{Syl}_2(G)$. Show $T \trianglelefteq G$ or G has a normal 2-complement or $|G : N_G(T)| = 5$ and $\ker_{N_G(T)}(G) = 1$.)

14

The geometry of groups of Lie type

Chapters 4 and 7 introduced geometries preserved by the classical groups. Chapter 14 considers these geometries (and related geometries preserved by Coxeter groups) in detail, and uses the representations of the classical groups on their geometries to establish various group theoretical results.

For example we'll see that the finite classical groups $L_n(q)$, $U_n(q)$, $PSp_n(q)$, and $P\Omega_n^\varepsilon(q)$ are simple, with a few exceptions when n and q are small. Also $L_n(F)$ and $PSp_n(F)$ are simple for infinite fields F, as are $U_n(F)$ and $P\Omega_n(F)$ under suitable restrictions on F. If F is finite of characteristic p, it will develop that the stabilizer B of a maximal flag of the geometry of a classical group G over F is the normalizer of a Sylow p-group of G, and the subgroups of G containing B are precisely the stabilizers of flags fixed by B. These subgroups and their conjugates are the *parabolic subgroups* of G. We say B is the *Borel group* of G.

It also turns out that to each classical group G there is associated a Coxeter group called the *Weyl group* of G. The Weyl groups of the classical groups are of type A_n, C_n, or D_n. The structure of G is controlled to a large extent by that of its Weyl group: see for example lemma 43.7 and Exercise 14.6.

41 Complexes

Before beginning this section the reader may wish to review the discussion of geometries in section 3. Section 41 is devoted to a related class of objects: complexes. A *complex* is a pair $\mathscr{C} = (\Gamma, \mathscr{C})$ where Γ is a geometry over some finite index set I and $\mathscr{C}$ is a collection of flags of Γ of type I. The members of $\mathscr{C}$ are called *chambers*. Subflags of chambers are called *simplices*. Simplices of corank 1 are called *walls*. A complex is *thin* if each wall is contained in exactly two chambers. A complex is *thick* if each wall is contained in at least three chambers.

Define the *chamber graph* of $\mathscr{C}$ to be the graph on $\mathscr{C}$ obtained by joining chambers which have a common wall. A path in the chamber graph is called a *gallery*. A complex is said to be *connected* if its chamber graph is connected. A connected complex $(\Gamma, \mathscr{C})$ in which every flag of Γ of rank 1 or 2 is a simplex is called a *chamber complex*.

A morphism $\alpha: (\Gamma, \mathscr{C}) \to (\Delta, \mathscr{D})$ of complexes is a morphism $\Gamma \to \Delta$ of geometries with $\mathscr{C}\alpha$ contained in $\mathscr{D}$. A *subcomplex* of $(\Gamma, \mathscr{C})$ is a complex

$(\Delta, \mathscr{D})$ with Δ a subgeometry of Γ and $\mathscr{D} = \mathscr{C} \cap \Delta$ the set of chambers contained in Δ.

The notions of complex, building, and Tits system (which are the subject of this chapter) come from Tits [Ti]. However the definition of a complex I've just given is somewhat less general than that of Tits [Ti] in that under my definition there is a type function defined on simplices inherited from the associated geometry. However Tits shows how to associate type functions to Coxeter complexes and buildings, so in the end the class of object considered is the same.

The treatment of complexes, buildings, and Tits systems given here is extracted from Tits [Ti] and Bourbaki [Bo], modulo remarks above.

Let G be a group and $\mathscr{F} = (G_i : i \in I)$ a family of subgroups of G. The coset geometry $\Gamma(G, \mathscr{F})$ determined by G and $\mathscr{F}$ is defined in section 3. Let $\mathscr{C}(G, \mathscr{F})$ be the complex on $\Gamma(G, \mathscr{F})$ whose chamber set (also denoted by $\mathscr{C}(G, \mathscr{F})$) consists of the flags $S_{I,x}, x \in G$, where $S_{I,x} = \{G_i x : i \in I\}$.

(41.1) Let G be a group, $\mathscr{F} = (G_i : i \in I)$ a family of subgroups of G, $\Gamma = \Gamma(G, \mathscr{F})$, and $\mathscr{C} = \mathscr{C}(G, \mathscr{F})$. Then

(1) G is represented as a group of automorphisms of $\mathscr{C}$ by right multiplication on the cosets in Γ.

(2) G is transitive on the simplices of $\mathscr{C}$ of type J for each subset J of I. In particular G is transitive on chambers.

(3) G_J is the stabilizer of the simplex S_J of type J.

(4) The walls of chambers of $\mathscr{C}$ are the conjugates of $S_{i'}, i \in I$, under G. $S_{i'}$ is contained in exactly $|G_{i'} : G_I|$ chambers.

(5) Every flag in Γ of rank 1 or 2 is a simplex of $\mathscr{C}$.

The proof is straightforward; the notation is explained in section 3.

(41.2) Let C be a chamber in a chamber complex $\mathscr{C}$ and α a morphism of $\mathscr{C}$ fixing C. Then

(1) α fixes each simplex contained in C.

(2) If $\mathscr{C}$ is thin and α is a bijection on $\mathscr{C}$, then $\alpha = 1$.

Proof. α is a morphism of geometries so it preserves type. Hence as C contains a unique simplex of each type, (1) holds.

To prove (2) it suffices to show α fixes each chamber adjacent to C, since $\mathscr{C}$ is connected. Let D be such a chamber; then $D \cap C = W$ is a wall of C, so α fixes W by (1). As $\mathscr{C}$ is thin, D and C are the only chambers containing W, so, as $W = W\alpha = D\alpha \cap C\alpha$, $\{C\alpha, D\alpha\} = \{C, D\}$. Now, as $C = C\alpha$, also $D = D\alpha$.

Let $\mathscr{C}$ be a thin chamber complex. A *folding* of $\mathscr{C}$ is an idempotent morphism whose fibres on chambers are all of order 2. $\mathscr{C}$ is a *Coxeter complex* if for each pair C, D of adjacent chambers there exists a folding mapping C to D.

In the remainder of this section assume $\mathscr{C} = (\Gamma, \mathscr{C})$ is a Coxeter complex.

(41.3) Let ϕ be a folding of $\mathscr{C}$. Then
  (1) ϕ induces a bijection $\phi \colon (\mathscr{C} - \mathscr{C}\phi) \to \mathscr{C}\phi$.
  (2) ϕ fixes each member of $\Gamma\phi$ and $\mathscr{C}\phi$.

Proof. Part (2) is just a restatement of the hypothesis that ϕ is idempotent. Let $C \in \mathscr{C}\phi$. By hypothesis there is a unique chamber D distinct from C with $D\phi = C$. By (2), D is not in $\mathscr{C}\phi$.

For C, D in $\mathscr{C}$ let $d(C, D)$ be the distance between C and D in the chamber graph.

(41.4) Let ϕ be a folding of $\mathscr{C}$, C in $\mathscr{C}\phi$, and D in $\mathscr{C} - \mathscr{C}\phi$. Then
  (1) If $C = C_0, \ldots, C_n = D$ is a gallery, there exists $0 \le i < n$ with $C_i \in \mathscr{C}\phi$ and $C_{i+1} \notin \mathscr{C}\phi$.
  (2) If C is adjacent to D then $D\phi = C$.
  (3) Let $C_i' = C_i$ if C_i is not in $\mathscr{C}\phi$ and $C_i' = \phi^{-1}(C_i) - \{C_i\}$ if C_i is in $\mathscr{C}\phi$. Then C_i' is adjacent to C_{i+1}'.

Proof. Part (1) is clear. Assume $W = C \cap D$ is a wall. By 41.3, ϕ fixes C and W. Thus $W = W\phi \subseteq D\phi$, so, as $\mathscr{C}$ is thin, $D\phi = C$ or D. As D is not in $\mathscr{C}\phi$, $D\phi = C$, so (2) holds.

Part (3) is clear if neither C_i nor C_{i+1} is in $\mathscr{C}\phi$, so let C_i be in $\mathscr{C}\phi$. Let $U = C_i \cap C_{i+1}$, $V = \phi^{-1}(U) \cap C_i'$, and E the chamber through V distinct from C_i'. Then $U = V\phi \subseteq E\phi$, so $E\phi = C_i$ or C_{i+1}. If $E\phi = C_{i+1}$ then $E = C_{i+1}'$ is adjacent to C_i'. If $E\phi = C_i$ then, as $\phi^{-1}(C_i) = \{C_i', C_i\}$, we have $E = C_i$. Hence $V = V\phi = U$ by 41.2.1, so $C_i' = C_{i+1}$. Therefore $C_{i+1}\phi = C_i \ne C_{i+1}$, so $C_{i+1} \notin \mathscr{C}\phi$, and hence $(C_{i+1})' = C_{i+1}$. Thus $C_i' = C_{i+1}$ is adjacent to $C_{i+1} = C_{i+1}'$.

(41.5) Let C and C' be distinct adjacent chambers and let ϕ and ϕ' be foldings with $C'\phi = C$ and $C\phi' = C'$. Then
  (1) If D is in $\mathscr{C}\phi$ then $d(C', D) = d(C, D) + 1$.
  (2) If D is not in $\mathscr{C}\phi$ then $d(C', D) = d(C, D) - 1$.
  (3) $\mathscr{C}\phi' = \mathscr{C} - \mathscr{C}\phi$.

Proof. As C is adjacent to C', $|d(C, D) - d(C', D)| \le 1$ for each $D \in \mathscr{C}$.

Assume D is in $\mathscr{C}\phi$ and $C' = C_0, \ldots, C_n = D$ is a gallery of length $d(C', D)$. Then $(C_i\phi: 0 \le i \le n)$ is a gallery between $C = C'\phi$ and $D\phi = D$ as ϕ is a morphism. By 41.4 there is $0 \le i < n$ with $C_i\phi = C_{i+1} = C_{i+1}\phi$. Hence $C = C_0\phi, \ldots, C_{i-1}\phi, C_{i+1}\phi, \ldots, C_n\phi = D$ is a gallery of length at most $n - 1$ from C to D. Therefore $d(C, D) < n = d(C', D)$, so (1) follows from the first paragraph of the proof.

Similarly if $D \notin \mathscr{C}\phi$ let $C = C_0, \ldots, C_n = D$ be a gallery of length $d(C, D)$. Define C_i' as in 41.4.3. By 41.4.3, $C' = C_0', \ldots, C_n' = D$ is a gallery, while by 41.4.1 and 41.4.2 there is $0 \le i < n$ with C_i in $\mathscr{C}\phi$, C_{i+1} not in $\mathscr{C}\phi$ and $C_{i+1}\phi = C_i$. Hence $C_i' = C_{i+1}' = C_{i+1}$, so, as in the preceding paragraph, $d(C', D) < n$ and then (2) holds.

Finally let $D \in \mathscr{C}$. If D is in $\mathscr{C}\phi'$, then, by (1) applied to ϕ' and ϕ, D is not in $\mathscr{C}\phi$. So $\mathscr{C}\phi' \subseteq \mathscr{C} - \mathscr{C}\phi$. If D is not in $\mathscr{C}\phi$ then, by (2) applied to ϕ and ϕ', D is in $\mathscr{C}\phi'$. Thus $\mathscr{C} - \mathscr{C}\phi \subseteq \mathscr{C}\phi'$. So (3) is established.

Foldings ϕ and ϕ' are defined to be *opposite* if $\mathscr{C}\phi' = \mathscr{C} - \mathscr{C}\phi$.

(41.6) Let ϕ and ϕ' be opposite foldings and define $\alpha = \alpha(\phi, \phi'): \Gamma \to \Gamma$ by $v\alpha = v\phi$ if $v \in \Gamma\phi'$ and $v\alpha = v\phi'$ if $v \in \Gamma\phi$. Then α is an automorphism of $\mathscr{C}$ of order 2 whose orbits on $\mathscr{C}$ are the fibres of ϕ.

Proof. By hypothesis $\mathscr{C} = \mathscr{C}\phi \cup \mathscr{C}\phi'$. By definition of chamber complex, each member of Γ is a simplex, so $\Gamma = \Gamma\phi \cup \Gamma\phi'$. By 41.3.2, ϕ and ϕ' agree on $\Gamma\phi \cap \Gamma\phi'$, so α is well defined. By 41.3.1, α is bijective on $\mathscr{C}$. If u and v are incident members of Γ, then as $\mathscr{C}$ is a chamber complex there is C in $\mathscr{C}$ with u and v in C. Then $u\alpha, v\alpha \in C\alpha \in \mathscr{C}$, so $u\alpha * v\alpha$. Hence α is morphism. By 41.4.1 there exist adjacent chambers C and D with C in $\mathscr{C}\phi$ and D not in $\mathscr{C}\phi$. By 41.4.2, $D\phi = C$ and $C\phi' = D$. Then $C\alpha^2 = C\phi'\phi = D\phi = C$. Hence, by 41.2.2, $\alpha^2 = 1$. Therefore $\alpha^{-1} = \alpha$ is a morphism, so α is an automorphism of order 2.

The element $\alpha(\phi, \phi')$ of 41.6 is called a *reflection* and is said to be a reflection through $C \cap C'$ for each pair C, C' of adjacent chambers with $C \in \mathscr{C}\phi$ and $C' \notin \mathscr{C}\phi$.

Fix a chamber C in $\mathscr{C}$. $\mathscr{C}$ is defined over some finite index set I; let $m = |I|$ for the moment. Notice each of the m subsets of I of order $m - 1$ determines a wall of C, and, as $\mathscr{C}$ is thin, each wall X determines a unique chamber C' with $C' \cap C = X$. Let $\Delta(C)$ denote the set of these chambers; then $\Delta(C)$ is the set of chambers distinct from C and adjacent to C in the chamber graph. Further $\Delta(C)$ is of order m. As $\mathscr{C}$ is a Coxeter complex, for each C' in $\Delta(C)$

there exist foldings ϕ and ϕ' of $\mathscr{C}$ with $C\phi = C'$ and $C'\phi' = C$. By 41.5, ϕ and ϕ' are opposite, so by 41.6 they determine an involutory automorphism $\alpha(\phi, \phi')$ of $\mathscr{C}$, called a *reflection through* $C \cap C'$. Let S denote the set of all reflections through $C \cap C'$ as C' varies over $\Delta(C)$, and let W be the subgroup of $\text{Aut}(\mathscr{C})$ generated by S. It will develop that (W, S) is a Coxeter system and $W = \text{Aut}(\mathscr{C})$.

For $w \in W$ let $l(w)$ be the length of w with respect to the generating set S.

(41.7) Let $w = s_1 \ldots s_n \in W$ with $s_i \in S$. Then

(1) $C, Cs_n, Cs_{n-1}s_n, \ldots, Cs_1 \ldots s_n$ is a gallery from C to Cw.

(2) W is transitive on $\mathscr{C}$.

(3) $d(C, Cw) = l(w)$.

Proof. Let $s \in S$. By definition of S there exists $C' \in \Delta(C)$ and foldings ϕ, ϕ' through $C \cap C'$ with $s = \alpha(\phi, \phi')$, and, by 41.6, (C, C') is a cycle of s.

Let $w_i = s_{n-i+1} \ldots s_n$. We've just seen C is adjacent to Cs_{n-i+1} so $Cw_i = Cs_{n-i+1}w_{i-1}$ is adjacent to Cw_{i-1}. Hence $C, Cw_1, \ldots, Cw_n = Cw$ is a gallery from C to Cw of length n, so $d(C, Cw) \le n$. In particular $d(C, Cw) \le l(w)$.

Conversely let $C = C_0, C_1, \ldots, C_m$ be a gallery of length $d(C, C_m) = m$. I'll show there exist $r_i \in S$, $1 \le i \le m$, such that $C_k = Cu_k$, where $u_k = r_{m-k+1} \ldots r_m$. Notice that this establishes the transitivity of W on $\mathscr{C}$. Proceed by induction on m; the case $m = 1$ has been handled in the first paragraph of this proof, so take $m > 1$. By induction $C_k = Cu_k$ for $k < m$. Then C_m is adjacent to $C_{m-1} = Cu_{m-1}$ so $C_m(u_{m-1})^{-1}$ is adjacent to C. Hence there is $r_1 \in S$ with $Cr_1 = C_m(u_{m-1})^{-1}$, so $Cu_m = Cr_1u_{m-1} = C_m$.

Finally let $u = r_1 \ldots r_m$ and assume $C_m = Cw$. Then $Cu = C_m = Cw$ so uw^{-1} fixes C and hence, by 41.2, $u = w$. So $l(w) = l(u) \le m = d(C, Cw)$, completing the proof.

(41.8) (1) $W = \text{Aut}(\mathscr{C})$.

(2) W is regular on $\mathscr{C}$.

(3) $|S| = |I|$; that is there exists a unique reflection through each wall of C.

Proof. By 41.7, W is transitive on $\mathscr{C}$, so $G = \text{Aut}(\mathscr{C}) = WG_C$ where G_C is the stabilizer in G of C. But, by 41.2.2, $G_C = 1$, so (1) and (2) hold. For each $C' \in \Delta(C)$ there is $s \in S$ with cycle (C, C'), so $|S| \ge |\Delta(C)| = |I|$. Further if t is a member of S with cycle (C, C') then $st \in W_C = 1$, so $s = t$.

(41.9) (W, S) is a Coxeter system.

Proof. It suffices to establish the exchange condition of 29.4. So let $w = s_1 \ldots s_n \in W$ with $l(w) = n$ and $s \in S$ with $l(ws) \le l(w)$. Let $C_k = Cs_{n-k+1} \ldots s_n$ for $1 \le k \le n$. By 41.7,

$$\mathscr{G} = (C = C_0, C_1, \ldots, C_n = Cw)$$

is a gallery and $d(C, Cw) = l(w) \ge l(ws) = d(C, Cws)$. Let ϕ be a folding with $Cs\phi = C$. Now $d(Cs, Cw) = d(C, Cws) \le d(C, Cw)$, so, by 41.5, $Cw \notin \mathscr{C}\phi$. Then by 41.4 applied to $\mathscr{G}$ there exists i, $1 \le i \le n$ with C_i in $\mathscr{C}\phi$, C_{i+1} not in $\mathscr{C}\phi$, and $C_{i+1}\phi = C_i$. By definition of s, $C_i s = C_{i+1}$, so $Cs_{n-i+1} \ldots s_n s = C_i s = C_{i+1} = Cs_{n-i} \ldots s_n$, and hence, by 41.8.2, $s_{n-i+1} \ldots s_n s = s_{n-i} \ldots s_n$, so the exchange condition is verified.

(41.10) Let (G, R) be a Coxeter system, $R = (r_i : i \in I)$, $G_i = \langle r_i : j \ne i \rangle$, and $\mathscr{F} = \mathscr{F}(G, R) = (G_i : i \in I)$. Then $\mathscr{C}(G, \mathscr{F})$ is a Coxeter complex, $G = \mathrm{Aut}(\mathscr{C}(G, \mathscr{F}))$, and R is the set of reflections through the walls of the chamber $C = \{G_i : i \in I\}$.

Proof. Adopt the notation of section 3. By 29.13.3, $G_{i'} = \langle r_i \rangle$ (recall $i' = I - \{i\}$) and $G_I = 1$. So, by 41.1.4, $\mathscr{D} = \mathscr{C}(G, \mathscr{F})$ is thin. Also $G = \langle R \rangle = \langle G_{i'} : i \in I \rangle$, so, by Exercise 14.3, $\mathscr{D}$ is a chamber complex. By 41.1.2, G is transitive on $\mathscr{D}$ and, by 41.1.3, G_I is the stabilizer of the chamber C, so, as $G_I = 1$, G is regular on $\mathscr{D}$. In particular G is faithful on $\mathscr{D}$ so, by 41.1.1, G is a subgroup of $\mathrm{Aut}(\mathscr{D})$. Also G is transitive on $\mathscr{D}$ so it remains only to show for each wall $S_{i'}$ of C that there is a folding ϕ through $S_{i'}$ with C in $\mathscr{D}\phi$.

Let $r = r_i$ and define $\phi : G \to G$ by

$$g\phi = g \quad \text{if } l(gr) = l(g) + 1$$
$$g\phi = gr \quad \text{if } l(gr) = 1(g) - 1.$$

As $G_I = 1$ we can identify G with $\mathscr{D}$ via $g \mapsto Cg$ and regard ϕ as a function from $\mathscr{D}$ into $\mathscr{D}$. Claim that if $D, D' \in \mathscr{D}$ and $D \cap D'$ is a wall of type j' then $D\phi \cap D'\phi$ is also a wall of type j'. Now $D = Cg$ and $D' = Cr_j g$ for some $g \in G$, so we must show $Cg\phi \cap C(r_j g)\phi$ is a wall of type j', so it suffices to show $(r_j g)\phi = r_j(g\phi)$ for each $g \in G$. This is clear if $l(r_j gr) - l(r_j g) = l(gr) - l(g)$. So, by symmetry between $r_j g$ and g, we may assume $l(gr) = l(g) - 1$ and $l(r_j gr) = l(r_j g) + 1$. Thus $g\phi = gr$ and $(r_j g)\phi = r_j g$. Let $g = s_n \ldots s_1$ with $l(g) = n$ and $s_k = r_{i_k}$. As $l(gr) \le l(g)$, the Exchange Condition says $s_m \ldots s_1 = s_{m-1} \ldots s_1 r$ for some m. Thus $g = s_n \ldots s_{m+1} s_{m-1} \ldots s_1 r$, so without loss $r = s_1$, Hence $g\phi = gr = s_n \ldots s_2$. Next $r_j gr = r_j s_n \ldots s_2$ is of length at most $n = l(g)$ so $l(r_j g) = l(r_j gr) - 1 \le l(g)$. Then, by the Exchange Condition, $r_j s_n \ldots s_{k+1} = s_n \ldots s_k$ for some k. If $k \ne 1$ we may take $s_n = r_j$.

But then $r_j g r = s_{n-1} \ldots s_2$ is of length $n - 2 < n - 1 = l(r_j g)$. So $k = 1$ and $r_j s_n \ldots s_2 = s_n \ldots s_1 = g$. Thus $(r_j g)\phi = r_j g = s_n \ldots s_2 = g\phi$, as desired.

So the claim is established. By 29.13.3,

$$G_{k'} = \langle G_{\{k,j\}'} : j \in k' \rangle$$

for each $k \in I$, if $|I| > 2$. Hence by Exercise 14.8, the map $G_k g \mapsto G_k(g\phi)$ is a morphism of the complex $(\Gamma(G, \mathscr{F}). \mathscr{D})$. I also write this map as ϕ.

Next observe that for each $g \in G$, $g\phi^2 = g\phi$. So ϕ is an idempotent morphism. Also if $Cg \in \mathscr{D}\phi$ then $\phi^{-1}(Cg) = \{Cg, Cgr\}$ so ϕ is a folding. Finally C and Cr are the chambers through $S_{i'}$ and $Cr\phi = C$ so ϕ is a folding through $S_{i'}$. By construction r is the reflection through $S_{i'}$; that is the fibres of ϕ on $\mathscr{D}$ are the orbits of r on $\mathscr{D}$. So R is the set of reflection through the walls of C and hence $\text{Aut}(\mathscr{D}) = \langle R \rangle = G$.

(41.11) The map $(G, R) \mapsto \mathscr{C}(G, \mathscr{F}(G, R))$ is a bijection between the set of all Coxeter systems and the set of all Coxeter complexes (up to isomorphism). $\mathscr{F}(G, R)$ is the family of maximal parabolics defined in 41.12 and the inverse of the correspondence is $\mathscr{C} \mapsto (\text{Aut}(\mathscr{C}), R)$ where R is the set of reflections through the walls of some fixed chamber of $\mathscr{C}$.

Proof. If (G, R) is a Coxeter system then by 41.10, $(G, R)\varphi = \mathscr{C}(G, \mathscr{F}(G, R))$ is a Coxeter complex, while if $\mathscr{C}$ is a Coxeter complex then by 41.8 and 41.9, $\mathscr{C}\psi = (\text{Aut}(\mathscr{C}), R)$ is a Coxeter system. By Exercise 14.2.3, $\mathscr{C}\psi\varphi \cong \mathscr{C}$, while by 41.10, $(G, R)\varphi\psi = (G, R)$, so φ and ψ are inverses of each other and the lemma holds.

42 Buildings

A *building* is a thick chamber complex $\mathscr{B} = (\Gamma, \mathscr{C})$ together with a set $\mathscr{A}$ of subcomplexes of $\mathscr{B}$, called *apartments*, such that the following axioms are satisfied:

(B1) The apartments are thin chamber complexes.

(B2) Each pair of chambers of $\mathscr{B}$ is contained in an apartment.

(B3) If A and B are simplices of $\mathscr{B}$ contained in apartments Σ and Σ', then there exists an isomorphism of Σ with Σ' which is the identity on $A \cup B$.

In sections 13 and 22 a geometry Γ was associated to each classical group G (occasionally subject to restrictions on the field), and it was shown that G is flag transitive on Γ. Now $\mathscr{C}(\Gamma)$ is a thick chamber complex and in Exercise 14.5 a set of apartments $\mathscr{A}$ is defined which admits the transitive action of G, and it is shown that $(\mathscr{C}(\Gamma), \mathscr{A})$ is a building admitting G as a group of automorphisms. This representation is then used to study G in section 43.

But before all that let's take a closer look at buildings. So for the rest of this section let $\mathscr{B} = (\Gamma, \mathscr{C})$ be a building with apartment set $\mathscr{A}$.

(42.1) Each pair of apartments is isomorphic.

Proof. Let $\Sigma_i, i = 1, 2$, be apartments and pick chambers $C_i \in \Sigma_i$. By axiom B2, there is an apartment Σ containing C_1 and C_2, and by axiom B3, $\Sigma_1 \cong \Sigma \cong \Sigma_2$.

Given a pair (Σ, C) with Σ an apartment and C a chamber in Σ define a map $\rho = \rho(\Sigma, C): \Gamma \to \Sigma$ by $v\rho = v\phi$, where C, v are contained in some apartment Σ' and $\phi: \Sigma' \to \Sigma$ is an isomorphism which is the identity on C. There are a number of points to be made here. First by B2 there does indeed exist an apartment Σ' containing C and v, and by B3 the map ϕ exists. Moreover if v, C are contained in an apartment θ and $\psi: \theta \to \Sigma$ is an isomorphism trivial on C then by B2 there exists an isomorphism $\alpha: \Sigma' \to \theta$ trivial on $C \cup \{v\}$. Then $(\phi^{-1})\alpha\psi = \beta \in \mathrm{Aut}(\Sigma)$ is trivial on C, so, by 41.2.2, $\beta = 1$. Thus $\alpha\psi = \phi$, so $v\phi = v\alpha\psi = v\psi$. What all this shows is that $\rho(\Sigma, C)$ is independent of the choice of Σ' and ϕ, so in particular $\rho(\Sigma, C)$ is well defined. As a matter of fact it shows there exists a unique isomorphism $\phi: \Sigma' \to \Sigma$ trivial on C. Let's see next that:

(42.2) $\rho = \rho(\Sigma, C)$ is a morphism of $\mathscr{B}$ onto Σ with $\rho^{-1}(C) = C$. Moreover if Σ' is an apartment containing C then $\rho: \Sigma' \to \Sigma$ is the unique isomorphism of Σ' with Σ trivial on C. In particular ρ is trivial on Σ.

I've already made the last observation of 42.2. The remaining parts of the lemma are an easy consequence of this observation and the building axioms.

Given simplices A and B of $\mathscr{B}$ define $d(A, B)$ to be the minimal integer n such that there exists a gallery $\mathscr{G} = (C_i: 0 \le i \le n)$ of $\mathscr{B}$ with A contained in C_0 and B contained in C_n. Then $\mathscr{G}$ is said to be a gallery between A and B of length n.

(42.3) (Rainy Day Lemma) Let Σ be an apartment, C a chamber of Σ, and X a simplex of Σ. Then Σ contains every gallery of $\mathscr{B}$ between X and C of length $d(X, C)$.

Proof. Let $\mathscr{G} = (C_i: 0 \le i \le n)$ be such a gallery and assume $\mathscr{G}$ is not contained in Σ. Then there exists an i with C_i not contained in Σ and C_{i+1} contained in Σ. Then $W = C_i \cap C_{i+1}$ is a wall of C_{i+1} so there is a chamber $T \ne C_{i+1}$ of Σ with $W \subseteq T$. Let $\rho = \rho(\Sigma, T)$. As ρ is a morphism trivial on Σ, $C_i\rho$ is a chamber of Σ containing W, so, as Σ is thin, $C_i\rho = T$ or C_{i+1}. As $\rho^{-1}(T) = T$, $C_i\rho = C_{i+1}$. But $\mathscr{G}\rho$ is a gallery between $X = X\rho$ and $C\rho = C$ of length $n = d(X, C)$, so, as $C_i\rho = C_{i+1} = C_{i+1}\rho, C_0\rho, \dots$, $C_{i-1}\rho, C_{i+1}\rho, \dots, C_n\rho$ is a gallery between X and C of length $n - 1$, contradicting $n = d(X, C)$.

(42.4) Let C and D be chambers, X a subset of C, and Σ an apartment containing C. Let $\rho = \rho(\Sigma, C)$ and $\mathscr{G}$ a gallery of length $d(X, D)$ between X and D in $\mathscr{B}$. Then $d(X\rho, D\rho) = d(X, D)$ and $\mathscr{G}\rho$ is a gallery of length $d(X\rho, D\rho)$ between $X = X\rho$ and $D\rho$.

Proof. By B2, $C \cup D$ is contained in an apartment Σ' and, by 42.3, $\mathscr{G}$ is contained in Σ'. By 42.2, $\rho : \Sigma' \to \Sigma$ is an isomorphism so the lemma holds.

(42.5) Each apartment is a Coxeter complex.

Proof. Let C and C' be adjacent chambers in an apartment Σ. We must show there exist opposite foldings ϕ and ϕ' of Σ through $B = C \cap C'$ with $C'\phi = C$ and $C\phi' = C'$. For the first time the hypothesis that $\mathscr{B}$ is thick is used. Namely as $\mathscr{B}$ is thick there is a chamber C^* distinct from C and C' through B. Let Σ_1 be an apartment containing C and C^*, $\rho_1 = \rho(\Sigma_1, C)$, $\rho_2 = \rho(\Sigma, C')$, and $\phi = \rho_1\rho_2 : \Sigma \to \Sigma$. As the composition of morphisms trivial on C, ϕ is a morphism trivial on C. Moreover applying 42.4 to ρ_1 and ρ_2 we obtain:

(42.5.1) If $\mathscr{G}$ is a gallery of length $d(B, T) = n$ between B and a chamber T of Σ, then $d(B, T\phi) = n$ and $\mathscr{G}\phi$ is a gallery of length n between B and $T\phi$.

Next $C'\phi$ is a chamber containing B, so $C'\phi = C$ or C'. As $\rho_1 : \Sigma \to \Sigma_1$ is an isomorphism trivial on C, $C'\rho_1 = C^*$. Then $C'\phi = C^*\rho_2 \neq C'$ as $(\rho_2)^{-1}(C') = C'$. So $C'\phi = C$.

Similarly there is a morphism ϕ' of Σ trivial on C', with $C\phi' = C'$, and satisfying 42.5.1.

Let D be a chamber in Σ and $(C_i : 0 \le i \le n) = \mathscr{G}$ a gallery of length $d(B, D)$ in Σ from B to D. Notice $C_0 = C$ or C'. Claim

(42.5.2) If $C_0 = C$ then

(i) ϕ and $\phi'\phi$ are trivial on D, and
(ii) $D\phi' \neq D$.

Assume otherwise and choose a counterexample with n minimal. As $C\phi' = C'$, $C'\phi = C$, and ϕ is trivial on C, it follows that $n > 0$. Now $(C_i : 0 \le i < n)$ is a gallery from B to $C_{n-1} = E$ of length $d(B, E) = n - 1$, so (i) and (ii) hold for E by minimality of n. Let $A = D \cap E$. As $A = A\phi \subseteq D\phi$, $D\phi = D$ or E. By 42.5.1, $n = d(B, D\phi)$, so, as $d(B, E) = n - 1$, it follows that $D\phi = D$. Similarly, as ϕ and ϕ' both satisfy 42.5.1, so does $\phi'\phi$, and then as $\phi'\phi$ is trivial on E the same argument shows $D\phi'\phi = D$.

So (i) is established. Notice $\mathscr{G}' = \mathscr{G}\phi'$ is gallery of length $n = d(B, D\phi')$ with $C_0\phi' = C\phi' = C'$, so, by symmetry between C and C', ϕ' is trivial on $D\phi'$. Thus, if $D = D\phi'$, $A = A\phi' \subseteq D \cap E\phi'$, so $E\phi' = D$ since $E\phi' \neq E$. This is impossible as $n - 1 = d(B, E) = d(B, E\phi')$ while $d(B, D) = n$. This completes the proof of 42.5.2.

Now, by 42.5.2, either $C_0 = C$ and $D = D\phi$, or $C_0 = C'$, in which case $C_0\phi = C$ and, applying 42.5.2 to $\mathscr{G}\phi$, we get $(D\phi)\phi = D\phi$. In either case $D\phi = D\phi^2$, so ϕ is idempotent. Also if $D = D\phi$ then by 42.5.2, $C_0 \neq C'$, so $C_0 = C$, and then again by 42.5.2, $D \neq D\phi'$ and $D = D\phi'\phi$. Moreover if $T \neq D$ is a chamber of Σ with $T\phi = D$ then $T \neq T\phi$, so $T = T\phi' = T\phi\phi'$. Thus $T = D\phi'$ and so $\{D, D\phi'\}$ is the fibre of D under ϕ. Hence ϕ is a folding of Σ through B with $C'\phi = C$, and ϕ' is the opposite folding. The proof of 42.5 is complete.

43 BN-pairs and Tits systems

A *Tits system* is a quadruple (G, B, N, S) such that G is a group, B and N are subgroups of G, S is a finite collection of cosets of $B \cap N$ in N, and the following axioms are satisfied:

(BN1) $G = \langle B, N \rangle$ and $H = B \cap N \trianglelefteq N$.
(BN2) $W = N/H$ is generated by S, and the members of S are involutions in W.
(BN3) For each $s \in S$ and $w \in W$, $sBw \subseteq BwB \cup BswB$.
(BN4) For each $s \in S$, $B \neq B^s$.

By convention, $wB = nB$ and $B^w = B^n$, for n a representative of the coset $w = Hn \in W$. As $H \trianglelefteq N$ and $H \leq B$, nB and B^n are independent of the choice of coset representative n, so the notation is well defined. Axioms BN3 and BN4 should be read subject to this convention.

B and N are also said to be a (B, N)-*pair* for G.

(43.1) Let $\mathscr{B} = (\Gamma, \mathscr{C})$ be a building with apartment set $\mathscr{A}$ and assume G is a group of automorphisms of $\mathscr{B}$ transitive on

$$\Omega = \{(C, \Sigma): C \in \mathscr{C}, \Sigma \in \mathscr{A}, C \subseteq \Sigma\}.$$

Let $(C, \Sigma) \in \Omega$, $B = G_C$, and $N = N_G(\Sigma)$. Then

(1) The representation of N on Σ maps N surjectively onto $\text{Aut}(\Sigma)$ with kernel $H = B \cap N$.

(2) (W, S) is a Coxeter system, where $W = N/H$ and S is the set of reflections in W through the walls of C.

(3) (G, B, N, S) is a Tits system.

Proof. By 42.5, Σ is a Coxeter complex, while by hypothesis N is transitive on the chambers of Σ, so (1) follows from 41.8. Then (2) follows from (1) and 41.9.

Let $s \in S$ and $w \in W$. Then C and Cs are adjacent and $E = C \cap Cs$ is a wall of C. Each $b \in B$ fixes E, so $E \subseteq Csb$ and $Ew \subseteq Csbw$. Let $\mathcal{G} = (C_i : 0 \leq i \leq n)$ be a gallery of length $d(C, Ew)$ from C to Ew, and Σ' an apartment containing C and $Csbw$. By the Rainy Day Lemma, 42.3, $\mathcal{G} \subseteq \Sigma \cap \Sigma'$. Let $a \in G$ with $(C, \Sigma')a = (C, \Sigma)$. Then $Ca = C$, so $a \in B$. C_1 and C are the only chambers of Σ or Σ' containing $C \cap C_1$ so, as a fixes C, it also fixes C_1. Proceeding by induction on k, a fixes C_k for each $0 \leq k \leq n$. Thus a fixes $Ew \subseteq C_n$. But $Ew \subseteq Csbw \subseteq \Sigma'$. So $Ew = Ewa \subseteq Csbwa \subseteq \Sigma'a = \Sigma$, so $Csbwa = Cw$ or Csw. Hence $sbw \in BwB$ or $BswB$, so BN3 is established.

As $\mathcal{B}$ is thick, there exists a chamber D through E distinct from C and Cs. Let θ be an apartment containing C and D, and let $g \in G$ with $(C, \Sigma)g = (C, \theta)$. Then $g \in B$ and C and Csg are the chambers through E in $\Sigma g = \theta$, so $Csg = D \neq Cs$. Thus $sgs \notin B$, so BN4 is established and the proof is complete.

It will develop later in this section that the converse of 43.1 also holds; that is each Tits system defines a building.

Notice that by 43.1 and Exercise 14.5 the classical groups possess a BN-pair. The BN-pair structure will be used to establish various facts about the classical groups.

In the remainder of this section assume (G, B, N, S) is a Tits system and let $H = B \cap N$ and $W = N/H$. We say B is the *Borel subgroup* of G, H is the *Cartan subgroup* of G, and W is the *Weyl group* of G. Let l be the length function on W defined by the generating set S.

(43.2) If $u, w \in W$ with $BwB = BuB$ then $u = w$.

Proof. Let $m = l(w) \leq l(u)$ and induct on m. If $m = 0$ then $w = 1$, so $u \in B \cap N = H$; that is $u = 1$ (remember the convention on W). So take $m > 0$. Then $w = sv$ with $s \in S$ and $l(v) = m - 1$. Now $svB \subseteq BuB$ by hypothesis, so $vB \subseteq sBuB \subseteq BuB \cup BsuB$ by BN3. Then $BvB = BuB$ or $BsuB$, so, by induction on m, $v = u$ or su. As $l(v) = m - 1$ and $m \le l(u)$, $v \ne u$, so $v = su$ and hence $w = sv = u$ as desired.

(43.3) Let $w \in W$ and $s \in S$.
  (1) If $l(sw) \ge l(w)$ then $sBw \subseteq BswB$.
  (2) If $l(sw) \le l(w)$ then $sBw \cap BwB$ is nonempty.

Proof. First the proof of (1), which is again by induction on $m = l(w)$. If $m = 0$ then $w = 1$ and the result is trivial, so take $m > 0$. Then $w = ur$, with $r \in S$ and $l(u) = m - 1$. If $l(su) < m - 1$ then $l(sw) = l(sur) \le l(su) + 1 < m$, contrary to hypothesis, so $l(su) \ge m - 1 = l(u)$. Then, by induction on m, $sBu \subseteq BsuB$. Hence $sBw = sBur \subseteq BsuBr \subseteq BsuB \cup BsurB$ by BN3. Also, by BN3, $sBw \subseteq BwB \cup BswB$. If $BwB = BsuB$ then, by 43.2, $w = su$. Hence $u = sw$ so $m - 1 = l(u) = l(sw) \ge l(w) = m$, a contradiction. Similarly if $BwB = BsurB$ then $w = sur = sw$, a contradiction. It follows that $sBw \subseteq BswB$, as desired.

So (1) is established and it's on to (2). By BN3 and BN4, $B^s \cap BsB$ is nonempty, so $sBsu \cap BsBu$ is nonempty for each $u \in W$. Take $u = sw$. Then $l(su) = l(w) \ge l(sw) = l(u)$, so, by (1), $sBu \subseteq BsuB = BwB$. Therefore, as $sBsu \cap BsBu$ is nonempty and $w = su$, (2) is established.

(43.4) Let $w = s_1 \ldots s_n \in W$ with $s_i \in S$. Then
  (1) If $n = l(w)$ then $s_i B \subseteq \langle B, B^{w^{-1}} \rangle \supseteq BwB$ for $1 \le i \le n$.
  (2) For each $u \in W$, $Bw Bu B \subseteq \bigcup_{i \in \Delta} Bs_{i_1} \ldots s_{i_r} u B$, where Δ consists of the sequences $i = i_1, \ldots, i_r$ with $i_j \in \{1, \ldots, n\}$ and $i_1 < i_2 < \cdots < i_r$.
  (3) If $s_0 \in S$ with $l(s_0 w) \le l(w) = n$ then there exists $1 \le k \le n$ with $s_0 s_1 \ldots s_{k-1} = s_1 \ldots s_k$.

Proof. Part (2) follows from BN3 by induction on n. Next part (1). Let $w_i = s_i \ldots s_1 w$ for $1 \le i \le n$, $n = l(w)$, and $w_0 = w$. By hypothesis $l(w_{i+1}) < l(w_i)$ so, by 43.3.2, $s_{i+1} Bw_i \cap Bw_i B$ is nonempty. Hence

$$ s_{i+1} B \subseteq Bw_i Bw_i^{-1} B \subseteq \langle B, B^{w_i^{-1}} \rangle. $$

By induction on i, $s_j \in \langle B, B^{w^{-1}} \rangle = X$ for each $j \le i$, so $s_{i+1} \in \langle B, B^{w_i^{-1}} \rangle \le X$. Hence $w = s_1 \ldots s_n \in X$ and then $BwB \subseteq X$. So (1) holds.

Similarly if $l(s_0 w) \le l(w)$ then, by 43.3.2, $s_0 B \subseteq Bw Bw^{-1} B$ so, by (2), $Bs_0 B = Bxw^{-1}B$ for $x = s_{i_1} \ldots s_{i_r}$ and some $i \in \Delta$. Hence, by 43.2, $s_0 = xw^{-1}$. But $l(xw^{-1}) \ge l(w^{-1}) - l(x) \ge n - r$, so, as $l(s_0) = 1$, $r = n - 1$ and $x = s_1 \ldots s_{k-1} s_{k+1} \ldots s_n$ for some $1 \le k \le n$. Therefore (3) holds.

(43.5) (W, S) is a Coxeter system.

Proof. This is immediate from 43.4.3 and 29.4.

(43.6) S is the set of $w \in W^{\#}$ such that $B \cup BwB$ is a group.

Proof. By BN3, $B \cup BsB$ is a group for $s \in S$. Conversely if $w = s_1 \ldots s_n \in W$ with $s_i \in S$ and $n = l(w) > 0$ then, by 43.4.1, $s_i \in \langle B, w \rangle$, so if $B \cup BwB$ is a group then, by 43.2, $s_i = w$ for each i.

Let $S = (s_i : i \in I)$ and for $J \subseteq I$ let $P_J = \langle B, s_j : j \in J \rangle$. The conjugates of the subgroups P_J, $J \subseteq I$, are called *parabolic subgroups* of G. Recall $W_J = \langle s_j : j \in J \rangle$ is a parabolic of W.

(43.7) (1) $P_J = BW_J B$. In particular $G = BNB$.
   (2) The map $J \mapsto P_J$ is a bijection of the power set of I with the set of all subgroups of G containing B.
   (3) $P_{J \cup K} = \langle P_J, P_K \rangle$.
   (4) $P_{J \cap K} = P_J \cap P_K$.
   (5) If $g \in G$ with $B^g \leq P_J$ then $g \in P_J$. In particular $N_G(P_J) = P_J$ is conjugate in G to P_K only if $J = K$.
   (6) $W_J = W \cap P_J$.

Proof. To prove (1) it suffices to show $uBw \subseteq BW_J B$ for each $u, w \in W_J$. Proceeding by induction on $l(u)$ it suffices to show this for $u = s_j$, $j \in J$. But this is just BN3. So (1) is established.
   Let $B \leq X \leq G$. By (1), $X = \bigcup_{y \in Y} ByB$ for some $Y \subseteq W$. Let $y = r_1 \ldots r_n \in Y$ with $r_i \in S$ and $l(y) = n$. By 43.4, $r_i \in \langle B, B^{y^{-1}} \rangle \leq X$. Let J be the set of $j \in I$ such that $s_j = r_i$ for some $y \in Y$ and some such expression for y. I've shown $P_J \leq X$, while by definition certainly $X \leq P_J$. Thus $X = P_J$.
   Suppose $P_J = P_K$; to complete the proof of (2) it remains to show $K = J$. By (1) and 43.2, $W_J = W \cap P_J = W \cap P_K = W_K$. Hence, by 29.13.4, $J = K$. So (2) and (6) are established.
   Evidently the bijection of (1) preserves inclusion, and hence also least upper bounds and greatest lower bounds. Hence (3) and (4) hold.
   Let $B^g \leq P_J . g^{-1} \in BwB$ for some $w \in W$, and, by 43.4.1, $w \in \langle B, B^{w^{-1}} \rangle = \langle B, B^g \rangle \leq P_J$. So $g \in \langle B, g \rangle = \langle B, w \rangle \leq P_J$.

Exercises 4.9 and 7.8 show that, if G is a classical group over a finite field of characteristic p, then the Cartan group H is a Hall p'-group of the Borel group B and B possesses a normal p-complement U with $U \in \mathrm{Syl}_p(G)$. Further $B = N_G(U)$, so the parabolics are the subgroups of G containing the normalizer of a Sylow p-group of G. Indeed it can be shown that the maximal parabolics $P_{i'}, i \in I$, are the maximal subgroups of G containing the Sylow p-group U, and further every p-local is contained in some maximal parabolic (cf. section 47).

Some more notation. For $i \in I$ let $G_i = P_{i'}$ (recall for $J \subseteq I$ that $J' = I - J$) and let $\mathscr{F} = (G_i : i \in I)$ be the set of maximal parabolics containing B. Form the geometry $\Gamma(G, \mathscr{F})$, the subgeometry $\Sigma = \{G_i w : i \in I, w \in W\}$, and the complex $\mathscr{B} = \mathscr{C}(G, \mathscr{F})$. Let C be the chamber $\{G_i : i \in I\}$; then $\mathscr{C} = \{Cg : g \in G\}$ is the set of chambers of $\mathscr{B}$, and we can also regard Σ as the complex $(\Sigma, \mathscr{C} \cap \Sigma)$. Let $\mathscr{A} = \{\Sigma g : g \in G\}$, so that $\mathscr{A}$ is a collection of subcomplexes of $\mathscr{B}$. Write $\mathscr{B}(G, B, N, S)$ for the pair $(\mathscr{B}, \mathscr{A})$. Finally let $U_i = W_{i'}$, $\mathscr{E} = (U_i : i \in I)$ and form the complex $\mathscr{C}(W, \mathscr{E})$.

(43.8) (1) The map $G_i w \mapsto U_i w$ is an isomorphism of the complex $(\Sigma, \Sigma \cap \mathscr{C})$ with $\mathscr{C}(W, \mathscr{E})$.

(2) G is represented as a group of automorphisms of $\mathscr{B}$ by right multiplication; the kernel of this representation is $\ker_B(G)$.

(3) $\mathscr{B}(G, B, N, S)$ is a building.

(4) G is transitive on $\Omega = \{(D, \theta): D \in \mathscr{C}, \theta \in \mathscr{A}, D \subseteq \theta\}$.

(5) $B = G_C$ and $(B_\Sigma)N = N_G(\Sigma)$ are the stabilizers in G of C and Σ, respectively. $N^\Sigma = \mathrm{Aut}(\Sigma) \cong W$ and $H = N_\Sigma$.

Proof. By 29.13.3, $W_{J \cap K} = W_J \cap W_K$ for $J, K \subseteq I$. Thus $U_J = W_{J'}$, where $U_J = \bigcap_{j \in J} U_j$. In particular, as (W, S) is a Coxeter system, $U_I = W_\emptyset = 1$ and $U_{i'} = W_i = \langle s_i \rangle$ is of order 2, so $\mathscr{C}(W, \mathscr{E})$ is a thin chamber complex by 41.1.4 and Exercise 14.3.

Next 43.7.6 says the map $\pi: G_i w \mapsto U_i w$ of (1) is well defined, after which π is evidently a bijection and π^{-1} is a morphism of geometries. So to prove (1) it remains to show that if $G_k w \cap G_j v$ is nonempty then $U_k w \cap U_j v$ is too. But $G_k u \cap G_j v$ is nonempty if and only if $1 \in G_k u v^{-1} G_j$, in which case $U_k w \cap U_j v$ is nonempty by the following observation, which is an easy consequence of 43.2 and 43.4.2:

(43.9) For each $J, K \subseteq I$ and each $w \in W$, $(G_J w G_K) \cap W = U_J w U_K$.

Hence (1) holds and therefore by the first paragraph of this proof, $\mathscr{B}$ satisfies B1.

Notice (2) is immediate from 41.1.1. Exercise 14.3 and 43.7 show $\mathscr{B}$ is a chamber complex. To show $\mathscr{B}$ is thick, by 41.1.4 and 43.7 we must show $|P_i : B| > 2$ for each $i \in I$. But, by 43.6, $P_i = B \cup Bs_i B$ so $|P_i : B| = 2$ if and only if $s_i \in N_G(B)$. Thus BN4 says $\mathscr{B}$ is thick.

For $x \in G$ and $J \subseteq I$ let $C_{J,x} = \{G_j x : j \in J\}$ be the simplex of type J in Cx. Let $C_{J,x}$ and $C_{K,y}$ be simplices. Then there exist $a, b \in B$ and $n \in N$ with $xy^{-1} = anb$. Now $C_{K,y} = C_{K,by}$ and $C_{J,x} = C_{J,nby}$ are in Σby, so $\mathcal{B}$ satisfies B2.

Suppose $C_{J,x}, C_{K,y} \subseteq \Sigma \cap \Sigma g$. Then we can choose $x, y \in N$ and there exist $n, m \in N$ with $C_{J,x} = C_{J,ng}$ and $C_{K,y} = C_{K,mg}$. Notice $h = mgy^{-1} \in G_K$. Define $u = nm^{-1}$ and $v = xy^{-1}$. Then $(G_J)uh = (G_J)ngy^{-1} = (G_J)xy^{-1} = (G_J)v$, so $u \in (G_J)v(G_K)$. So, by 43.9, $u = rvs$ for some $r \in G_J \cap N$ and $s \in G_K \cap N$. Set $l = y^{-1}shy$. Then $G_J xl = G_J vyl = G_J vshy = G_J r^{-1} uhy = G_J ng = G_J x$, and $G_K yl = G_K shy = G_K hy = G_K y$, and $\Sigma l = \Sigma yl = \Sigma shy = \Sigma hy = \Sigma mg = \Sigma g$. So $\Sigma l = \Sigma g$, $(C_{J,x})l = C_{J,x}$, and $(C_{K,y})l = C_{K,y}$. That is l induces an isomorphism of Σ with Σg trivial on $C_{J,x}$ and $C_{K,y}$. Therefore $\mathcal{B}$ satisfies B3. This completes the proof of (3).

As W is transitive on the chambers of $\mathscr{C}(W, \mathscr{E})$, (1) says N is transitive on $\mathscr{C} \cap \Sigma$. So, as G is transitive on $\mathscr{A}$ by construction, (4) holds. Also $N_G(\Sigma) = N N_B(\Sigma)$ as $N_B(\Sigma)$ is the stabilizer of C in $N_G(\Sigma)$. As $\mathscr{E}$ is thin, $N_B(\Sigma) = B_\Sigma$ by 41.2, so (5) holds.

(43.10) Let $G^* = G/\ker_B(G)$. Then (G^*, B^*, N^*, S^*) is a Tits system.

Proof. This is clear.

Because of 43.10 it does little harm to assume $\ker_B(G) = 1$. By 43.8 this has the effect of insuring that G is faithful on its building.

Recall that the Coxeter system (W, S) is irreducible if the graph of its Coxeter diagram is connected.

(43.11) Assume (W, S) is an irreducible Coxeter system and $\ker_B(G) = 1$. Then
  (1) If $X \trianglelefteq G$ then $G = XB$, and
  (2) if G is perfect and B is solvable then G is simple.

Proof. Let $X \trianglelefteq G$. Then $XB \leq G$ so, by 43.7, $XB = P_J$ for some $J \subseteq I$. Let $J_0 = \{i \in I : (Bs_i B) \cap X \neq \phi\}$. If $i \in J_0$ then $s_i \in Bs_i B \subseteq XB = P_J$ so $i \in J$ by 43.7.6. Conversely, as $P_J = XB$, X intersects each coset of B in P_J nontrivially so $J \subseteq J_0$. Thus $J = J_0$.

Claim $J = I$. If not as (W, S) is irreducible there exists $i \in I - J$ and $j \in J$ with $[s_i, s_j] \neq 1$. Therefore, as $\langle s_i, s_j \rangle$ is dihedral, $l(s_i s_j s_i) > l(s_j s_i) > l(s_i)$, so, by 43.3.1, $Bs_i Bs_j Bs_i B = Bs_i s_j s_i B$. As $X \trianglelefteq G$ and $J = J_0$ it follows that $s_i s_j s_i \in W \cap P_J = W_J$. But then $s_i s_j s_i \in W_J \cap W_{\{i,j\}} = W_j = \langle s_j \rangle$, contradicting $[s_i, s_j] \neq 1$.

Thus $I = J$, so $XB = P_J = P_I = G$. Hence (1) is established and of course (1) implies (2).

(43.12) Let F be a field and $1 < n$ and integer. Then

(1) $\mathrm{SL}_n(F)$ is quasisimple and $L_n(F)$ is simple unless $(n, |F|) = (2, 2)$ or $(2, 3)$.

(2) $\mathrm{Sp}_n(F)$ is quasisimple and $\mathrm{PSp}_n(F)$ is simple unless $(n, |F|) = (2, 2)$, $(2, 3)$, or $(4, 2)$.

(3) If F is finite then $\mathrm{SU}_n(F)$ is quasisimple and $U_n(F)$ is simple unless $(n, |F|) = (2, 4), (2, 9)$ or $(3, 4)$.

(4) If F is finite or algebraically closed, and $n \geq 6$, then $\Omega_n^\varepsilon(F)$ is quasisimple and $P\Omega_n^\varepsilon(F)$ is simple.

Proof. Let $G = \mathrm{SL}_n(F), \mathrm{Sp}_n(F), \mathrm{SU}_n(F)$, or $\Omega_n^\varepsilon(F)$, in the respective case. By Exercise 14.5, parts 1, 2, and 6, and 43.1, G possesses a BN-pair (B, N). Observe next that B is solvable. This follows from Exercise 4.9 and Exercise 7.8. Indeed these exercises show B is the semidirect product of a nilpotent group by a solvable group. In applying Exercise 7.8 observe $n - 2m \leq 2$ as $Z^\perp$ has no singular points; thus $\mathrm{O}(Z^\perp, f)$ is solvable by Exercise 7.2.

Next, by Exercise 14.5.5, $\ker_B(G) = Z(G)$. Further, except in the special cases listed in the lemma, G is perfect. This follows from 13.8, 22.3.4, 22.4, and Exercise 7.6. Finally, to complete the proof, 43.11.2 says $G/Z(G)$ is simple, since by Exercise 14.5.3 the Coxeter system of the Weyl group of G is irreducible.

Remarks. The material in chapter 14 comes from Tits [Ti] and Bourbaki [Bo]. Already in this chapter we begin to see the power of the Tits system building approach to the study of groups of Lie type. The proof of the simplicity of various classical groups in 43.12 probably provides the best example. Further results are established in section 47, where groups with a BN-pair generated by root subgroups and satisfying a weak version of the Chevalley commutator relations are investigated. These extra axioms facilitate the proof of a number of interesting results.

In [Ti], Tits classifies all buildings of rank at least 3 with a finite Weyl group, and hence also all groups with a Tits system of rank at least 3 and finite Weyl group. The *rank* of a building is $|I|$ while the rank of a Tits system (G, B, N, S) is $|S|$.

Exercises for chapter 14

1. If $\mathscr{A} = (\Delta, \mathscr{A})$ is a complex let $\Sigma(\mathscr{A})$ be the geometry whose objects of type i are the simplices of $\mathscr{A}$ of type i and whose edges are the simplices of $\mathscr{A}$ of rank 2. If Γ is a geometry let $\mathscr{C}(\Gamma)$ be the complex $(\Gamma, \mathscr{C}(\Gamma))$

where $\mathscr{C}(\Gamma)$ is the set of flags of Γ of type I. Prove

(1) The inclusion map π is an injective morphism of $\Sigma(\mathscr{A})$ into Δ; π is an isomorphism if and only if each rank 2 flag of Δ is a simplex of $\mathscr{A}$; π is an isomorphism of $\mathscr{C}(\Sigma(\mathscr{A}))$ with $\mathscr{A}$ if and only if every flag of Δ is a simplex of $\mathscr{A}$.

(2) π is an isomorphism of $\Sigma(\mathscr{C}(\Gamma))$ with Γ if and only if every rank 2 flag of Γ is contained in a flag of type I.

(3) For each simplex T of $\mathscr{A}$ let $\mathscr{A}_T = \{C - T : T \leq C \in \mathscr{A}\}$ and $\Sigma_T = \Sigma(\Delta_T, \mathscr{A}_T)$. Prove the map $\pi_T : \Sigma_T \to \Delta_T$ is an isomorphism of Δ_T with Σ_T for each simplex T of $\mathscr{A}$ if and only if every flag of Δ is a simplex of $\mathscr{A}$.

(4) $\mathrm{Aut}(\Gamma) = \mathrm{Aut}(\mathscr{C}(\Gamma))$.

2. Let $\mathscr{C} = (\Gamma, \mathscr{C})$ be a Coxeter complex over $I = \{1, \ldots, n\}$, let $W = \mathrm{Aut}(\mathscr{C})$, C in $\mathscr{C}$, and for $J \subseteq I$ let $J' = I - J$ and T_J the subflag of C of type J. For $i \in I$ let r_i be the reflection through the wall $T_{i'}$, let $R = \{r_1, \ldots, r_n\}$, and $R_J = \{r_i : i \in J'\}$. Prove

(1) If $J \subset I$ then $(\Sigma_{T_J}, \mathscr{C}_{T_J})$ is a Coxeter complex and $W_{T_J} = \mathrm{Aut}(\Sigma_{T_J}, \mathscr{C}_{T_J})$ (in the notation of Exercise 14.1).

(2) $W_{T_J} = \langle R_J \rangle$ and if $I \neq J$ then (W_{T_J}, R_J) is a Coxeter system.

(3) Let $T_i = \{v_i\}$, $W_i = W_{T_i}$, and $\mathscr{F} = \mathscr{F}(W, R) = (W_i : i \in I)$. Prove the map

$$v_i w \mapsto W_i w \quad i \in I, w \in W$$

is an isomorphism of $\mathscr{C}$ with $\mathscr{C}(W, \mathscr{F})$.

3. Let G be a group and $\mathscr{F} = (G_i : i \in I)$ a family of subgroups of G. Prove the complex $\mathscr{C}(G, \mathscr{F})$ is a chamber complex if and only if $G = \langle G_{i'} : i \in I \rangle$, where $i' = I - \{i\}$. Prove $\mathscr{C}(G, \mathscr{F}) = \mathscr{C}(\Gamma(G, \mathscr{F}))$ if and only if G is flag trasitive on $\Gamma(G, \mathscr{F})$.

4. Let $\mathscr{C} = (\Gamma, \mathscr{C})$ be a Coxeter complex and $W = \mathrm{Aut}(\mathscr{C})$. Prove

(1) Let A and B be simplices. Prove $A \cup B$ is a simplex if and only if $A \cup B$ is contained in $\Gamma\phi$ or $\Gamma\phi'$ for each pair ϕ, ϕ' of opposite foldings.

(2) Let A_i, $1 \leq i \leq 3$, be simplices such that $A_i \cup A_j$ is a simplex for each i, j. Then $\bigcup_{i=1}^{3} A_i$ is a simplex.

(3) Every flag of Γ is a simplex.

(4) W is flag transitive on Γ.

(5) Prove (in the notation of Exercise 14.1) that $(\Gamma_{T_J}, \mathscr{C}_{T_J})$ is a Coxeter complex for each $J \subset I$.

(6) $W = \mathrm{Aut}(\Gamma)$.

5. Let F be a field, V a finite dimensional vector space over F, and assume one of the following holds:

(A) Γ is the projective geometry on V, $Y = \{\langle x_i \rangle : 1 \leq i \leq n\}$ for some basis $X = (x_i : 1 \leq i \leq m)$ of V, $m = \dim(V)$, and $G = \mathrm{GL}(V)$.

(C) (V, f) is a symplectic or unitary space and $G = O(V, f)$ or (V, Q) is an orthogonal space and $G = O(V, Q)$. If (V, Q) is orthogonal assume it is not hyperbolic. $0 < m$ is the Witt index of the space and Γ its polar geometry. $X = (x_i : 1 \le i \le 2m)$ is a hyperbolic basis for a maximal hyperbolic subspace of V and $Y = \{\langle x \rangle : x \in X\}$.

(D) (V, Q) is a $2m$-dimensional hyperbolic orthogonal space and G is the subgroup of $O(V, Q)$ preserving the equivalence relation of 22.8, Γ is the oriflamme geometry of (V, Q), $X = (x_i : 1 \le i \le 2m)$ is a hyperbolic basis of V, and $Y = \{\langle x \rangle : x \in X\}$.

In each case let $\mathscr{B} = \mathscr{C}(\Gamma) = (\Gamma, \mathscr{C})$ be the complex on Γ defined in Exercise 14.1, let Σ_Y be the subgeometry consisting of the objects of Γ generated by members of Y, identify Σ_Y with the subcomplex $(\Sigma_Y, \mathscr{C} \cap \Sigma_Y)$, and let $\mathscr{A} = ((\Sigma_Y)g : g \in G)$. Prove

(1) $(\mathscr{B}\mathscr{A})$ is a building.

(2) G is transitive on $\Omega = \{(C, \Sigma) : C \in \mathscr{C}, \Sigma \in \mathscr{A}\}$.

(3) The Weyl group of $\mathscr{B}$ is of type A_m, C_m, or D_m in case A, C, and D, respectively.

(4) Let C be a chamber in Σ_Y, $B = G_C$, and $N = N_G(\Sigma_Y)$. Prove G is the semidirect product of a nilpotent group U by $H = B \cap N$, and H is abelian if A or D holds, or if (V, f) is symplectic or unitary, or if F is finite or algebraically closed.

(5) $\ker_B(G) = \ker_H(G)$ is the group of scalar transformations of V in G.

(6) $SL(V)$ and $Sp(V)$ are transitive on Ω and if F is finite or algebraically closed then $\Omega(V, Q)$ and $SU(V)$ are transitive on Ω.

6. Let (G, B, N, S) be a Tits system with Weyl group $W = N/H$ and $S = (s_i : i \in I)$. Prove

(1) For each $J, K \subseteq I$, the map

$$(P_J)w(P_K) \mapsto (W_J)w(W_K)$$

is a bijection of the set of orbits of the parabolic W_K on the coset space B/P_J with the orbits of the parabolic W_K on the coset space W/W_J.

(2) Assume the Coxeter diagram Δ of W is one of the following:

Prove:

    (a) If Δ is of type A_n then G is 2-transitive on $G/P_{1'}$, rank 3 on $G/P_{2'}$ if $n > 2$, and rank 6 on $P_{\{1,n\}'}$ if $n > 1$.

    (c) If Δ is of type C_n and $n > 1$, then G is rank $n + 1$ on $G/P_{n'}$, rank 3 on $G/P_{1'}$, and rank 6 on $G/P_{2'}$ if $n > 2$.

    (d) If Δ is of type D_n, $n \geq 4$, then G is rank 3 on $G/P_{1'}$ and rank 8 or 7 on $G/P_{2'}$ for $n = 4$ or $n > 4$, respectively.

7. (1) Every flag in a building is a simplex.

    (2) If (G, B, N, S) is a Tits system then G is flag transitive on its building $\mathcal{B}(G, B, N, S)$.

    (3) If $\mathcal{B} = (\Gamma, \mathscr{C})$ is a building then $\mathrm{Aut}(\mathcal{B}) = \mathrm{Aut}(\Gamma)$.

8. Let $(\Gamma, \mathscr{C})$ be a chamber complex and for $x \in \Gamma$ let $\mathscr{C}_x = \mathscr{C} \cap \Gamma_x$.

    (1) If $|I| > 2$ assume for each $x \in \Gamma$ that $(\Gamma_x, \mathscr{C}_x)$ is a connected complex. Assume $\phi: \mathscr{C} \mapsto \mathscr{C}$ is a function such that, for all $C, D \in \mathscr{C}$ and all $i \in I$, if $C \cap D$ is a wall of type i' then $C\phi \cap D\phi$ is also a wall of type i'. For $x \in \Gamma$ define $x\alpha$ to be the element of $C\phi$ of the same type as x, where $x \in C \in \mathscr{C}$. Prove $\alpha: (\Gamma, \mathscr{C}) \to (\Gamma, \mathscr{C})$ is a well-defined morphism.

    (2) Assume $|I| > 2$, $\Gamma = \Gamma(G, \mathscr{F})$, and $\mathscr{C} = C(G, \mathscr{F})$ for some group G and some family $\mathscr{F} = (G_i : i \in I)$ of subgroups. Assume for each $i \in I$ that $G_{i'} = \langle G_{\{i,j\}'} : j \in i' \rangle$. Prove $(\Gamma_x, \mathscr{C}_x)$ is a connected complex.

9. Let $T = (G, B, N, S)$ be a Tits system with finite Weyl group $W = N/H$. Let $\mathcal{B} = \mathcal{B}(T)$ be the building of T, C the chamber of $\mathcal{B}$ fixed by B, and $\Sigma = (xw : x \in C, w \in W)$ the apartment of $\mathcal{B}$ stabilized by N (cf. the discussion before 43.8). Let Φ be a root system for W and π a simple system for W with $S = (r_\alpha : \alpha \in \pi)$ (cf. 29.12 and 30.1). Let $\leq$ be the partial order on W defined in Exercise 10.6. Prove

    (1) If $u, w \in W$ with $u \leq w$ then there exists a gallery $(C_i : 0 \leq i \leq n)$ of length $n = l(w)$ from C to Cw with $C_{l(u)} = Cu$.

    (2) If $s \in S$ and $w \in W$ with $l(sw) < l(w)$ then $B \cap B^{sw} B^w \subseteq B^{sw}$.

    (3) If $s \in S$ and $w \in W$ with $l(w) \leq l(ws)$ then $B^s \cap B^w \leq B$ and $B^w \cap B^s B \subseteq B$.

    (4) If $u, w \in W$ with $u \leq w$ then $B \cap B^w \leq B^u$.

    (5) Define T to be *saturated* if $H = \bigcap_{w \in W} B^w$. Prove the following are equivalent:

        (a) T is saturated.

        (b) H is the pointwise stabilizer in G of Σ.

        (c) $B \cap B^{w_0} = H$, where w_0 is the element of W of maximal length in the alphabet S (cf. Exercise 10.5).

    (6) Let $\tilde{H} = \bigcap_{w \in W} B^w$, $\tilde{N} = \tilde{H}N$, and $\tilde{S} = \{\tilde{N}s : s \in S\}$. Prove $(G, B, \tilde{N}, \tilde{S})$ is a saturated Tits system.

(Hints: Use 41.7 in (1). To prove (2) observe $B \cap B^{sw} B^w \subseteq N_B$ $(Csw \cap Cw)$ and use the Rainy Day Lemma and (1) to prove N_B $(Csw \cap Cw) \leq B^{sw}$. Use (2) to prove (3) and (4) and use (4) and Exercise 10.6 to prove (5).)

10. (Richen [Ri]) Assume the hypothesis of Exercise 14.9 and assume further that T is saturated. Let $\alpha \in \pi$ and $s = r_\alpha \in S$. For $w \in W$ define $B_w = B \cap B^{w_0 w^{-1}}$, where w_0 is the maximal element of W. For $\beta \in \pi$ define $B_\beta = B_{r_\beta}$. Prove

   (1) If $w \in W$ with $\alpha w > 0$ then $B_\alpha \leq B^{w^{-1}}$.
   (2) If $w \in W$ with $\alpha w < 0$ then $B_\alpha \cap B^{w^{-1}} = H$.
   (3) If $w \in W$ with $l(w) \leq l(ws)$ then $B = (B \cap B^s)(B \cap B^w)$.
   (4) $B = B_\alpha(B \cap B^s)$ with $B^s \cap B_\alpha = H$.
   (5) $B_\alpha \neq H$.
   (6) For $w \in W$, $B_\alpha \leq B^{w^{-1}}$ if and only if $\alpha w > 0$.
   (7) Let $\Delta = \{(B_\alpha)^w : \alpha \in \pi, w \in W\}$. Prove the map $(B_\alpha)^w \mapsto \alpha w$ is a well-defined permutation equivalence of the representations of W on Δ and Φ. In particular we may define $B_\gamma = (B_\alpha)^w$ for each $\gamma \in \Phi$, where $\gamma = \alpha w$.
   (8) If $w \in W$ with $\alpha w < 0$ then $B_w = B_\alpha(B_{sw})^s$ and $B_\alpha \cap (B_{sw})^s = H$.
   (9) Let $w_0 = r_1 \ldots r_n$ with $n = l(w_0) = |\Phi^+|$ and $r_i = r_{\alpha_i}, \alpha_i \in \pi$. Let $w_i = r_i r_{i-1} \ldots r_1$. Prove $B = B_{\alpha_1} B_{\alpha_2 w_1} \ldots B_{\alpha_n w_{n-1}}$ with $B_{\alpha_i w_{i-1}} \cap \prod_{j=i+1}^n B_{\alpha_j w_{j-1}} = H$. Further the map $i \mapsto \alpha_i w_{i-1}$ is a bijection of $\{1, \ldots, n\}$ with Φ^+.

   (Hints: In (1) use Exercise 10.6.4 to show $w^{-1} < w_0 s$ and then appeal to Exercise 14.9.4. In (2) use parts (2) and (5) of Exercise 14.9 to conclude $B^{w^{-1}s} \cap B^s \leq B$ and $B \cap B^{w_0} = H$. In (3) use 43.3.1 to get $B \subseteq B^s B^w$ and then appeal to Exercise 14.9.3. Use (4) and BN4 to prove (5). In (7) use (6) to show for $\beta \in \pi$, $(B_\alpha)^w = B_\beta$ if and only if $\alpha w = \beta$. In (8) use (1) to show $B_\alpha \leq B_w$. Then use (4) to show $B_w = B_\alpha(B \cap B^s)$ and use Exercise 14.9.3 to show $B^s_{sw} \leq B$. Use (8) to prove (9).)

15

Signalizer functors

Let r be a prime, G a finite group, and A an abelian r-subgroup of G. An A-*signalizer functor* on G is a map θ from $A^{\#}$ into the set of A-invariant r'-subgroups of G such that, for each $a, b \in A^{\#}$, $\theta(a) \leq C_G(a)$ and $\theta(a) \cap C_G(b) \leq \theta(b)$. The signalizer functor θ is said to be *complete* if there is an A-invariant r'-subgroup $\theta(G)$ such that $\theta(a) = C_{\theta(G)}(a)$ for each $a \in A^{\#}$.

Notice that one way to construct an A-signalizer functor is to select some A-invariant r'-subgroup X of G and define $\theta(a) = C_X(a)$ for $a \in A^{\#}$. By construction this signalizer functor is complete. If $m(A) \geq 3$ it turns out that this is the only way to construct signalizer functors. That is, if $m(A) \geq 3$ then every A-signalizer functor is complete. This result is called the Signalizer Functor Theorem. It's one of the fundamental theorems in the classification of the finite simple groups. Unfortunately the proof of the Signalizer Functor Theorem is beyond the scope of this book. However, chapter 15 does contain a proof of a special case: the so-called Solvable 2-Signalizer Functor Theorem. It turns out that the Solvable Signalizer Functor Theorem suffices for many applications of signalizer functors.

An A-signalizer functor θ on G is said to be *solvable* if $\theta(a)$ is solvable for each $a \in A^{\#}$. We say θ is *solvably complete* if θ is complete and $\theta(G)$ is solvable. The main result of chapter 15 is:

Solvable 2-Signalizer Functor Theorem. Let A be an abelian 2-subgroup of a finite group G with $m(A) \geq 3$. Then each solvable A-signalizer functor on G is solvably complete.

Chapter 16 contains a discussion of the Classification Theorem which illustrates how the Signalizer Functor Theorem is used. Observe that the condition that $m(A) \geq 3$ is necessary in the Signalizer Functor Theorem by Exercise 15.1.

44 Solvable signalizer functors

In section 44, r is a prime, G is a finite group, A is an abelian r-subgroup of G of r-rank at least 3, and θ is a solvable A-signalizer functor on G.

Let $\mathcal{Q}(G)$ denote the set of solvable A-invariant r'-subgroups X of G with $C_X(a) = X \cap \theta(a)$ for each $a \in A^{\#}$. For $A \leq H \leq G$, let $\mathcal{Q}(H) = H \cap \mathcal{Q}(G)$ and

$$\theta(H) = \langle H \cap \theta(a): a \in A^{\#} \rangle.$$

If π is a set of primes then $\mathcal{Q}(H, \pi)$ denotes the set of π-groups in $\mathcal{Q}(H)$. Write $\mathcal{Q}^*(H)$ and $\mathcal{Q}^*(H, \pi)$ for the maximal members of $\mathcal{Q}(H)$ and $\mathcal{Q}(H, \pi)$, respectively, under the partial order of inclusion. For $X \in \mathcal{Q}(G)$, write $\mathcal{Q}_X(G, \pi)$ for the set of $H \in \mathcal{Q}(G, \pi)$ with $X \leq H$, and define

$$\pi(\theta) = \bigcup_{a \in A^\#} \pi(\theta(a)).$$

(44.1) If $X \in \mathcal{Q}(G)$ and Y is an A-invariant subgroup of X then $Y \in \mathcal{Q}(G)$.

(44.2) If $X, Y \in \mathcal{Q}(G)$ with $X \leq N_G(Y)$ then $XY \in \mathcal{Q}(G)$.

Proof. This follows from Exercise 6.1.

(44.3) Let $H \in \mathcal{Q}(G)$ with $H \trianglelefteq G$, set $G^* = G/H$, and define $\theta^*(a^*) = \theta(a)^*$ for $a^* \in A^{*\#}$. Then
   (1) θ^* is an A^*-signalizer functor.
   (2) $\mathcal{Q}(G^*) = \mathcal{Q}_H(G)^* = \{X^*: X \in \mathcal{Q}(G) \text{ and } H \leq X\}$.
   (3) $\mathcal{Q}(G^*) = \mathcal{Q}(G)^* = \{X^*: X \in \mathcal{Q}(G)\}$.

Proof. Let $a, b \in A^\#$. To prove (1) we must show $\theta^*(a^*) \cap C_{G^*}(b^*) \leq \theta^*(b^*)$. By coprime action, 18.7, $C_{G^*}(b^*) = C_G(b)^*$, so it suffices to show $C_{H\theta(a)}(b) \leq H\theta(b)$. But by Exercise 6.1, $C_{H\theta(a)}(b) = C_H(b)C_{\theta(a)}(b)$, so as $C_{\theta(a)}(b) \leq \theta(b)$, (1) is established.

Let $X \in \mathcal{Q}_H(G)$. Then X is a solvable A-invariant r'-group, so its image X^* has the same properties. Further for $a \in A^\#$, $C_X(a) \leq \theta(a)$, and by coprime action, $C_{X^*}(a^*) = C_X(a)^*$, so $C_{X^*}(a^*) \leq \theta(a)^* = \theta^*(a^*)$. That is, $X^* \in \mathcal{Q}(G^*)$. Conversely, suppose $Y^* \in \mathcal{Q}(G^*)$ and let Y be the full preimage of Y^* in G; to complete the proof of (2), we must show $Y \in \mathcal{Q}(G)$. As $Y/H = Y^*$ and H are solvable r'-groups, so is Y by 9.3. As $Y^* \in \mathcal{Q}(G^*)$, $C_{Y^*}(a^*) \leq \theta^*(a^*)$, so $C_Y(a) \leq \theta(a)H$. But $H \in \mathcal{Q}(G)$, so $C_H(a) \leq \theta(a)$, and then

$$C_Y(a) \leq C_G(a) \cap \theta(a)H = \theta(a)C_H(a) \leq \theta(a)$$

by the modular property of groups 1.14. So (2) is established.

By 44.2, $HX \in \mathcal{Q}_H(G)$ for each $X \in \mathcal{Q}(G)$, so $\mathcal{Q}_H(G)^* = \mathcal{Q}(G)^*$. Hence (2) implies (3).

(44.4) For each $1 \neq B \leq A$, $\theta(C_G(B)) \in \mathcal{Q}(G)$.

Proof. Recall $\theta(C_G(B)) = \langle C_G(B) \cap \theta(a): a \in A^\# \rangle$. But for $b \in B^\#$, $C_G(B) \cap \theta(a) \leq C_G(b) \cap \theta(a) \leq \theta(b)$, so $\theta(C_G(B)) \leq \theta(b)$, and hence the lemma follows from 44.1.

(44.5) For each $X \in \mathcal{Q}(G)$ and $\pi \subseteq \pi(\theta)$, $C_X(A) \leq \theta(C_G(A))$ and $C_X(A)$ is transitive on $\mathcal{Q}^*(X, \pi)$. Indeed, $\mathcal{Q}^*(X, \pi)$ is the set of A-invariant Hall π-subgroups of X.

Proof. Let T be the set of A-invariant π-subgroups of X and T^* the set of maximal members of T under inclusion. By 44.1, $T = \mathcal{Q}(X, \pi)$, so $T^* = \mathcal{Q}^*(X, \pi)$. By Exercise 6.2, T^* is the set of A-invariant Hall π-subgroups of X and $C_X(A)$ is transitive on T^*. Finally, as $X \in \mathcal{Q}(G)$, we have $C_X(A) \leq \theta(a)$ for each $a \in A^\#$, so $C_X(A) \leq \theta(C_G(A))$, completing the proof.

(44.6) Let $p \in \pi(\theta)$. Then
(1) For each $a \in A^\#$ there is a unique maximal $\theta(C_G(A))$-invariant member $\Delta(a)$ of $\mathcal{Q}(\theta(a), p')$.
(2) Δ is an A-signalizer functor.

Proof. By 44.4, $Y = \theta(C_G(A)) \leq X = \theta(a)$. Let $\mathcal{R}$ be the set of Y-invariant members of $\mathcal{Q}(X, p')$ and $\mathcal{S} = \mathcal{Q}^*(X, p')$. Then for $R \in \mathcal{R}$, $R \leq S \in \mathcal{S}$, so as Y is transitive on $\mathcal{S}$ by 45.5, R is contained in each member of $\mathcal{S}$. Therefore $M = \langle \mathcal{R} \rangle \leq S$, so M is the unique maximal member of $\mathcal{R}$ by 44.1. This establishes (1). Next for $b \in B^\#$, $\Delta(a) \cap C_G(b) = C_M(b) \leq \theta(b)$ as θ is a signalizer functor. Also $C_M(b)$ is an A-invariant p'-group, so $C_M(b) \leq \Delta(b)$, completing the proof of (2).

(44.7) (Transitivity Theorem) $\theta(C_G(A))$ is transitive on $\mathcal{Q}^*(G, p)$ for each $p \in \pi(\theta)$.

Proof. The proof is left as Exercise 15.2.

(44.8) Let $X \in \mathcal{Q}(G)$. Then
(1) If B is a noncyclic elementary abelian subgroup of A then
$$X = \langle \theta(C_G(D)) \cap X : |B : D| = r \rangle.$$
(2) If $1 \neq T \leq A$ then
$$\theta(C_G(T)) = \langle \theta(C_G(B)) : T \leq B \leq A \text{ and } B \text{ is noncyclic} \rangle.$$
(3) $X = \theta(X)$.

Proof. If $1 \neq T \leq A$ then by Exercise 6.5,
$$X = \langle C_X(S) : T/S \text{ is cyclic} \rangle.$$

Further if $S \neq 1$ then $C_X(S) \leq \theta(C_G(S))$ by 44.5, so (1) follows. Similarly by 44.4, $Y = \theta(C_G(T)) \in \mathcal{Q}(G)$ and $\bar{A} = A/T$ is abelian and acts on Y, so by Exercise 6.5,

$$Y = \langle C_Y(B) : \bar{A}/\bar{B} \text{ is cyclic} \rangle$$

with $C_Y(B) = \theta(C_G(B))$. If T is noncyclic then (2) is trivial, so we may assume T is cyclic. Then as $m(A) \geq 3$, B is noncyclic, so (2) holds.

For $H \leq G$, $\theta(H) \leq H$ by definition of $\theta(H)$, while $X \leq \theta(X)$ by (1).

This about exhausts the results on solvable signalizer functors which can be established outside of an inductive setting. To proceed further we must work inside a minimal counterexample to the Solvable 2-Signalizer Functor Theorem.

So in the remainder of chapter 15 assume (A, G, θ) is a counterexample to the Solvable 2-Signalizer Functor Theorem with $|G| + |\pi(\theta)|$ minimal. In particular now $r = 2$. Let $I = \theta(C_G(A))$.

(44.9) (1) If $A \leq H < G$ then $\theta(H) \in \mathcal{Q}(G)$.

(2) $G = A\langle \theta(b) : b \in B^{\#} \rangle$ for each E_4-subgroup B of A.

(3) If $1 \neq H \in \mathcal{Q}(G)$ then $G \neq N_G(H)$.

Proof. In each case we use the minimality of the counterexample. Under the hypotheses of (1), $\theta_H(a) = \theta(a) \cap H$ is an A-signalizer functor on H, so (1) follows from minimality of $|G|$.

Let B be a 4-subgroup of A and $K = \langle \theta(b) : b \in B^{\#} \rangle$. If $G \neq KA$ then by (1), $Y = \theta(KA) \in \mathcal{Q}(G)$. But now for $a \in A^{\#}$,

$$\theta(a) = \langle \theta(a) \cap \theta(b) : b \in B^{\#} \rangle \leq C_Y(a) = \langle \theta(b) \cap C_Y(a) : b \in B^{\#} \rangle \leq \theta(a)$$

by 44.8.1, so θ is solvably complete, contrary to the choice of (A, G, θ) as a counterexample. Thus (2) holds.

Suppose $1 \neq H \in \mathcal{Q}(G)$ and $H \trianglelefteq G$. Define G^* and θ^* as in 44.3. By 44.3.1, θ^* is an A^*-signalizer functor, so by minimality of $|G|$, θ^* is solvably complete. Thus there is a unique maximal member X^* of $\mathcal{Q}(G^*)$. By 44.3.2, $X \in \mathcal{Q}(G)$. As $\theta(a)^* = \theta^*(a^*) \leq X^*$, $\theta(a) \leq X$, so $\theta(a) = C_X(a)$ as $X \in \mathcal{Q}(G)$. Therefore θ is solvably complete, contrary to the choice of (A, G, θ) as a counterexample.

(44.10) Let $\pi \subseteq \pi(\theta)$, $M \in \mathcal{Q}^*(G, \pi)$, and $1 \neq X \in \mathcal{Q}(G)$ with $X \trianglelefteq M$. Then

(1) M is a Hall π-subgroup of $\theta(N_G(X))$.

(2) $O_p(M) \leq O_{p^\pi}(\theta(N_G(X)))$ for each $p \in \pi$.

(3) If $\pi = \pi(\theta)$, so that $M \in \mathcal{Q}^*(G)$, then $\theta(N_G(X)) = M$. Hence if $X \leq Y \in \mathcal{Q}(G)$, then $N_Y(X) \leq M$.

Proof. As $X \trianglelefteq M$, $M \leq H = \theta(N_G(X))$ by 44.8.3. By 44.9.3, $G \neq N_G(X)$, so $H \in \mathcal{Q}(G)$ by 44.9.1. As $M \in \mathcal{Q}^*(G, \pi)$, $M \in \mathcal{Q}^*(H, \pi)$, so (1) follows from 44.5. Part (2) follows from (1) and Exercise 11.1. If $M \in \mathcal{Q}^*(G)$ then $H = M$ by maximality of M, so (3) holds.

(44.11) $C_A(P) = 1$ for each $p \in \pi(\theta)$ and each $P \in \mathcal{Q}^*(G, p)$.

Proof. Suppose $T = C_A(P) \neq 1$. Then by 44.7, $\mathcal{Q}^*(G, p) \subseteq C_G(T)$, so, by Exercise 11.1.3, $[H, T] \leq O_{p'}(H)$ for each $H \in \mathcal{Q}(G)$.

Form the signalizer functor Δ of 44.6 with respect to p. By minimality of $|\pi(\theta)|$, Δ is solvably complete. Hence $K = \Delta(G) \in \mathcal{Q}(G)$. Observe

$$[\theta(a), T] \leq O_{p'}(\theta(a)) \leq \Delta(a) = C_K(a)$$

by paragraph one of this proof. Recall that, by 24.5, $[\theta(a), T, T] = [\theta(a), T]$, so

$$[C_K(a), T] = [\theta(a), T].$$

Then by 8.5.6, $[C_K(a), T] \trianglelefteq \theta(a)$.

Let S be the set of noncyclic subgroups of A containing T. By Exercise 8.9,

$$[K, T] = \langle [C_K(b), T] : b \in B^{\#} \rangle$$

for each $B \in S$. Therefore

$$\theta(C_G(B)) = \bigcap_{b \in B^{\#}} \theta(b)$$

acts on $[K, T]$. Hence $\theta(C_G(T))$ acts on $[K, T]$ by 44.8.2, so by 44.2 and 44.4, $H = [K, T]\theta(C_G(T)) \in \mathcal{Q}(G)$. But by 24.4, $\theta(a) \leq \theta(C_G(T))[\theta(a), T]$, so, by the second paragraph of this proof, $\theta(a) \leq H$ for each $a \in A^{\#}$. This contradicts 44.9.2.

(44.12) For each $\pi \subseteq \pi(\theta)$, $a \in A^{\#}$, and $M \in \mathcal{Q}^*(G, \pi)$, $[a, F(M)] \neq 1$.

Proof. Assume otherwise. Then by Exercise 11.1.2, a centralizes M.

Let $p \in \pi(F(M))$ and $O_p(M) = X \leq P \in \mathcal{Q}^*(G, p)$. By 44.11, $[a, P] \neq 1$, so by the Thompson $A \times B$ Lemma, $[N_P(X), a] \neq 1$. But $N_P(X)$ is contained in a Hall π-subgroup of $H = \theta(N_G(X))$, so by 44.10.1 and 44.5, $N_P(X)^h \leq M$ for some $h \in C_H(A)$. This is impossible, as a centralizes M but $[N_P(X)^h, a] \neq 1$.

For $\pi \subseteq \pi(\theta)$ and $a \in A^{\#}$, let $\mathcal{M}(a, \pi)$ denote those $M \in \mathcal{Q}^*(G, \pi)$ such that $C_M(a) \in \mathcal{Q}^*(C_G(a), \pi)$. Recall that for π a set of primes and $p \in \pi$, $p^{\pi} = \pi' \cup \{p\}$.

(44.13) Let $\pi \subseteq \pi(\theta)$, $p \in \pi$, $a \in A^{\#}$, $M \in \mathcal{M}(a, \pi)$, $V \in \mathcal{Q}(O_p(M))$, and $N \in \mathcal{Q}(G)$ with $N = \langle V, C_N(a) \rangle$. Then $V \leq O_{p^\pi}(N)$.

Proof. Assume $V \not\leq O_{p^\pi}(N)$. By 31.20.2, there is a 4-subgroup B of A with $C_V(B) \not\leq O_{p^\pi}(\langle C_N(a), C_V(B) \rangle)$, so replacing V by $C_V(B)$, we may assume $[V, B] = 1$. Next by 31.20.3 there is $b \in B^{\#}$ with $V \not\leq O_{p^\pi}(\langle C_N(\langle a, b \rangle), V \rangle)$, so we may assume $N \leq \theta(b)$. However as $Y = C_M(a) \in \mathcal{Q}^*(C_G(a), \pi)$, $C_Y(b) \in \mathcal{Q}^*(C_{\theta(b)}(a), \pi)$. Further $C_Y(b)$ acts on $C_{O_p(M)}(b)$ and $C_V(b) \leq C_{O_p(M)}(b)$, so 31.20.1 supplies a contradiction, completing the proof.

For $\pi \subseteq \pi(\theta)$ and $a \in A^{\#}$, let $\mathcal{U}(a, \pi)$ consist of the those $U \in \mathcal{Q}(G, \pi)$ such that $[U, a] \neq 1$ and U is invariant under some member of $\mathcal{Q}^*(\theta(a), \pi)$. Let $\mathcal{U}^*(a, \pi)$ be the minimal members of $\mathcal{U}(a, \pi)$ under inclusion. Given $p \in \pi$, let $\mathcal{V}(p, \pi)$ consist of those $V \in \mathcal{Q}(G, p)$ such that $V \leq O_{p^\pi}(H)$ for each $H \in \mathcal{Q}_V(G)$.

(44.14) For each $\pi \subseteq \pi(\theta)$ and $a \in A^{\#}$:
   (1) $\mathcal{U}(a, \pi) \neq \varnothing$.
   (2) For $U \in \mathcal{U}^*(a, \pi)$, $U = [U, a]$ is a p-group for some prime p.
   (3) Let $Y \in \mathcal{Q}^*(\theta(a), \pi)$ with $Y \leq N_G(U)$ and $M \in \mathcal{Q}^*_{UY}(G, \pi)$. Then $U \leq O_p(M)$ and U centralizes $C_{F(M)}(a)$.
   (4) $U \in \mathcal{V}(p, \pi)$.

Proof. Let $Y \in \mathcal{Q}^*(\theta(a), \pi)$ and $Y \leq X \in \mathcal{Q}^*(G, \pi)$. By 44.12, $[a, F(X)] \neq 1$, so $X \in \mathcal{U}(a, \pi)$, and (1) is established. Let $U \in \mathcal{U}^*(a, \pi)$; without loss Y acts on U. By Exercise 11.1.2, $[O_p(U), a] \neq 1$ for some prime p, and by 24.5, $[O_p(U), a, a] = [O_p(U), a]$, so (2) follows from minimality of U.

Let $M \in \mathcal{Q}^*_{UY}(G, \pi)$. Then $Y = C_M(a)$, so U is $C_M(a)$-invariant. Thus $U \leq O_p(M)$ by 36.3. Hence U centralizes $O^p(F(M))$, while by the Thompson Lemma and minimality of U, U centralizes $C_{O_p(M)}(a)$. This establishes (3).

Finally let $H \in \mathcal{Q}_U(G)$ and $N = \langle C_H(a), U \rangle$. By 44.13, $U \leq O_{p^\pi}(N)$, so $U = [U, a] \leq [O_{p^\pi}(N), a] \leq O_{p^\pi}(H)$ by 36.3, establishing (4).

(44.15) Let $p \in \pi \subseteq \pi(\theta)$, B a hyperplane of A, $P \in \mathcal{Q}^*(G, p)$, and for $b \in B^{\#}$, let $M(b) \in \mathcal{M}(b, \pi)$. Then
   (1) If $F(M(b)) = O_p(M(b))$ and $P \cap M(b) \in \mathcal{Q}^*(M(b), p)$ for each $b \in B^{\#}$, then

$$Z(P) \leq \bigcap_{b \in B^{\#}} O_p(M(b)).$$

(2) We have

$$Q\left(\bigcap_{b \in B^{\#}} O_p(M(b))\right) \subseteq V(p, \pi).$$

Proof. Assume the hypotheses of (1) and let $Q(b) = O_p(M(b))$ and $N(b) = \theta(N_G(Q(b)))$. By 44.10.1, $M(b)$ is a Hall π-subgroup of $N(b)$, so as $P(b) = P \cap M(b) \in \mathrm{Syl}_p(M(b))$ by the hypotheses of (1), $P(b) \in \mathrm{Syl}_p(N(b))$. Thus $Q(b) \le P(b) \le P$, so $Z(P) \le \theta(C_G(Q(b))) \le N(b)$. Hence $Z(P) \le P(b)$, so as $C_{M(b)}$ $(F(M(b))) \le F(M(b))$, (1) holds. Let

$$1 \ne X \in Q\left(\bigcap_{b \in B^{\#}} O_p(M(b))\right)$$

and $N \in Q_X(G)$. By 44.13, $X \le O_{p^\pi}(\langle C_N(b), X\rangle)$ for each $b \in B^{\#}$, so by 31.20.3, $X \le O_{p^\pi}(N)$, establishing (2).

(44.16) Let $M \in Q^*(G, \pi)$, $\alpha \subseteq \pi(F(M))$, and $X \le F(M)$ with $\pi(X) = \alpha$ and $Z(O_\alpha(F(M))) \le X$. Then for each $N \in Q_X(G, \pi)$:
   (1) $[O^p(X), O_p(N)] = 1$ for each $p \in \alpha$.
   (2) If $\pi(F(N)) \subseteq \alpha$ then $\alpha = \pi(F(N))$ and $[O_p(X), O^p(F(N))] = 1$ for each $p \in \alpha$.
   (3) If $\pi(F(N)) \subseteq \alpha$ and $|\alpha| > 1$ then $X \le F(N)$.

Proof. For $p \in \alpha$, let $X_p = O_p(X)$ and $Z_p = Z(O_p(M))$. By 44.10.2,

$$X_q \le O_{q^\pi}(\theta(N_G(Z_p))) \quad \text{for each } p, q \in \alpha,$$

so $X_q \le O_q(N_N(Z_p))$. Thus $O^p(X) \le O_{p'}(N_N(Z_p))$, so by 31.15, $O^p(X) \le O_{p'}(N) \le C_N(O_p(N))$, establishing (1).

Suppose $\pi(F(N)) \subseteq \alpha$. Then by (1), $1 \ne O_p(X) \le C_N(O^p(F(N)))$, so by 31.10, $p \in \pi(F(N))$, establishing (2). Assume in addition that $|\alpha| > 1$ but $O_p(X) \not\le F(N)$. Pick $p \ne q \in \alpha$ and let $H = C_N(O^p(F(N)))$ and $K = HZ_q$. By (2), $X \le H$, so as $H \trianglelefteq N$, we have $X_p \not\le O_p(H)$. But if $Z_q \le O_q(K)$, then by 31.14, $X_p \le O_p(N_H(Z_q)) = O_p(H)$, so $Z_q \not\le O_q(K)$. However, $F(H) = O_p(N)Z(F(N))$ and $[Z_q, O_p(N)] = 1$, so $[Z_q, H] \le C_H(O_p(N)) = C_H(F(H)) = Z(F(H))$. As $[Z_q, O^p(F(H))] = 1$, we have $Z_q \le O_q(Z_q Z (F(H)))$, so $[Z_q, H] \le O_q(H)$, contradicting $Z_q \not\le O_q(K)$. This completes the proof of (3).

(44.17) Let $\pi \subseteq \pi(\theta)$, $H \in Q^*(G, \pi)$, $X, L \in Q(G, \pi)$, and $C_{F(H)}(X) \le X \le F(H) \cap L$. Then
   (1) $[O^p(X), O_p(L)] = 1$ for each $p \in \pi(F(H))$.
   (2) $\pi(F(L) \cap H) \subseteq \pi(F(H))$.

(3) If $\pi(F(L)) \subseteq \pi(F(H))$ then $\pi(F(L)) = \pi(F(H))$ and either $|\pi(F(H))| = 1$ or $X \leq F(L)$.

(4) Suppose $1 \neq V \in \mathcal{Q}(O_p(L) \cap O_p(H))$ with $N_H(V) \in \mathcal{Q}^*(N_G(V), \pi)$ and $L \in \mathcal{Q}^*(G, \pi)$. Then $\pi(F(H)) = \pi(F(L))$, and either $\pi(F(H)) = \{p\}$ or $X \leq F(L)$, $L \in H^I$, and $L = H$ if $\pi = \pi(\theta)$.

Proof. Part (1) follows from 44.16.1, and (3) follows from 44.16.2 and 44.16.3. As

$$[O_p(L) \cap H, O^p(X)] = 1 \text{ by (1)} \quad \text{and} \quad C_{F(H)}(X) \leq X,$$

(2) is a consequence of the $A \times B$ Lemma and 31.10.

Assume the hypotheses of (4) and let $Y = N_{F(L)}(V)$. By hypothesis, $N_H(V) \in \mathcal{Q}^*(N_G(V), \pi)$, so by 44.5, $Y \leq H^i$ for some $i \in N_I(V)$. Thus (L, Y, H^i) satisfies the hypotheses of (H, X, L). Also, by Exercise 11.1, $O_p(X) \leq O_{p^\pi}(\theta(N_G(V)))$, so $X \leq F(N_L(V))$, and hence $O^p(X) \leq F(L)$ by 31.14.2. Therefore $\pi(F(H)) \subseteq \pi(F(L))$, so applying (3) to (L, Y, H^i), we conclude $\pi(F(H)) = \pi(F(L))$ and either $Y \leq F(H^i)$ or $\pi(F(H)) = \{p\}$. So to complete the proof of (4) we may assume $|\pi(F(H))| > 1$ and $Y \leq F(H^i)$. By (3), $X \leq F(L)$. As $O^p(Y) = O^p(F(L)) \leq O^p(F(H^i))$, $|O^p(F(L))| \leq |O^p(F(H))|$. As $(Z(O_p(L)), L, H^i, Y)$ satisfy the hypotheses of (V, H, L, X), by symmetry, $|O^p(F(H))| \leq |O^p(F(L))|$, so $O^p(F(L)) = O^p(Y) = O^p(F(H^i))$. Then by 44.10.1 and 44.5, H^i is conjugate to L in $N_I(O^p(Y))$, so $L \in H^I$. Finally, if $\pi = \pi(\theta)$ then $L = \theta(N_G(O^p(Y))) = H^i = H$ by 44.10.3.

(44.18) Let $\pi \subseteq \pi(\theta)$, $U \in \mathcal{V}(p, \pi)$, and $H, L \in \mathcal{Q}_U^*(G, \pi)$ with $N_L(U) \in \mathcal{Q}^*(N_G(U), \pi)$.

Then

(1) If $F(L) = O_p(L)$ then $F(H) = O_p(H)$.

(2) If $q \in \pi$, $V \in \mathcal{V}(q, \pi)$, $N_H(V) \in \mathcal{Q}^*(N_G(V), \pi)$, and $[U, V] \leq U \cap V$ then either

(i) $p = q$ and $F(H)$ and $F(L)$ are p-groups, or

(ii) $L \in H^I$ and if $\pi = \pi(\theta)$ then $H = L$.

Proof. As $U \in \mathcal{V}(p, \pi)$, $U \leq O_p(L) \cap O_p(H)$. As $N_L(U) \in \mathcal{Q}^*(N_G(U), \pi)$, $X = N_{F(H)}(U) \leq L^i$ for some $i \in N_I(U)$, and we may assume $i = 1$. Thus we have the hypotheses of 44.17. If $F^*(L) = O_p(L)$ then $\pi(F(L)) \subseteq \pi(F(H))$, so (1) holds by 44.17.3. Assume the hypotheses of (2). Then $V \leq O_q(H)$ because $V \in \mathcal{V}(p, \pi)$, so as $[V, U] \leq U$, we have $V \leq X \leq L$ and then $V \leq O_q(L)$ because $V \in \mathcal{V}(q, \pi)$. So we have the hypotheses of 44.17.4, so that lemma completes the proof of (2).

For $\pi \subseteq \pi(\theta)$ let $\mathcal{H}(\pi)$ consist of those $H \in \mathcal{Q}(G, \pi)$ such that $\mathcal{Q}^*(H, p) \subseteq \mathcal{Q}^*(G, p)$ for each $p \in \pi$.

(44.19) If B is a hyperplane of A and $H \in \mathcal{M}(b, \pi)$ for each $b \in B^{\#}$ then $H \in \mathcal{H}(\pi)$.

Proof. Let $p \in \pi$, $Q \in \mathrm{Syl}_p(H)$, and $Q \leq P \in \mathcal{Q}^*(G, p)$. As $H \in \mathcal{M}(b, \pi)$, we have $C_H(b) \in \mathcal{Q}^*(C_G(b), \pi)$, so $C_Q(b) = C_P(b)$. Thus $P = Q$ by 44.8.1.

(44.20) Let $\pi \subseteq \pi(\theta)$. Then either
  (1) $\mathcal{H}(\pi) \neq \varnothing$, or
  (2) There exists a prime p such that $F^*(M) = O_p(M)$ for each $M \in \mathcal{M}(a, \pi)$ and each $a \in A^{\#}$.

Proof. Let $a \in A^{\#}$. By 44.14.1 there is $U \in \mathcal{U}^*(a, \pi)$, and by 44.14.2, U is a p-group for some prime p. By minimality of U and Exercise 8.9, $C_A(U) = B$ is a hyperplane of A. Thus $U \leq M(b) \in \mathcal{M}(b, \pi)$ for each $b \in B^{\#}$. By 44.14.4, $U \in \mathcal{V}(p, \pi)$, so $U \leq O_p(M(b))$. By 44.12 there is $U(b) \in \mathcal{U}^*(b, \pi)$ with $U(b) \leq F(M(b))$, and by 44.14.3, $[U, U(b)] = 1$.

Let $H \in \mathcal{Q}^*(G, \pi)$ with $N_H(U) \in \mathcal{Q}^*(N_G(U), \pi)$. If $F(H)$ is a p-group then by 44.18.1, so is $F^*(M)$ for each $M \in \mathcal{Q}_U^*(G, \pi)$. In particular $F(M(b)) = O_p(M(b))$. Further, for $c \in A^{\#}$ and $M \in \mathcal{M}(c, \pi)$, from the proof of 44.14 there is $U(c) \in \mathcal{U}(a, \pi)$ with $U(c) \leq M$. By symmetry there is a hyperplane $B(c)$ of A centralizing $U(c)$, and as $B \cap B(c) \neq 1$, we have $U(c) \leq M(b)^i$ for some $i \in I$. As $F(M(b))$ is a p-group, so is $F(H(c))$ for $H(c) \in \mathcal{Q}^*(G, \pi)$ with $N_{H(c)}(U(c)) \in \mathcal{Q}^*(N_G(U(c)), \pi)$ by 44.18.2. Hence $F^*(M) = O_p(M)$, as $U(c) \leq M$. Thus (2) holds in this case, so we may assume $F^*(H)$ is not a p-group. Hence by 44.18.2, $M(b) \in H^l$ for each $b \in B^{\#}$. Thus (1) holds by 44.19.

(44.21) There is a prime p such that for each $a \in A^{\#}$ and each $M \in \mathcal{M}(a, \pi(\theta))$, $F(M) = O_p(M)$.

Proof. If not, by 44.20 there is $H \in \mathcal{H}(\pi(\theta))$. But then for $p \in \pi(\theta)$ and $P \in \mathrm{Syl}_p(H)$, $P \in \mathcal{Q}^*(G, p)$, so for $a \in A^{\#}$, $C_P(a) \in \mathrm{Syl}_p(\theta(a))$, and hence $\theta(a) = \langle C_P(a) : p \in \pi(\theta) \rangle \leq H$, contrary to 44.9.2.

In the remainder of this section pick p as in 44.21 and let $\pi = \pi(\theta) - \{p\}$. If $H \in \mathcal{H}(\pi)$, pick $q \in \pi(F(H))$, $Q \in \mathcal{Q}^*(G, q)$, and B a hyperplane of A with $Z = Z(Q) \cap O_q(H) \cap C_H(B) \neq 1$. As $H \in \mathcal{H}(\pi)$, $H \in \mathcal{M}(b, \pi)$ for each $b \in B^{\#}$. In this case let $M(b) = H$ for $b \in B^{\#}$.

If $\mathcal{H}(\pi) = \varnothing$ then by 44.20 there is a prime q such that $F^*(M) = O_q(M)$ for each $M \in \mathcal{M}(a, \pi)$ and each $a \in A^{\#}$. Pick $Q \in \mathcal{Q}^*(G, q)$, B a hyperplane of A with $Z = C_{Z(Q)}(B) \neq 1$, and for $b \in B^{\#}$, $M(b) \in \mathcal{M}(b, \pi)$ with $Q \cap M(b) \in \mathcal{Q}^* (M(b), q)$. By 44.15.1, $Z \leq O_q(M(b))$.

So in any case, $Z \le O_q(M(b))$ for each $b \in B^\#$. Hence by 44.15.2, $Z \in \mathcal{V}(q, \pi)$.

For $a \in A^\#$, pick $M_a \in \mathcal{M}(a, \pi(\theta))$ and let $P \in \mathcal{Q}^*(G, p)$. As $I \le M_a$ and I is transitive on $\mathcal{Q}^*(G, p)$, $P \cap M_a \in \mathcal{Q}^*(M_a, p)$. Thus $Z(P) \le O_p(M_a)$ by 44.15.1 and 44.21. Therefore

$$Z(P) \le P_0 = \bigcap_{b \in B^\#} O_p(M_b).$$

Next, $\langle Z, I \rangle \le \theta(b) \le M_b$ for each $b \in B^\#$, so $\langle Z, I \rangle$ acts on P_0. Let $U = \langle Z^I \rangle$, and $\mathcal{W}$ the set of $W \in \mathcal{Q}(G)$ with $I \le N_G(W)$, $W = O_p(W)U$ and $P_0 \le O_p(W)$.

(44.22) $P_0 \in \mathcal{V}(p, \pi(\theta))$, and $F(M) = O_p(M)$ for each $M \in \mathcal{Q}^*_{P_0}(G)$.

Proof. First $P_0 \in \mathcal{V}(p, \pi(\theta))$ by 44.15.2. Let $L \in \mathcal{Q}^*(G)$ with $\theta(N_G(P_0)) \le L$. As $P_0 \in \mathcal{V}(P, \pi(\theta))$, $P_0 \le O_p(L)$. If $1 \ne O_r(L)$ for some prime $r \ne p$, then as B is noncyclic, $1 \ne C_{O_r(L)}(b)$ for some $b \in B^\#$. But $X = C_{F(M_b)}(P_0) \le L$, so by 44.17.2, $r \in \pi(F(M_b))$, contrary to 44.21.

Thus $F(L) = O_p(L)$, so the lemma follows from 44.18.1.

(44.23) Let $M, L \in \mathcal{Q}^*_{P_0}(G)$ and $O_p(M) \le R \in \mathcal{Q}(M, p)$ with $\theta(N_G(C)) \le L$ for each $1 \ne C$ char R. Then $M = L$.

Proof. Let $X = RO_{p,F}(M)$. As $Q = O_p(M) \le R$, $R \in \mathrm{Syl}_p(X)$, and by 44.22, $Q = F(M)$, so $Q = F(X)$. Thus by Thompson Factorization, 32.6, we have $X = N_X(J(R))C_X(Z(R))$, so $X \le L$ by hypothesis. As $Q = O_p(X)$, $Q = O_p(L \cap M)$. By 44.10.3, $\theta(N_G(C)) \le M$ for each $1 \ne C$ char Q. In particular $N_{O_p(L)}(Q) \le O_p(L \cap M) = Q$, so $O_p(L) \le Q$ by Exercise 3.2.1. But now (M, Q, L) satisfies the hypotheses of (L, R, M), so by symmetry, $Q \le O_p(L)$. Therefore $Q = O_p(L)$, so $M = L$ by 44.10.3.

(44.24) Let $W \in \mathcal{W}$, $S = O_p(W)$, $Y = O^p(W)$, $L \in \mathcal{Q}^*(G)$ with $\theta(N_G(Y)) \le L$, and $M \in \mathcal{Q}^*_{IW}(G)$. Then
  (1) $Y \le O_{p,q}(M)$.
  (2) If $O_p(M) \le S$ then $M = L$.

Proof. Recall $Z \in \mathcal{V}(q, \pi)$, so $Z \le O_{q^\pi}(M) = O_{\{p,q\}}(M)$ by definition of $\mathcal{V}(q, \pi)$. By construction $Z \le Z(Q)$ for some $Q \in \mathcal{Q}^*(G, q)$, and as $I \le M$, $Q \cap M \in \mathrm{Syl}_q(M)$. Thus as $Z \le O_{\{p,q\}}(M)$, $Z \le O_{p,q}(M)$ by 31.10. Hence $U = \langle Z^I \rangle \le O_{p,q}(M)$. But $W = UO_p(W)$, so $Y = O^p(W) = \langle U^W \rangle \le O_{p,q}(M)$, establishing (1).

Suppose $R = O_p(M) \le S$ and let $M^* = M/R$ and $X^* = F(M^*)$. By (1), $Y^* \le O_q(X^*)$ and as $R \le S \le N_G(Y)$, $Y = O^p(YR) \trianglelefteq N_X(Y^*)$, so $N_X(Y^*) =$

$N_X(Y) \leq L$. Let $T = O_p(L)$. Then $N_T(R) \leq M$ by 44.10.3, so

$$[N_T(R)^*, N_{X^*}(Y^*)] \leq N_T(R)^* \cap N_{X^*}(Y^*) = 1.$$

Therefore by the $A \times B$ Lemma, $N_T(R)^* \leq C_{M^*}(X^*) \leq X^*$, so $N_T(R) \leq R$ and hence $T \leq R$. Finally by 44.10.3, $\theta(N_G(C)) \leq M$ for each $1 \neq C$ char R, so $M = L$ by 44.23.

If $M_b = M_{b'}$ for all $b, b' \in B^\#$, then the argument establishing 44.21 supplies a contradiction. Therefore as $M_b \in \mathcal{Q}^*_{P_0UI}(G)$ for each $b \in B^\#$, $|\mathcal{Q}^*_{P_0UI}(G)| > 1$. Pick $W \in \mathcal{W}$ with $S = O_p(W)$ maximal subject to $|\mathcal{Q}^*_{WI}(G)| > 1$. Set $Y = O^p(W)$ and pick $M \in \mathcal{Q}^*(G)$ with $\theta(N_G(S)) \leq M$.

(44.25) For each $N \in \mathcal{Q}^*_{WI}(G) - \{M\}$, S is the unique maximal WI-invariant member of $\mathcal{Q}(N, p)$.

Proof. Suppose $S < S_0 \in \mathcal{Q}(N, p)$ with $WI \leq N_G(S_0)$. Then $S < N_{S_0}(S)$, so without loss, $S \trianglelefteq S_0$. Therefore $S_0 \leq \theta(N_G(S)) \leq M$. Let $W_0 = WS_0$. Then $W_0 \in \mathcal{W}$ and $M, N \in \mathcal{Q}^*_{W_0I}(G)$, contradicting the maximal choice of S.

(44.26) (1) $\mathcal{Q}^*_{\theta(N_G(Y))}(G)$ contains a unique member L.
(2) $\mathcal{Q}^*_{\theta(N_G(S))}(G) = \{M\}$.
(3) $\mathcal{Q}^*_{WI}(G) = \{M, L\}$.

Proof. Let $L \in \mathcal{Q}^*_{\theta(N_G(Y))}(G)$ and $N \in \mathcal{Q}^*_{WI}(G) - \{M\}$. By 44.25, $O_2(N) \leq S$, so by 44.24.2, $N = L$. Thus (3) holds and as $|\mathcal{Q}^*_{WI}(G)| > 1$, $M \neq L$, so (1) and (2) hold.

We are now in a position to obtain a contradiction and hence establish the Solvable 2-Signalizer Functor Theorem. By 44.26.2, $\theta(N_G(C)) \leq M$ for each $1 \neq C$ char S, while by 44.25, $O_p(L) \leq S$. Hence $M = L$ by 44.23, contradicting 44.26.3 and the choice of W with $|\mathcal{Q}^*_{WI}(G)| > 1$.

This completes the proof of the Solvable 2-Signalizer Functor Theorem.

Remarks. Gorenstein introduced the concept of the signalizer functor and initiated the study of these objects [Gor 1]. He was motivated by earlier work of Thompson. Goldschmidt [Gol 1, Gol 2] simplified some of Gorenstein's definitions and proved the Solvable 2-Signalizer Functor Theorem and the Solvable Signalizer Functor Theorem for r odd when $m_r(A) > 3$. Glauberman [Gl 1] established the general theorem for solvable functors. Bender gave a new short proof of the Solvable 2-Signalizer Functor Theorem [Be 2]; the proof given here is based on his proof, although our proof is longer and more complicated than Bender's for reasons I'll explain in a moment. Finally, McBride [Mc] proved the Signalizer Functor Theorem subject to the hypothesis that the composition factors of $\theta(a)$ are known simple groups. The Classification Theorem allows this hypothesis to be removed.

Our proof is longer than Bender's because Bender uses the ZJ-Theorem, whereas we use Thompson Factorization, since the ZJ-Theorem is not proved in this text. There are at least two advantages to this approach: First, the increase in the length of the proof caused by not appealing to the ZJ-Theorem is probably less than the length of the proof of the ZJ-Theorem, although of course the ZJ-Theorem is of interest in its own right. Second, the change gives some insight into how Thompson Factorization is used in the literature, and Thompson Factorization is used often, while the ZJ-Theorem is not.

In the first edition of the book, we gave a proof of the Solvable Signalizer Functor Theorem for all primes which was similar to the proof given here but still more complicated. In this edition we have opted for less generality in the hope that a simpler proof will better expose the underlying concepts.

Exercises for chapter 15

1. Let $G = A_7$, $A = \langle (1, 2)(3, 4), (3, 4)(5, 6) \rangle$, and $\theta(a) = O_3(C_G(a))$ for $a \in A^\#$. Prove θ is a solvable A-signalizer functor on G which is not complete.

2. Prove lemma 44.7. (Hint: Assume the lemma is false and choose $P, Q \in Q^*(G, p)$ not conjugate under $\theta(C_G(A))$ with $P \cap Q$ maximal subject to these constraints. Prove $P \cap Q \neq 1$ by observing $C_P(a) \neq 1 \neq C_Q(a)$ for some $a \in A^\#$, and considering $\theta(a)$. Then consider $N_G(P \cap Q)/(P \cap Q)$ and use 44.3.)

3. Let A be a noncyclic abelian r-subgroup of a finite group G and θ an A-signalizer functor on G. For $B \leq A$ let $W_B = \langle \theta(b) : b \in B^\# \rangle$ and let $W = W_A$. Prove
   (1) If B is a noncyclic subgroup of A then $W = W_B$.
   (2) If $\theta(a^g) = \theta(a)^g$ for each $g \in G$ and $a \in A^\#$, then $\Gamma_{2,A}(G) \leq N_G(W)$, where $\Gamma_{2,A}(G) = \langle N_G(B) : B \leq A, m(B) \geq 2 \rangle$.
   (3) Assume the hypothesis of (2) and let $\mathcal{V}_G(A, r')$ denote the set of A-invariant r'-subgroups of G. Prove $\mathcal{V}_G(A, r') \subseteq N_G(W)$. If θ is complete, $G = \Gamma_{2,A}(G)$ and $O_{r'}(G) = 1$, prove $\theta(a) = 1$ for all $a \in A^\#$.

4. Let A be an elementary abelian r-subgroup of a finite group G with $m(A) \geq 3$. Prove
   (1) If $a \in A^\#$ and $C_G(a)$ is solvable then $C_G(a)$ is balanced for the prime r. (See section 31 for the definition of balance.)
   (2) If $C_G(a)$ is balanced for the prime r for each $a \in A^\#$ then θ is an A-signalizer functor on G, where $\theta(a) = O_{r'}(C_G(a))$.

(3) Assume $C_G(a)$ is solvable for each $a \in A^\#$ and $G = \Gamma_{2,A}(G)$. Prove $O_{r'}(C_G(a)) \leq O_{r'}(G)$ for each $a \in A^\#$, and if $O_{r'}(G) = 1$ then $F^*(C_G(a)) = O_r(C_G(a))$.

5. Let r be a prime, G a finite group, $E_{r^3} \cong A \leq G$, and π a set of primes with $r \notin \pi$. For $E_{r^2} \cong B \leq G$ define

$$\alpha(G) = \alpha_G(B) = \langle\langle[O_\pi(C_G(B)), b]: b \in B\rangle\rangle \left(\bigcap_{b \in B^\#} O_\pi(C_G(b))\right).$$

Assume $\alpha_{C_G(c)}(B) \leq O_\pi(C_G(c))$ for each $c \in A^\#$ and each hyperplane B of A. Also either assume the Signalizer Functor Theorem or assume $O_\pi(C_G(c))$ is solvable for each $c \in A^\#$. Prove $\alpha(B)$ is a π-group for each hyperplane B of A and $\alpha(B)$ is independent of the choice of B. (Hint: Define

$$\gamma_B(a) = [O_\pi(C_G(a)), B](O_\pi(C_G(B)) \cap C(a))$$

for $E_{p^2} \cong B \leq A$ and $a \in A^\#$. Prove γ_B is an A-signalizer functor and $\alpha(B) \leq \alpha(D)$ for each pair B, D of hyperplanes of A.)

16

Finite simple groups

To my mind the theorem classifying the finite simple groups is the most important result in finite group theory. As I indicated in the preface, the Classification Theorem is the foundation for a powerful theory of finite groups which proceeds by reducing suitable group theoretical questions to questions about representations of simple groups. The final chapter of this book is devoted primarily to a discussion of the Classification Theorem and the finite simple groups themselves.

Sections 45 and 46 introduce two classes of techniques useful in the study of simple groups. Section 45 investigates consequences of the fact that each pair of involutions generates a dihedral group. The two principal results of the section are the Thompson Order Formula and the Brauer–Fowler Theorem. The Thompson Order Formula supplies a formula for the order of a finite group with at least two conjugacy classes of involutions in terms of the fusion of those involutions in the centralizers of involutions. The Brauer–Fowler Theorem shows that there are at most a finite number of finite simple groups possessing an involution whose centralizer is isomorphic to any given group.

Section 46 considers the commuting graph on the set of elementary abelian p-subgroups of p-rank at least k in a group G. The determination of the groups for which this graph is disconnected for small k plays a crucial role in the Classification Theorem.

Section 47 contains a brief description of the finite simple groups. The groups of Lie type are described as groups with a split BN-pair and generated by root groups satisfying a weak version of the Chevalley commutator relations. The last portion of section 47 explores consequences of these axioms. In particular the existence of Levi complements is derived, it is shown that the maximal parabolics are the maximal overgroups of Sylow p-groups of groups in characteristic p, and the Borel–Tits Theorem is (essentially) proved for finite groups of Lie type. This last result says that in a finite group of Lie type and characteristic p, each p-local is contained in a maximal parabolic.

Finally section 48 consists of a sketchy outline of the Classification Theorem. This discussion provides a nice illustration of many of the techniques developed in earlier chapters.

45 Involutions in finite groups

Section 45 seeks to exploit the following property of involutions established in Exercise 10.1:

(45.1) Let x and y be distinct involutions of a group G. Then $\langle x, y \rangle$ is a dihedral group of order $2|xy|$.

Throughout section 45, G will be assumed to be a finite group. To begin let's look more closely at 45.1:

(45.2) Let x and y be distinct involutions in G, $n = |xy|$, and $D = \langle x, y \rangle$. Then
  (1) Each element in $D - \langle xy \rangle$ is an involution.
  (2) If n is odd then D is transitive on its involutions, so in particular x is conjugate to y in D.
  (3) If n is even then each involution in D is conjugate to exactly one of x, y, or z, where z is the unique involution in $\langle xy \rangle$. Further $z \in Z(D)$.
  (4) If n is even and z is the involution in $\langle xy \rangle$ then xz is conjugate to x in D if and only if $n \equiv 0 \mod 4$.

Proof. Let $u = xy$ and $U = \langle u \rangle$. Then $u^x = u^{-1}$ so $v^x = v^{-1}$ for each $v \in U$. Then $(vx)^2 = vv^x = vv^{-1} = 1$, so (1) holds. Further for $w \in U$, $(vx)^w = vx^w = vw^{-1}w^x x = vw^{-2}x$, so $x^D = \{v^2x : v \in U\}$. In particular if n is odd then $x^D = D - U$ and U contains no involutions, so (2) holds.

So take n even and let z be the involution in U. Then

$$D - U = \{v^2x : v \in U\} \cup \{v^2ux : v \in U\}$$

with $ux = y$, so $D - U = x^D \cup y^D$ and (3) holds. Finally $zx \in x^D$ precisely when z is a square in U; that is when $n \equiv 0 \mod 4$.

(45.3) Let G be of even order with $Z(G) = 1$, let m be the number of involutions in G, and $n = |G|/m$. Then G possesses a proper subgroup of index at most $2n^2$.

Proof. Let I be the set of involutions of G, R the set of elements of G inverted by a member of I, and $(x_i : 0 \le i \le k)$ a set of representatives for the conjugacy classes of G in R. Pick $x_0 = 1$ and let $m_i = |x_i^G|$, let B_i be the set of pairs (u, v) with $u, v \in I$ and $uv = x_i$, and let $b_i = |B_i|$.

Observe first that if $u, v \in I$ then either $u = v$ and $uv = 1$ or $\langle u, v \rangle$ is dihedral. In either case u inverts uv, so $uv \in R$. Thus counting $I \times I$ in two different ways, we obtain:

(a) $$m^2 = |I \times I| = \sum_{i=0}^{k} m_i b_i.$$

Moreover there is an involution t_i inverting x_i and if $(u, v) \in B_i$ then u inverts x_i and $v = ux_i$. Hence the map $(u, v) \mapsto u$ is an injection of B_i into $t_i C_G(x_i)$ so $b_i \leq |t_i C_G(x_i)| = |C_G(x_i)|$. Also $m_i = |G : C_G(x_i)|$, so $m_i b_i \leq |G|$. If $i = 0$ we can be more precise: $m_0 = 1$ and $b_0 = m$. Combining these remarks with (a) gives:

(b) $$m^2 \leq m + k|G|.$$

Let s be the minimal index of a proper subgroup of G. If $i > 0$ then by hypothesis $x_i \notin Z(G)$, so $m_i = |G : C_G(x_i)| \geq s$. Hence $|G| \geq \sum_{i=0}^{k} m_i \geq 1 + ks$, which I record as:

(c) $$k \leq (|G| - 1)/s.$$

Combining (b) and (c) gives:

(d) $$m^2 \leq (|G|(|G| - 1)/s) + m.$$

Then, as $n = |G|/m$, it follows from (d) that $s \leq n(n - m^{-1})/(1 - m^{-1})$, and, as $m \geq 2$, $s \leq 2n^2$.

(45.4) Let G be a finite simple group of even order, t an involution in G, and $n = |C_G(t)|$. Then $|G| \leq (2n^2)!$.

Proof. By 45.3, G possesses a proper subgroup H of index at most $2n_0^2$, where $n_0 = |G|/m$ and m is the number of involutions in G. Then $m \geq |t^G| = |G : C_G(t)|$, so $n_0 \leq |G|/|G : C_G(t)| = n$.

Represent G as a permutation group on the coset space G/H. Then $k = |G/H| \leq 2n^2$, and as G is simple the representation is faithful. So G is isomorphic to a subgroup of the symmetric group S_k, and hence $|G| \leq k! \leq (2n^2)!$.

As an immediate corollary to 45.4 we have the following theorem of Brauer and Fowler:

(45.5) (Brauer–Fowler [BF]) Let H be a finite group. Then there exists at most a finite number of finite simple groups G with an involution t such that $C_G(t) \cong H$.

Recall that the Odd Order Theorem of Feit and Thompson says that nonabelian finite simple groups are of even order and hence possess involutions. The Brauer–Fowler Theorem suggests it is possible to classify finite simple groups in terms of the centralizers of involutions. This approach will be discussed in more detail in section 48.

If G has more than one class of involutions it is possible to make a much more precise statement than 45.5:

(45.6) (Thompson Order Formula) Assume G has $k \geq 2$ conjugacy classes of involutions x_i^G, $1 \leq i \leq k$, and define n_i to be the number of ordered pairs (u, v) with $u \in x_1^G$, $v \in x_2^G$, and $x_i \in \langle uv \rangle$. Then

$$|G| = |C_G(x_1)||C_G(x_2)| \left(\sum_{i=1}^{k} n_i / |C_G(x_i)| \right).$$

Proof. Let I be the set of involutions in G and $\Omega = x_1^G \times x_2^G$. Then

(*) $$|\Omega| = \left| x_1^G \right| \left| x_2^G \right| = |G : C_G(x_1)||G : C_G(x_2)|.$$

For $(u, v) \in \Omega$, $u \notin v^G$, so by 45.2 there is a unique involution $z(u, v)$ in $\langle uv \rangle$. Hence Ω is partitioned by the subsets Ω_z consisting of those pairs (u, v) with $z = z(u, v)$. Thus $|\Omega| = \sum_{z \in I} |\Omega_z|$. Further $|\Omega_z| = |\Omega_{x_i}|$ for $z \in x_i^G$, so

$$\sum_{z \in x_i^G} |\Omega_z| = \left| x_i^G \right| \left| \Omega_{x_i} \right| = |G : C_G(x_i)| n_i.$$

Therefore

(**) $$|\Omega| = \sum_{i=1}^{k} |G : C_G(x_i)| n_i.$$

Of course the lemma follows from (*) and (**).

Notice that the integer n_i can be calculated if $x_j^G \cap C_G(x_i)$, $j = 1, 2$, is known for each i. Hence the order of G is determined by the fusion of x_1 and x_2 in $C_G(x_i)$, $1 \leq i \leq k$. In particular the order of G can be determined from the fusion of involutions in local subgroups.

46 Connected groups

In this section G is a finite group and p is a prime.

If Ω is a collection of subgroups of G permuted by G via conjugation, define $\mathscr{D}(\Omega)$ to be the graph on Ω obtained by joining A to B if $[A, B] = 1$. Evidently G is represented as a group of automorphisms of $\mathscr{D}(\Omega)$ via conjugation.

(46.1) Let Δ be a G-invariant collection of subgroups of G and $H \leq G$. Then the following are equivalent:

(1) H controls fusion in $H \cap \Delta$ and $N_G(X) \leq H$ for each $X \in H \cap \Delta$.
(2) $H \cap H^g \cap \Delta$ is empty for $g \in G - H$.

(3) The members of $H \cap \Delta$ fix a unique point in the permutation representation of G on G/H by right multiplication.

(4) $H \cap \Delta$ is the union of a set Γ of connected components of $\mathscr{D}(\Delta)$, and $\Gamma \cap \Gamma^g$ is empty for $g \in G - H$.

Proof. Assume (1) and let $g \in G$ with $H \cap H^g \cap \Delta$ nonempty. Then there is $X \in H \cap \Delta$ with $X^g \leq H$ so, as H controls fusion in $H \cap \Delta$, $X^{gh} = X$ for some $h \in H$. Then $gh \in N_G(X) \leq H$, so $g \in H$. Thus (1) implies (2).

Assume (2) and consider the representation of G on G/H. Then $X \in H \cap \Delta$ fixes Hg if and only if $X \leq H^g$, which holds precisely when $g \in H$ by (2). Thus (2) implies (3).

Assume (3). Then, for $A \in H \cap \Delta$, $\{H\} = \mathrm{Fix}(A)$, so $N_G(A) \leq H$, as H is the stabilizer in G of the coset H. In particular if $B \in \Delta$ is incident to A in $\mathscr{D}(\Delta)$ then $B \leq C(A) \leq H$, so $B \in H \cap \Delta$. Thus $H \cap \Delta$ is the union of some set Γ of connected components of $\mathscr{D}(\Delta)$. Further if $A \in \theta \in \Gamma$ and $\theta^g \in \Gamma$ then $A^g \leq H$, so $\{H\} = \mathrm{Fix}(A^g) = \{Hg\}$ and hence $g \in H$. So (3) implies (4).

Finally assume (4). If $X \in H \cap \Delta$ and $g \in G$ with $X^g \leq H$, then $X \in \theta \in \Gamma$ and $X^g \in \theta' \in \Gamma$, so $\theta' = \theta^g \in \Gamma \cap \Gamma^g$. Hence, by (4), $g \in H$. So (4) implies (1).

If k is a positive integer write $\mathscr{E}_k^p(G)$ for the set of all elementary abelian p-subgroups of G of p-rank at least k. G is said to be *k-connected for the prime p* if $\mathscr{D}(\mathscr{E}_k^p(G))$ is connected.

(46.2) Let G be a p-group. Then

(1) G is 1-connected for the prime p.

(2) If $m(G) > 2$ then there is $E_{p^2} \cong U \trianglelefteq G$ and $X \in \mathscr{E}_2^p(G)$ is in the same connected component of $\mathscr{D}(\mathscr{E}_2^p(G))$ as U precisely when $m(C_G(X)) > 2$.

(3) If $p = 2$ and G has a normal E_8-subgroup then G is 2-connected for the prime 2.

(4) If $p = 3$ and $m(G) > 3$, then G is 2-connected for the prime 3.

Proof. Part (1) follows from the fact that $Z(G) \neq 1$. Assume $m(G) > 2$; by Exercise 8.4 there is $E_{p^2} \cong U \trianglelefteq G$. $G/C_G(U) \leq SL_2(p)$ and $SL_2(p)$ is of p-rank 1. So if $A \in \mathscr{E}_3^2(G)$ then $m(C_A(U)) \geq 2$, so $m(C_A(U)U) > 2$. Hence $m(C_G(U)) > 2$. Also, if $U \neq E \in \mathscr{E}_2^p(G)$ is in the same connected component as U, then there is $E \neq D \in \mathscr{E}_2^p(G)$ adjacent to E and $m(C_G(E)) \geq m(DE) \geq 3$. Thus to prove (2) it remains to show that each $A \in \mathscr{E}_3^p(G)$ is in the connected component of U. But $A, C_A(U), C_A(U)U, U$ is a path in $\mathscr{D}(\mathscr{E}_2^p(G))$, so the proof of (2) is complete.

Assume $p = 2$ and $E_8 \cong V \trianglelefteq G$. Let $E \in \mathscr{E}_2^p(G)$; to prove (3) we must exhibit a path from E to V. For $e \in E$, $m(C_V(e)) \geq 2$ by Exercise 9.8, and we may assume there is $e \in E - V$, so $E, \langle e, C_V(E) \rangle, C_V(e), V$ is a path in $\mathscr{D}(\mathscr{E}_2^p(G))$.

Similarly under the hypothesis of (4) there is $E_{81} \cong V \trianglelefteq G$ by Exercise 8.11, after which we can use Exercise 9.8 and the argument of the last paragraph to establish (4).

Given a p-group P acting as a group of automorphisms of G, and given a positive integer k, define

$$\Gamma_{P,k}(G) = \langle N_G(X): X \le P, m(X) \ge k \rangle$$
$$\Gamma_{P,k}^0(G) = \langle N_G(X): X \le P, m(X) \ge k, m(XC_P(X)) > k \rangle.$$

The following observation is an easy consequence of Exercise 3.2.1:

(46.3) Let $H \le G$, $P \in \mathrm{Syl}_p(H)$, and k a positive integer. Then either of the following implies that $P \in \mathrm{Syl}_p(G)$:

(1) $m(P) \ge k$ and $\Gamma_{P,k}(G) \le H$
(2) $m(P) > k$ and $\Gamma_{P,k}^0(G) \le H$.

(46.4) Let $H \le G$, $P \in \mathrm{Syl}_p(H)$, and $m(P) \ge k$. Then the following are equivalent:

(1) $\Gamma_{P,k}(G) \le H$.
(2) H controls fusion in $\mathscr{E}_k^P(H)$ and $N_G(X) \le H$ for each $X \in \mathscr{E}_k^P(H)$.
(3) $m_p(H \cap H^g) < k$ for each $g \in G - H$.
(4) Each member of $\mathscr{E}_k^P(G)$ fixes a unique point in the permutation representation of G by right multiplication on G/H.

Proof. Parts (2), (3), and (4) are equivalent by 46.1, except in (4), $\mathscr{E}_k^P(G)$ should be $\mathscr{E}_k^P(H)$. This weaker version of (4) evidently implies (1) and (1) and 46.3 show $P \in \mathrm{Syl}_p(G)$, from which the strong version of (4) follows. It remains to show (1) implies (2).

Assume $\Gamma_{P,k}(G) \le H$ and let $X \in \mathscr{E}_k^P(H)$. It suffices to show that if $g \in G$ with $X^g \le H$, then $g \in H$. By Sylow's Theorem we may assume $\langle X, X^g \rangle \le P$. By 46.3, $P \in \mathrm{Syl}_p(G)$. By Alperin's Fusion Theorem there exists $P_i \in \mathrm{Syl}_p(G)$, $1 \le i \le n$, and $g_i \in N_G(P \cap P_i)$ with $g = g_1 \ldots g_n$, $X \le P_1$, and $X^{g_1 \cdots g_i} \le P \cap P_i$. Then $m(P \cap P_i) \ge k$, so $g_i \in N_G(P \cap P_i) \le \Gamma_{P,k}(G) \le H$. Hence $g = g_1 \ldots g_n \in H$, so that (2) holds.

If $P \in \mathrm{Syl}_p(G)$ then $\Gamma_{P,k}(G)$ is called the *k-generated p-core* of G. By Sylow the k-generated p-core of G is determined up to conjugacy in G. The hypothesis of 46.4 are equivalent to the assertion that H contains the k-generated p-core of G, and hence, if H is a proper subgroup of G, that p-core is a proper subgroup of G. By 46.1, if the k-generated p-core is proper then G is k-disconnected for the prime p, although the converse need not be true. However in the following special case the converse is valid:

(46.5) Let $P \in \mathrm{Syl}_p(G)$ and assume P is k-connected. Then

(1) $\mathscr{E}_k^p(\Gamma_{P,k}(G))$ is a connected component of $\mathscr{D}(\mathscr{E}_k^p(G))$ and $\Gamma_{P,k}(G)$ is the stabilizer in G of that connected component.

(2) G is k-disconnected for the prime p if and only if G has a proper k-generated p-core.

Proof. Evidently (1) implies (2). Let $H = \Gamma_{P,k}(G)$. By 46.1 and 46.4 $\mathscr{E}_k^p(H)$ is the union of connected components of $\mathscr{D}(\mathscr{E}_k^p(G))$, while, as P is k-connected, $\mathscr{E}_k^p(P)$ is contained in some component Δ and of course $H \leq N_G(\Delta)$. Hence $\mathscr{E}_k^p(H) \subseteq \Delta$ by Sylow's Theorem. So $\Delta = \mathscr{E}_k^p(H)$ and $H = N_G(\Delta)$ by 46.1.4. That is (1) holds.

If H is a proper subgroup of G satisfying any of the equivalent conditions of 46.4 with $k = 1$, then we say H is a *strongly p-embedded subgroup* of G. As a direct consequence of 46.5 and 46.2.1 we have:

(46.6) G possesses a strongly p-embedded subgroup if and only if G is 1-disconnected for the prime p.

(46.7) Let $m_p(G) > 2$, $P \in \mathrm{Syl}_p(G)$, and let $\mathscr{E}_2^p(G)^0$ denote the set of subgroups $X \in \mathscr{E}_2^p(G)$ with $m_p(XC_G(X)) > 2$. Then

(1) $\mathscr{E}_2^p(G)^0$ is the set of points of $\mathscr{E}_2^p(G)$ which are not isolated in $\mathscr{D}(\mathscr{E}_2^p(G))$.

(2) $\mathscr{E}_2^p(\Gamma_{P,2}^0(G))^0$ is a connected component of $\mathscr{D}(\mathscr{E}_2^p(G))$ and $\Gamma_{P,2}^0(G)$ is the stabilizer in G of this connected component.

(3) $\mathscr{D}(\mathscr{E}_2^p(G)^0)$ is connected if and only if $G = \Gamma_{P,2}^0(G)$.

Proof. Part (1) is trivial, as is (3) given (2). So it remains to establish (2). By 46.2.2, $\mathscr{E}_2^p(P)^0$ is contained in a connected component Δ of $\mathscr{D}(\mathscr{E}_2^p(G))$. Thus $H = \Gamma_{P,2}^0(G) \leq N_G(\Delta)$. Let $\Gamma = \mathscr{E}_2^p(H)^0$. As $\mathscr{E}_2^p(P)^0 \subseteq \Delta$ and H acts on Δ, $\Gamma \subseteq \Delta$ by Sylow's Theorem. If $\Delta \neq \Gamma$ there is $x \in \Gamma$, $Y \in \Delta - \Gamma$ with X and Y adjacent in $\mathscr{D}(\Delta)$. Without loss $X \in \mathscr{E}_2^p(P)^0$, so $Y \leq C_G(X) \leq H$. Hence, as $m_p(C_H(Y)) \geq m_p(XY) \geq 3$, $Y \in \Gamma$, a contradiction.

So $\Delta = \Gamma$. Thus, to complete the proof of (2), it remains to show that if $X, X^g \in \Gamma$ then $g \in H$. Suppose $X, X^g \in \Gamma$. Then $N_G(X) \leq H \geq N_G(X^g)$ so there is $E_{p^3} \cong A \leq C_G(X) \leq H$ and $A^g \leq H$. By Sylow we may take $A, A^g \leq P$. Now apply Alperin's Fusion Theorem as in the proof of 46.4, using $\Gamma_{P,3}(G) \leq \Gamma_{P,2}^0(G) \leq H$, to conclude $g \in H$.

Lemma 46.7 says that if $m_p(G) > 2$ and $P \in \mathrm{Syl}_p(G)$, then, neglecting the isolated points of $\mathcal{D}(\mathcal{E}_2^p(G))$, G is 2-connected for the prime p precisely when $G = \Gamma_{P,2}^0(G)$.

These observations are very useful in conjunction with the Signalizer Functor Theorem, as the next two lemmas and Exercise 16.1 indicate.

(46.8) Let Γ be the set of elements a of G of order p with $m_p(C_G(a)) > 2$ and let θ be a map from Γ into the set of p'-subgroups of G such that, for all $a, b \in \Gamma$ with $[a, b] = 1$ and all $g \in G$, $\theta(a^g) = \theta(a)^g$ and $\theta(a) \cap C_G(b) \leq \theta(b)$. Let $P \in \mathrm{Syl}_p(G)$, assume $G = \Gamma_{P,2}^0(G)$, $O_{p'}(G) = 1$, and either

(1) $\theta(a)$ is solvable for each $a \in \Gamma$, or
(2) the Signalizer Functor Theorem holds on G.

Then $\theta(a) = 1$ for each $a \in \Gamma$.

Proof. For $A \in \mathcal{E}_3^p(G)$, θ is an A-signalizer functor by hypothesis. For $B \in \mathcal{E}_2^p(G)^0$ define $W_B = \langle \theta(b): b \in B^\# \rangle$. Then there exists $A \in \mathcal{E}_3^p(G)$ with $B \leq A$ and, by Exercise 15.3, $W_B = W_A$ and $\Gamma_{A,2}(G) \leq N_G(W_A)$. In particular if B and D are distinct members of $\mathcal{E}_2^p(G)$ adjacent in $\mathcal{D}(\mathcal{E}_2^p(G))$ then $BD \in \mathcal{E}_3^p(G)$ so $W_B = W_{BD} = W_D$. Thus, as $\Gamma_{P,2}^0(G) = G$, 46.7 says $W_B = W$ is independent of the choice of $B \in \mathcal{E}_2^p(G)^0$. Thus $G = \Gamma_{P,2}^0(G) \leq N_G(W)$ as $N_G(B) \leq \Gamma_{A,2}(G) \leq N_G(W)$. But, by (1) and the Solvable Signalizer Functor Theorem or by (2), W is a p'-group, so $W \leq O_{p'}(G)$. As $O_{p'}(G) = 1$ by hypothesis, and as $\theta(a) \leq W$ for each $a \in \Gamma$, the lemma is established.

(46.9) Let Γ be the set of elements a of G of order p with $m_p(C_G(a)) > 2$, let $P \in \mathrm{Syl}_p(G)$, and assume $\Gamma_{P,2}^0(G) = G$, $O_{p'}(G) = 1$, $C_G(a)$ is balanced for the prime p for each $a \in \Gamma$, and either

(1) $O_{p'}(C_G(a))$ is solvable for each $a \in \Gamma$, or
(2) the Signalizer Functor Theorem holds on G.

Then $O_{p'}(C_G(a)) = 1$ for each $a \in \Gamma$.

Proof. For $a \in \Gamma$ let $\theta(a) = O_{p'}(C_G(a))$. By Exercise 15.4.2, $\theta(a) \cap C_G(b) \leq \theta(b)$ for each $a, b \in \Gamma$ with $[a, b] = 1$. Hence 46.8 completes the proof.

47 The finite simple groups

Section 47 describes a list $\mathcal{K}$ of finite simple groups. The Classification Theorem says that any finite simple group is isomorphic to a member of $\mathcal{K}$. The proof of the Classification Theorem is far beyond the scope of this book, but there is a brief outline of that proof in the final section of this chapter.

$\mathcal{K}$ can be described as the following collection of groups:

(1) The groups of prime order.
(2) The alternating groups A_n of degree $n \geq 5$.
(3) The finite simple groups of Lie type listed in Table 16.1.
(4) The 26 sporadic simple groups listed in Table 16.3.

The groups of prime order are the abelian simple groups (cf. 8.4.1). They are the simplest of the simple groups.

By 15.16, the alternating groups $A_n, n \geq 5$, are simple. The permutation representation of A_n of degree n is an excellent tool for investigating the group, and, as in 15.3 and Exercise 5.3, to determine its conjugacy classes. Lemma 33.15 describes the covering group and Schur multiplier of A_n. Exercise 16.2 says $\mathrm{Aut}(A_n) = S_n$, unless $n = 6$.

The groups of Lie type are the analogues of the semisimple Lie groups. This class of groups is extremely interesting; for one thing by some measure most finite simple groups are of Lie type. We've already encountered examples of these groups; the classical groups are of Lie type.

The groups of Lie type can be described in terms of various representations: as groups of automorphisms of certain Lie algebras or fixed points of suitable automorphisms of such groups; as fixed points of suitable endomorphisms of semisimple algebraic groups; as groups of automorphisms of buildings with finite Weyl groups; as groups with a BN-pair and a finite Weyl group. I'll take a modified version of this last point of view here to avoid complications.

A finite group of Lie type satisfies the following conditions:

(L1) G possesses a Tits system (G, B, N, S) with a finite Weyl group $W = N/(B \cap N)$. The *Lie rank* of G is the integer $|S|$.

(L2) $H = B \cap N$ possesses a normal complement U in B.

(L3) There is a root system Σ for W and a simple system π for Σ with $S = (s_\alpha : \alpha \in \pi)$ where s_α is the reflection through α. Let Σ^+ be the positive system of π. There exists an injection $\alpha \mapsto U_\alpha$ of Σ into the set of subgroups of G such that for each $\alpha \in \Sigma$ and $w \in W$, $H \leq N_G(U_\alpha)$ and $(U_\alpha)^w = U_{\alpha w}$. Each member of U can be written uniquely as a product $\prod_{\alpha \in \Sigma^+} u_\alpha$, with $u_\alpha \in U_\alpha$, in some fixed ordering of Σ^+ respecting height.

(L4) The *root groups* $(U_\alpha : \alpha \in \Sigma^+)$ satisfy

$$[U_\alpha, U_\beta] \subseteq (Z(U_\alpha) \cap Z(U_\beta)) \prod U_{i\alpha + j\beta}, \quad \alpha, \beta \in \Sigma^+,$$

where the product is over all roots $i\alpha + j\beta$ with i and j positive integers.

We will be most interested in groups which also satisfy:

(L5) Let w_0 be the unique member of W of maximal length in W (cf. Exercise 10.3). Then $G = \langle U, U^{w_0} \rangle$, $B \cap B^{w_0} = H$, and W is irreducible.

Some observations about these axioms are in order. Define a Tits system $T = (G, B, N, S)$ to be *saturated* if $\bigcap_{w \in W} B^w = H$. By Exercise 14.9.5, T is saturated if and only if $B \cap B^{w_0} = H$. So the condition $U \cap U^{w_0} = H$ in L5 can be replaced by the hypothesis that T is saturated.

Next, from Exercise 10.3, $\Sigma^+ w_0 = \Sigma^-$. Hence, by L3, 47.1.1, and 30.7, $U_\alpha = U \cap U^{w_0 s_\alpha}$ for $\alpha \in \pi$. Moreover the proof of 47.1 depends on L1–L4 but not L5. By 30.9 each $\gamma \in \Sigma$ can be written $\gamma = \alpha w$ for some $\alpha \in \pi$, $w \in W$. So, by L3, $U_\gamma = U_{\alpha w} = (U_\alpha)^w = (U \cap U^{w_0 s_\alpha})^w$. Hence L1–L4 determine the root groups U_γ, $\gamma \in \Sigma$ uniquely.

Conversely if T is a saturated Tits system satisfying L2 then by Exercise 14.10 we can define $U_\gamma = (U \cap U^{w_0 s_\alpha})^w$, where $\gamma = \alpha w$, $\alpha \in \pi$, and $w \in W$. Then, by Exercise 14.10 parts (7) and (9), L3 is satisfied. Thus L3 can be dispensed with if we assume L_1 and L_2 with T saturated. Equivalently L1, L2, and L5 imply L3.

The properties L1 and L2 essentially characterize the finite groups of Lie type (cf. [FS], [HKS], and [Ti]). There are some rather trivial examples of groups of Lie rank 1 satisfying L1–L5 which are not of Lie type. The finite groups of Lie type satisfying L5 and possessing a trivial center are listed in Table 16.1. Column 1 lists G. The parameters q and n are an arbitrary prime power and positive integer, respectively, unless some restriction is listed explicitly. Column 2 lists the order of a certain central extension $\tilde{G}$ of G. Usually G is simple, $\tilde{G}$ is the universal covering group of G, and $d(G)$ is the order of the Schur multiplier of G. In any event $|G| = |\tilde{G}|/d(G)$. Finally, column 4 lists the Dynkin diagram of the root system Σ of G. Table 16.2 explains the notation A_n, B_n, etc. The nodes of the Dynkin diagram are indexed by the set π of simple roots. Distinct nodes α and β are joined by an edge of weight $2(\alpha, \beta)(\beta, \alpha)/(\alpha, \alpha)(\beta, \beta) = m_{\alpha\beta} - 2$. It turns out that $|s_\alpha s_\beta| = m_{\alpha\beta}$, so, if we neglect the arrows in the diagram, the Dynkin diagram of Σ is just the Coxeter diagram of the Weyl group W of Σ or G, as defined in section 29. By 30.9, each member of Σ is conjugate to a member of π under W, and it turns out that there are one or two orbits of W on Σ depending on whether the Dynkin diagram has no multiple bonds or has multiple bonds, respectively. Indeed if α and β are joined by an edge of weight 1 then α is conjugate to β under $\langle s_\alpha, s_\beta \rangle$. If the diagram has multiple bonds, roots in different orbits are of different lengths, and hence are called *long* or *short* roots. The arrow in the diagram points to the short root in the pair joined.

Let q be a power of the prime p. p is called the *characteristic* of G. Recall that G can be defined in terms of one of several representations. Consider for the moment the representation of G as a group of automorphisms of a Lie algebra L. L is a Lie algebra over a field of characteristic p obtained from a simple Lie algebra L' over the complex numbers. L' has an associated root

Table 16.1 Finite groups of Lie type with L5 and trivial center

G	$|G|d(G) = |\tilde{G}|$	$d(G)$	Σ
$A_n(q)$	$q^{n(n+1)/2} \prod_{i=1}^{n}(q^{i+1} - 1)$	$(n+1, q-1)$	A_n
$B_n(q)$, q odd	$q^{n^2} \prod_{i=1}^{n}(q^{2i} - 1)$	$(2, q-1)$	B_n
$C_n(q)$	$q^{n^2} \prod_{i=1}^{n}(q^{2i} - 1)$	$(2, q-1)$	C_n
$D_n(q)$, $n \geq 2$	$q^{n(n-1)}(q^n - 1) \prod_{i=1}^{n-1}(q^{2i} - 1)$	$(4, q^n - 1)$	D_n
$E_6(q)$	$q^{36}(q^{12} - 1)(q^9 - 1)(q^8 - 1)$ $(q^6 - 1)(q^5 - 1)(q^2 - 1)$	$(3, q-1)$	E_6
$E_7(q)$	$q^{63}(q^{18} - 1)(q^{14} - 1)$ $\cdot(q^{12} - 1)(q^{10} - 1)$ $\cdot(q^8 - 1)(q^6 - 1)$ $\cdot(q^2 - 1)$	$(2, q-1)$	E_7
$E_8(q)$	$q^{120}(q^{30} - 1)(q^{24} - 1)$ $\cdot(q^{20} - 1)(q^{18} - 1)$ $\cdot(q^{14} - 1)(q^{12} - 1)$ $\cdot(q^8 - 1)(q^2 - 1)$	1	E_8
$F_4(q)$	$q^{24}(q^{12} - 1)(q^8 - 1)$ $\cdot(q^6 - 1)(q^2 - 1)$	1	F_4
$G_2(q)$	$q^6(q^6 - 1)(q^2 - 1)$	1	G_2
$^2A_n(q)$	$q^{n(n+1)/2} \prod_{i=1}^{n}(q^{i+1} - (-1)^{i+1})$	$(n+1, q+1)$	$C_{[n+1/2]}$
$^2B_2(q)$, $q = 2^{2m+1}$	$q^2(q^2 + 1)(q - 1)$	1	A_1
$^2D_n(q)$, $n > 1$	$q^{n(n-1)}(q^n + 1) \prod_{i=1}^{n-1}(q^{2i} - 1)$	$(4, q^n + 1)$	C_{n-1}
$^3D_4(q)$	$q^{12}(q^8 + q^4 + 1)(q^6 - 1)$ $\cdot(q^2 - 1)$	1	C_2
$^2E_6(q)$	$q^{36}(q^{12} - 1)(q^9 + 1)(q^8 - 1)$ $\cdot(q^6 - 1)(q^5 + 1)(q^2 - 1)$	$(3, q+1)$	F_4
$^2F_4(q)$, $q = 2^{2m+1}$	$q^{12}(q^6 + 1)(q^4 - 1)$ $\cdot(q^3 + 1)(q - 1)$	1	dihedral 16
$^2G_2(q)$, $q = 3^{2m+1}$	$q^3(q^3 + 1)(q - 1)$	1	A_1

system Σ' with Dynkin diagram Σ'. If G is of type A, B, C, D, F, or G then G is called an *ordinary Chevalley group* and $\Sigma = \Sigma'$. On the other hand the groups of type ^mX are called *twisted Chevalley groups*. $^mX(q)$ is the fixed points on an ordinary Chevalley group of type $X(q^e)$ of a suitable automorphism of order m, and in this case Σ is not equal to Σ'.

It turns out also that U is a Sylow p-subgroup of G, where p is the characteristic of G (cf. 47.3).

Table 16.2 Some Dynkin diagrams

The classical groups are groups of Lie type; namely $A_n(q) = L_{n+1}(q)$, $B_n(q) = P\Omega_{2n+1}(q)$, for q odd, $C_n(q) = PSp_{2n}(q)$, $D_n(q) = P\Omega_{2n}^+(q)$, ${}^2A_n(q) = U_{n+1}(q)$, and ${}^2D_n(q) = P\Omega_{2n}^-(q)$. Also the groups ${}^2B_2(q)$ were first discovered as permutation groups by M. Suzuki and hence are called *Suzuki groups* and sometimes denoted by $Sz(q)$.

There are some isomorphisms among the groups of Lie type, and of groups of Lie type with alternating groups. Also some centerless groups of Lie type are not simple. Here's the list of such exceptions; we've already seen a number of them:

$$A_1(q) \cong B_1(q) \cong C_1(q) \cong {}^2A_1(q) \cong L_2(q),$$

$$B_2(q) \cong C_2(q) \cong PSp_4(q),$$

$$D_2(q) \cong P\Omega_4^+(q) \cong L_2(q) \times L_2(q),$$

$${}^2D_2(q) \cong P\Omega_4^-(q) \cong L_2(q^2),$$

$$D_3(q) \cong A_3(q), \quad {}^?D_3(q) \cong {}^2A_3(q).$$

In particular recall $L_2(2)$ and $L_2(3)$ are not simple, and of course neither is $L_2(q) \times L_2(q)$. Also $C_2(2) \cong S_6$, and ${}^2A_2(2) \cong PSU_3(2)$ are not simple. If

$G = G_2(2)$ or $^2F_4(2)$ then $|G : G^{(1)}| = 2$ with $G^{(1)}$ simple. $(G_2(2))^{(1)} \cong U_3(3)$. $(^2F_4(2))^{(1)}$ is called the *Tits group*. $^2B_2(2)$ is a Frobenius group of order 20, and hence solvable. $|^2G_2(3) : ^2G_2(3)^{(1)}| = 3$ with $^2G_2(3)^{(1)} \cong L_2(8)$. All other centerless groups of Lie type are simple. $L_2(4) \cong L_2(5) \cong A_5$, $L_2(9) \cong A_6$, $L_4(2) \cong A_8$, $U_4(2) \cong PSp_4(3)$, and $L_3(2) \cong L_2(7)$. This exhausts the isomorphisms among groups of Lie type and between groups of Lie type and alternating groups.

Later in this section we will explore further consequences of the properties L1–L5, using the theory of *BN*-pairs developed in chapter 14. But first let's take a look at the sporadic simple groups.

The sporadic simple groups and their orders are listed in Table 16.3. At present there is no nice class of representations available to describe the sporadic groups in a uniform manner, although some recent work on the

Table 16.3 The sporadic simple groups

Notation	Name	Order
M_{11}	Mathieu	$2^4 \cdot 3^2 \cdot 5 \cdot 11$
M_{12}		$2^6 \cdot 3^3 \cdot 5 \cdot 11$
M_{22}		$2^7 \cdot 3^2 \cdot 5 \cdot 7 \cdot 11$
M_{23}		$2^7 \cdot 3^2 \cdot 5 \cdot 7 \cdot 11 \cdot 23$
M_{24}		$2^{10} \cdot 3^3 \cdot 5 \cdot 7 \cdot 11 \cdot 23$
J_1	Janko	$2^3 \cdot 3 \cdot 5 \cdot 7 \cdot 11 \cdot 19$
$J_2 = HJ$	Hall–Janko	$2^7 \cdot 3^3 \cdot 5^2 \cdot 7$
$J_3 = HJM$	Higman–Janko–McKay	$2^7 \cdot 3^5 \cdot 5 \cdot 17 \cdot 19$
J_4	Janko	$2^{21} \cdot 3^3 \cdot 5 \cdot 7 \cdot 11^3 \cdot 23 \cdot 29 \cdot 31 \cdot 37 \cdot 43$
HS	Higman–Sims	$2^9 \cdot 3^2 \cdot 5^3 \cdot 7 \cdot 11$
Mc	McLaughlin	$2^7 \cdot 3^6 \cdot 5^3 \cdot 7 \cdot 11$
Sz	Suzuki	$2^{13} \cdot 3^7 \cdot 5^2 \cdot 7 \cdot 11 \cdot 13$
Ly = LyS	Lyons-Sims	$2^8 \cdot 3^7 \cdot 5^6 \cdot 7 \cdot 11 \cdot 31 \cdot 37 \cdot 67$
He = HHM	Held–Higman–McKay	$2^{10} \cdot 3^3 \cdot 5^2 \cdot 7^3 \cdot 17$
Ru	Rudvalis	$2^{14} \cdot 3^3 \cdot 5^3 \cdot 7 \cdot 13 \cdot 29$
O'N = O'NS	O'Nan–Sims	$2^9 \cdot 3^4 \cdot 5 \cdot 7^3 \cdot 11 \cdot 19 \cdot 31$
$Co_3 = .3$	Conway	$2^{10} \cdot 3^7 \cdot 5^3 \cdot 7 \cdot 11 \cdot 23$
$Co_2 = .2$		$2^{18} \cdot 3^6 \cdot 5^3 \cdot 7 \cdot 11 \cdot 23$
$Co_1 = .1$		$2^{21} \cdot 3^9 \cdot 5^4 \cdot 7^2 \cdot 11 \cdot 13 \cdot 23$
$M(22) = F_{22}$	Fischer	$2^{17} \cdot 3^9 \cdot 5^2 \cdot 7 \cdot 11 \cdot 13$
$M(23) = F_{23}$		$2^{18} \cdot 3^{13} \cdot 5^2 \cdot 7 \cdot 11 \cdot 13 \cdot 17 \cdot 23$
$M(24)' = F_{24}$		$2^{21} \cdot 3^{16} \cdot 5^2 \cdot 7^3 \cdot 11 \cdot 13 \cdot 17 \cdot 23 \cdot 29$
$F_3 = E$	Thompson	$2^{15} \cdot 3^{10} \cdot 5^3 \cdot 7^2 \cdot 13 \cdot 19 \cdot 31$
$F_5 = D$	Harada-Norton	$2^{14} \cdot 3^6 \cdot 5^6 \cdot 7 \cdot 11 \cdot 19$
$F_2 = B$	Baby monster	$2^{41} \cdot 3^{13} \cdot 5^6 \cdot 7^2 \cdot 11 \cdot 13 \cdot 17$ $\cdot 19 \cdot 23 \cdot 31 \cdot 47$
$F_1 = M$	Monster	$2^{46} \cdot 3^{20} \cdot 5^9 \cdot 7^6 \cdot 11^2 \cdot 13^3 \cdot 17$ $\cdot 19 \cdot 23 \cdot 29 \cdot 31 \cdot 41 \cdot 47 \cdot 59 \cdot 71$

representations of the sporadic groups on geometries may eventually lead to such a description. Instead the sporadic groups were discovered in various ways. The Mathieu groups all have multiply transitive permutation representations; specifically M_n is 3, 4, or 5-transitive of degree n. J_2, Mc, HS, Sz, Ru, $M(22)$, $M(23)$, and $M(24)$ all have rank-3 permutation representations. Hence each is a group of automorphisms of a strongly regular graph via the construction of section 16. The Conway groups act on the Leech lattice, a certain discrete integer submodule of 24-dimensional Euclidean space. The other sporadic groups were discovered through the study of centralizers of involutions (cf. section 48).

In the remainder of this section assume G satisfies the conditions L1–L5. We'll see that these conditions together with the theory in chapter 14 have a number of interesting consequences.

A subset Δ of Σ is *closed* if, for each $\alpha, \beta \in \Delta$ with $\alpha + \beta \in \Sigma$, we have $\alpha + \beta \in \Delta$. A subset of Γ of Δ is an *ideal* of Δ if, whenever $\alpha \in \Gamma, \beta \in \Delta$, and $\alpha + \beta \in \Sigma$, then $\alpha + \beta \in \Gamma$. Define $U_\Delta = \langle U_\alpha : \alpha \in \Delta \rangle$.

(47.1) Let Δ be a closed subset of Σ^+ and Γ an ideal of Δ. Then

(1) $U_\Delta = \Pi_{\alpha \in \Delta} U_\alpha$, with the product in any order, and each element in U_Δ can be written uniquely as a product $\Pi_{\alpha \in \Delta} u_\alpha, u_\alpha \in \Delta$ (for any fixed ordering of Δ).

(2) $U_\Gamma \trianglelefteq U_\Delta$.

(3) $H \leq N_G(U_\Delta)$.

Proof. H acts on U_α by L3, so (3) holds. Property L4 says $U_\Gamma \trianglelefteq U_\Delta$. Recall the definition of the height function h from section 30, and let $\alpha \in \Delta$ with $h(\alpha)$ minimal. Observe $\Delta' = \Delta - \{\alpha\}$ is an ideal of Δ. Thus $U_{\Delta'} \trianglelefteq U_\Delta$, so $U_\Delta = U_\alpha U_{\Delta'}$, and then, by induction on the order of Δ, $U = \Pi_{\beta \in \Delta} U_\beta$ with respect to some ordering of Δ. Now L3 and Exercise 16.3 complete the proof.

(47.2) Let $w \in W, n \in N$ with $Hn = w$, and $\Delta = \Sigma^+ \cap \Sigma^- w$. Then Δ is a closed subset of Σ and each element of BwB can be written uniquely in the form $bnu, b \in B, u \in U_\Delta$.

Proof. Observe Σ^+ and Σ^- are closed subsets of Σ, the image of a closed subset under an element of W is closed, and the intersection of closed subsets is closed. So Δ is closed. Similarly $\Gamma = \Sigma^+ - \Delta = \Sigma^+ \cap \Sigma^+ w$ is closed. By 47.1, $U = U_\Gamma U_\Delta$ so $BwB = BwHU_\Gamma U_\Delta = B((U_\Gamma)^{w^{-1}})wU_\Delta = BwU_\Delta$ as $(U_\Gamma)^{w^{-1}} = U_{\Gamma w^{-1}} \leq U$. Suppose $bnu = anv, a, b \in B, u, v \in U_\Delta$. Then

$a^{-1}b = (vu^{-1})^{n^{-1}} \in B \cap (U_\Delta)^{w^{-1}} \le B \cap U^{w_0}$ as $\Delta w^{-1} \subseteq \Sigma^- = \Sigma^+ w_0$, by Exercise 10.3. But by L5, $B \cap B^{w_0} = H$, so $B \cap U^{w_0} = 1$. Hence $a^{-1}b = 1$, so $a = b$ and $u = v$.

If G is a finite group of Lie type in characteristic p satisfying L5 then $U \in \mathrm{Syl}_p(B)$. Thus the next result shows $U \in \mathrm{Syl}_p(G)$.

(47.3) If G is finite and $U \in \mathrm{Syl}_p(B)$ then $U \in \mathrm{Syl}_p(G)$.

Proof. For $w \in W$ let $\Delta(w) = \Sigma^+ \cap \Sigma^- w$. Then, by 47.2,

$$|G| = |B| \sum_{w \in W} |U_{\Delta(w)}|.$$

As $U \in \mathrm{Syl}_p(B)$, $|B|_p = |U|$ and $|U_{\Delta(w)}| \equiv 0 \bmod p$ if $U_{\Delta(w)} \ne 1$. But $U_{\Delta(w)} = 1$ if and only if $\Delta(w) = \varnothing$, and, by 30.12, this happens precisely when $w = 1$. Hence $|G| \equiv |B| \bmod(p|U|)$, so $U \in \mathrm{Syl}_p(G)$.

Recall from section 43 that, for $J \subseteq \pi$, $S_J = (s_j : j \in J)$ and further that $W_J = \langle S_J \rangle$ and $P_J = \langle B, W_J \rangle$ are parabolics of W and G, respectively. Let Σ_J be the set of roots spanned by J and let $\psi_J = \Sigma^+ - \Sigma_J$. Observe ψ_J is an ideal of Σ^+ and let $V_J = U_{\psi_J}$. Similarly Σ_J^+ is a closed subset of Σ and set $U_J = U_{\Sigma_J^+}$. Finally set $L_J = \langle W_J, U_J \rangle$.

(47.4) (1) $P_J = N_G(V_J)$.
    (2) L_J is a complement to V_J in P_J.
    (3) (L_J, HU_J, N, S_J) is a Tits system satisfying L1–L4 with respect to the root system Σ_J and the simple system J.

Proof. By 30.20, Σ_J is a root system with simple system J and Weyl group W_J. As ψ_J is an ideal of Σ^+, $V_J \trianglelefteq U$ and $U = U_J V_J$ by 47.1. Then $P_J = \langle U, W_J \rangle = \langle L_J, V_J \rangle$. Finally, for $s_j \in S_J$ and $\alpha \in \psi_J$, $\alpha s_j = \alpha - 2(\alpha, j)j/(j, j) \in \psi_J$, because α has a positive projection on $\pi - J$, and hence αs_j does too. Thus $U_\alpha^{s_j} = U_{\alpha s_j} \le V_J$, so $W_J \le N_G(V_J)$. Therefore $P_J = \langle U, W_J \rangle \le N_G(V_J)$. So, by 43.7, $N_G(V_J) = P_K$ for some $K \subseteq \pi$ with $J \subseteq K$. If $k \in K - J$ then $k \in \psi_J$ and $ks_k = -k$, so $U_{-k} = U_k^{s_k} \le V_J$. But then $U_{-k} \le U^{w_0} \cap U = 1$. So $K = J$ and hence (1) holds.

Let $A = HU_J$, Claim $L_J = AW_J A$. It suffices to show $wU_J w' \subseteq AW_J A = Y$ for all $w, w' \in W_J$. Then by induction on the length of w', it suffices to show $wU_J s_j \subseteq Y$ for each $w \in W_J$ and $j \in J$. Let $\Delta = \Sigma_J - \{j\}$. By 30.7, $\Delta s_j \subseteq \Sigma_J^+$, so $U_\Delta^{s_j} = U_{\Delta s_j} \subseteq U_J$. Thus as $U_J = U_j U_\Delta$, it suffices to show $wU_J s_j \subseteq Y$. Similarly, if $l(ws_j) > l(w)$, then by 30.10, $jw^{-1} > 0$, so

$U_j^{w^{-1}} = U_{jw^{-1}} \le U_J$ and hence $wU_j s_j = U_j^{w^{-1}} w s_j \subseteq Y$. So assume $l(ws_j) < l(w)$. Then by induction on $l(w)$, $ws_j U_j s_j \subseteq Y$. Now if $AW_j A$ is a group then $U_J^{s_j} \subseteq AW_j A$, so

$$wU_J s_j = ws_j U_J^{s_j} \subseteq ws_j AW_j A = ws_j A \cup ws_j As_j A \subseteq Y.$$

Thus it suffices to show $AW_j A$ is a group.

By Exercise 10.3, $s_j^{w_0} = s_i$ for some $i \in \pi$, and w_0 is an involution. Let $X = P_i^{w_0}$. I'll show $X \cap P_j = AW_j A$, which will show $AW_j A$ is a group and hence establish the claim. By 43.7, $P_j = B \cup Bs_j B = V_j T$, where $T = AW_j A$. Similarly $X = B^{w_0} \cup B^{w_0} s_j B^{w_0} = B^{w_0} \cup s_j B^{w_0 s_j} B^{w_0}$, and, as $s_j \in X$, $s_j X = X$, so $X = s_j B^{w_0} \cup B^{w_0 s_j} B^{w_0}$. Let $\Omega = \Sigma^- - \{-j\}$. Then $U^{w_0} = U_{-j} U_\Omega$ so $U^{w_0 s_j} = U_j U_\Omega$ by 30.7, and hence $B^{w_0 s_j} B^{w_0} = U_j B^{w_0}$. Next $B \cap s_j B^{w_0} = (Bw_0 \cap s_j w_0 B)w_0$ and, as $w_0 \ne s_j w_0$, $Bw_0 B \cap Bs_j w_0 B = \emptyset$ by 43.2. So $B \cap s_j B^{w_0} = \emptyset$. Hence $B \cap X = B \cap B^{w_0 s_j} B^{w_0} = B \cap U_j B^{w_0} = U_j (B \cap B^{w_0}) = U_j H = A$. As s_j, $A \subseteq X$, $T \subseteq X$, so $X \cap P_j = TV_j \cap X = T(V_j \cap X) \subseteq T(B \cap X) = TA = T$. Thus the claim is at last established.

It's now easy to see that (L_J, A, N, S_J) is a Tits system. The only one of the BN-pair axioms which is not evident is BN3. But, for $s \in S_J$, $w \in W_J$,

$$sAw \subseteq (BwB \cup BswB) \cap AW_J A$$
$$= AwA \cup AswA$$

with the equality following from 47.2 and 43.2. Similarly it is clear that L2-L4 inherit to L_J. So (3) holds.

We saw $V_J \trianglelefteq P_J = \langle L_J, V_J \rangle$, so $P_J = L_J V_J$. By 43.2 and 47.1, $V_J \cap L_J = V_J \cap A = 1$, establishing (2).

The subgroup L_J is called a *Levi factor* of the parabolic P_J, and V_J is the *unipotent radical* of P_J.

(47.5) (1) $HU_J \cap U_J^{x_0} = 1$, where x_0 is the element of W_J of maximal length.
(2) $C_H(U_J) = \ker_{HU_J}(L_J)$.
(3) If G is finite, $Z(G) = 1$, and $U \in \mathrm{Syl}_p(B)$ for some prime p, then $V_J = O_p(P_J) = F^*(P_J)$ for each $J \subset \pi$.

Proof. Let $A = HU_J$ and $L = L_J$. By Exercise 10.3, $(\Sigma_J^+)x_0 = \Sigma_J^-$, so $U_J^{x_0} \cap B \le U^{w_0} \cap B = 1$. Thus (1) holds. Let $D = \ker_A(L)$. Then $D \le A \cap A^{x_0} = H$ by (1). So $[D, U_J] \le H \cap U_J = 1$, and hence $D \le C_H(U_J) = E$. I'll prove $E \le D$ by induction on the Lie rank l of L. Without loss of generality, $J = I$. (I'll only use (1), not L5.) As $U \trianglelefteq B$, $E = C_H(U) \trianglelefteq H$, so $E \trianglelefteq B = UH$. If $l > 1$ then, by induction on l, $E \le \ker_{HU_i}(L_i)$ for each $i \in \pi$.

Thus $E \leq C(U_i^x)$ for each $x \in L_i$, so $L_i = \langle H, U_i^{L_i} \rangle \leq N_G(E)$. Therefore $G = \langle L_i, B : i \in \pi \rangle \leq N_G(E)$, so $E \leq D$. So take $l = 1$, and consider the action of G on B^G by conjugation. By BN3, $G = B \cup BwB$, so B is transitive on $B^G - \{B\}$. By 47.2, $BwB = BwU$, so $B^G - \{B\} = B^{wU}$. Hence, as $E \leq B^w$ and $U \leq C(E)$, $E \leq \bigcap_{g \in G} B^g = D$. So (2) is established.

Let $X = C_G(V_J)$. By 47.4.1, $X \leq P_J$. Let $Y = BX$ and for $1 \leq m \in Z$ define $\Omega_m = \{\alpha \in \psi_J : h(\alpha) \geq m\}$. Observe Ω_m is an ideal in Σ^+, so $V_m = V_{\Omega_m} \trianglelefteq B$. Indeed by L4, $U_\alpha V_{m+1} \trianglelefteq B$ for each $\alpha \in \Omega_m$. So also $U_\alpha V_{m+1} \trianglelefteq Y$. Now for $y \in Y \cap W_J$, $U_\alpha^y = U_{\alpha y}$ so $U_{\alpha y} V_{m+1} = U_\alpha V_{m+1}$ and hence by 47.1.1, $\alpha y = \alpha$. Therefore y fixes each member of ψ_J. Now if $y = s_j$ for some $j \in J$ then $\psi_J \subseteq C_{\langle \Sigma \rangle}(y)$. Hence, by Exercise 10.4, y centralizes $\langle \Sigma \rangle$, a contradiction. Therefore $S_J \cap Y$ is empty. However, by 43.7, $Y = P_K$ for some $K \subseteq J$ so as $S_J \cap Y = 1$, $K = \emptyset$ and $Y = B$.

So $X \leq B$. Next, as L_J is a complement to V_J in P_J, $X V_J = V_J (X V_J \cap L_J)$ and, as $X V_J \leq B$, $X V_J \cap L_J \leq \ker_A(L_J) = C_H(U_J)$. Assume the hypothesis of (3). Then H is a p'-group and $H_0 = X V_J \cap L_J$ induces inner automorphisms on the p-group V_J, so $[H_0, V_J] = 1$. Thus $H_0 = C_H(U_J V_J) = C_H(U) = \ker_B(G)$. But, as $G = \langle U^G \rangle$, $\ker_B(G) = Z(G) = 1$.

I've shown $C_G(V_J) \leq V_J$ under the hypothesis of (3). Thus to complete the proof of (3) it remains only to observe that, by (2), $V_J = O_p(P_J)$.

(47.6) $U^G = \bigcup_{w \in W} (U^w)^U$.

Proof. By 43.7, $G = \bigcap_{w \in W} BwU$, so the remark holds.

(47.7) If $U \leq X = \langle U^G \cap X \rangle \leq G$ then $H \leq N_G(X)$ and $HX = P_J$ for some $J \subseteq \pi$. In particular $(P_{\pi - \{i\}} : i \in \pi)$ is the set of maximal subgroups of G containing U.

Proof. By 47.6, $X = \langle U^d : d \in D \rangle$ for some $D \subseteq W$. So, as H normalizes U^w for each $w \in W$, H normalizes X. So $HX \leq G$ and, as $B = HU \leq HX$, $HX = P_J$ for some $J \subseteq \pi$, by 43.7.1.

Let M be a maximal subgroup of G containing U and $X = \langle U^G \cap M \rangle$. Then $X \trianglelefteq M$, so by maximality of M either $M = N_G(X)$ or $X \trianglelefteq G$. In the latter case $G = \langle U, U^{w_0} \rangle = X$, a contradiction. So $HX \leq N_G(X) = M$ and hence $M = P_J$ for some $J \subseteq \pi$. By maximality of M, $J = \pi - \{i\}$ for some $i \in \pi$.

If G is a finite group of Lie type in characteristic p then, as a consequence of a theorem of Borel and Tits [BT], each p-local subgroup of G is contained in a maximal parabolic. This result can be derived using L1–L5 together with three extra properties of finite groups of Lie type listed as hypotheses in the next lemma:

(47.8) Assume the following hold:

(a) G is a finite and $U \in \mathrm{Syl}_p(B)$.
(b) Let Γ be the set of nontrivial p-subgroups R of G with $R = O_p(N_G(R))$.
   Assume for each $R \in \Gamma$ that $R = P \cap Q$ for some $P, Q \in \mathrm{Syl}_p(N_G(R))$.
(c) $U = \langle U_i : i \in \pi \rangle$.

Then the following hold:

(1) $(N_G(R): R \in \Gamma)$ is the set of proper parabolic subgroups of G.
(2) If Q is a nontrivial p-subgroup of G then there exists a proper parabolic M of G with $N_G(Q) \leq M$ and $Q \leq O_p(M)$.
(3) $F^*(X) = O_p(X)$ for each p-local X of G, if $Z(G) = 1$.

Proof. Let $R \in \Gamma$ and $M = N_G(R)$. By hypothesis there are $P, Q \in \mathrm{Syl}_p(M)$ with $R = P \cap Q$. By 47.3, $U \in \mathrm{Syl}_p(G)$, so without loss $P \leq U$. Similarly, by 47.6, $Q \leq U^{wu}$ for some $w \in W$, $u \in U$, and replacing R by $R^{u^{-1}}$ we may assume $Q \leq U^w$. Observe $R = U \cap U^w$. For if not, $R < U \cap U^w \cap M \leq P \cap Q = R$, a contradiction.

Let $\Delta = \Sigma^+ \cap \Sigma^+ w$ and $\Omega = \Sigma^- \cap \Sigma^+ w$. Then $U^w = U_\Delta U_\Omega$ with $U_\Delta \leq R$, so $R = U_\Delta(U \cap U_\Omega) \subseteq U_\Delta(U \cap U^{w_0}) = U_\Delta$. That is $R = U_\Delta$. I'll show $U_i^{v_i} \leq M$ for each $i \in \pi$ and some $v_i \in U$. Then, by (c) and Exercise 8.12, $U \leq M$. Also $H \leq N(U_\Delta) \leq M$, so M is a parabolic by 43.7, and (1) holds.

Let $i \in \pi$; it remains to show $U_i \leq M$. If not $U_i \not\leq R$, so $i \notin \Delta$. Hence $-i \in (\Sigma^+)w$ by Exercise 10.5, so $U_{-i} \leq U^w$. By 47.5, $U \cap U^{s_i} = U_\psi \trianglelefteq \langle U, U^{s_i} \rangle$, where $\psi = \Sigma^+ - \{i\}$. So $U_{-i} \leq U^{s_i} \leq N_G(U \cap U^{s_i})$ and of course $U_{-i} \leq U^w \leq N_G(U^w)$, so U_{-i} acts on $U \cap U^w \cap U^{s_i} = U_\psi \cap U_\Delta = U_\Delta = R$, as $\Delta \subseteq \psi$. That is $U_{-i} < M$.

Observe next that, as $P \in \mathrm{Syl}_p(M)$, $P = U \cap M$. If $P \leq U_\psi$ then even $P = U_\psi \cap M$ so, as U_{-i} acts on U_ψ, U_{-i} acts on P. But then, as $P \in \mathrm{Syl}_p(M)$, $U_{-i} \leq P$, contradicting $U \cap U^{w_0} = 1$. So $P \not\leq U_\psi$. Consider the parabolic P_i. By 43.7, B^{s_i} is a maximal subgroup of P_i, while, as $P \not\leq U_\psi$, $M \cap P_i \not\leq B^{s_i}$. Hence, as $B^{s_i} = HU_{-i}U_\psi \leq (M \cap P_i)U_\psi$, it follows that $P_i = (M \cap P_i)U_\psi$. Therefore $s_i v \in M$ for some $v \in U$, so $U_i^v = (U_{-i})^{s_i v} \leq P$.

This completes the proof of (1). Arguing by induction on the order of Q, there exists $R \in \Gamma$ with $Q \leq R$ and $N_G(Q) \leq N_G(R)$, so (1) implies (2). Finally (2), 47.5.3, and 31.16 imply (3).

48 An outline of the Classification Theorem

Let $\mathcal{K}$ be the list of finite simple groups appearing in section 47. Section 48 provides a brief outline of the Classification Theorem, which asserts:

Classification Theorem. Every finite simple group is isomorphic to a member of $\mathcal{K}$.

The usual procedure for classifying a collection of objects is to associate to each member of the collection some family of invariants, prove that each object is uniquely determined by its invariants, and determine which sets of invariants actually correspond to objects. The invariants used to classify the finite simple groups are certain local subgroups of the group, usually the normalizers of suitable subgroups of prime order, particularly centralizers of involutions.

A rationale for this approach can be obtained from the Odd Order Theorem of Feit and Thompson and the Brauer–Fowler Theorem (Theorem 45.5).

Odd Order Theorem. (Feit–Thompson [FT]) Groups of odd order are solvable.

The Odd Order Theorem says that every nonabelians simple group G is of even order and hence possesses an involution t. The Brauer–Fowler Theorem says there is only a finite number of finite simple groups G_0 possessing an involution t_0 with $C_{G_0}(t_0) \cong C_G(t)$. In practice, with a small number of exceptions, G is the unique simple group with such a centralizer. Even in the exceptional cases at most three simple groups possess the same centralizer (e.g. $L_5(2)$, M_{24}, and He all possess an involution with centralizer $L_3(2)/D_8^3$). Exercise 16.6 illustrates how the isomorphism type of a simple group can be recovered from the isomorphism type of the centralizers of involutions.

So centralizers of involutions provide a set of invariants upon which to base the Classification Theorem. For various reasons it turns out to be better to enlarge this set of invariants to include suitable normalizers of subgroups of odd prime order.

To be more precise, define a *standard subgroup* of a group G for the prime p to be a subgroup H of G such that $H = C_G(x)$ for some element x of order p, H has a unique component L, and $C_H(L)$ has cyclic Sylow p-groups. Standard subgroups provide the principal set of invariants for the classification of the finite simple groups.

However certain small groups either possess no standard subgroup or cannot be effectively characterized in terms of this invariant. Such groups are characterized by other methods. Hence we have our first partition of the simple groups for purposes of the classification: the partition into generic groups and small groups. I'll define the appropriate measure of size in a moment; but before that another partition.

When possible, we'd like to characterize a simple group G in terms of a standard subgroup for the prime 2. But often G possesses no such subgroup. G is said to be of *characteristic p-type* if $F^*(H) = O_p(H)$ for each p-local subgroup H of G. For example, if G is of Lie type and characteristic p, we saw in 47.8 that G is of characteristic p-type. In particular if G is of characteristic 2-type then it possesses no standard subgroup for the prime 2. Our second partition of the simple groups is the partition into groups of even and odd characteristic, where by definition G is of even characteristic if G is of characteristic 2-type and G is of odd characteristic otherwise.

Define the *2-local p-rank* of G to be

$$m_{2,p}(G) = \max\{m_p(H): H \text{ is a 2-local of } G\},$$

and define

$$e(G) = \max\{m_{2,p}(G): p \text{ odd}\}.$$

In a group of Lie type and characteristic 2, $e(G)$ is a good approximation of the Lie rank.

A group G of even characteristic is small if $e(G) \leq 2$ and generic otherwise. A group of odd characteristic is small if G is 2-disconnected for the prime 2 and generic otherwise. Thus we have a four part partition of the finite simple groups for purposes of the classification.

The proof of the Classification Theorem proceeds by induction on the order of the simple group to be classified. Thus we consider a minimal counter example G to the Classification Theorem; that is G is a finite simple group of minimal order subject to $G \notin \mathcal{K}$. Define a finite group H to be a $\mathcal{K}$-*group* if every simple section of H is in $\mathcal{K}$ (a *section* of H is a factor group A/B, where $B \trianglelefteq A \leq H$). Observe that every proper subgroup of our minimal counterexample is a $\mathcal{K}$-group. This property will be used repeatedly.

If G is a finite simple group with $m_2(G) \leq 2$, a moment's thought shows G to be 2-disconnected for the prime 2. If $m_2(G) > 2$ then, by 46.7, neglecting isolated points in the graph $\mathcal{D}(\mathcal{E}_2^2(G))$, G is 2-disconnected for the prime 2 precisely when $G \neq \Gamma_{P,2}^0(G)$, where $P \in \mathrm{Syl}_2(G)$. As a matter of fact, if G is simple and $m_2(G) > 2$, it can be shown that G is 2-disconnected for the prime 2 if and only if G has a proper 2-generated 2-core. Thus to classify the small simple groups of odd characteristic it suffices to prove the following two results:

Theorem 48.1. If G is a nonabelian finite simple group with $m_2(G) \leq 2$ then either:

(1) a Sylow 2-group of G is dihedral, semidihedral, or Z_{2^n} wr Z_2, and $G \cong L_2(q), L_3(q), U_3(q), q$ odd, or M_{11}, or
(2) $G \cong U_3(4)$.

Theorem 48.2. Let G be a nonabelian finite simple group with $m_2(G) > 2$ and assume G has a proper 2-generated 2-core. Then either G is a group of Lie type of characteristic 2 and Lie rank 1 (i.e. $L_2(2^n)$, $U_3(2^n)$, or $Sz(2^n)$) or $G \cong J_1$.

In brief, Theorem 48.1 is proved by using local theory to restrict the subgroup structure of our minimal counterexample G. At this point there is either enough information available to conclude $G \in \mathcal{K}$ or G possesses many *TI*-sets. *TI*-sets can be exploited using character theory (along the lines of 35.22 or Frobenius' Theorem) to derive a contradiction or to restrict the structure of G further and show $G \in \mathcal{K}$.

It's interesting to note that character theory plays an important role in the proof of Theorem 48.1, but is rarely used in the remainder of the classification. In essence character theory is used to deal with small groups such as the groups of Lie rank 1 and some groups of Lie rank 2. The generic groups of higher dimension can be identified using geometric or quasigeometric techniques.

The local theory used in the proof of Theorem 48.1 is of two sorts. Transfer and fusion techniques are used to pin down the structure of a Sylow 2-group of G and the fusion of 2-elements; see for example Exercise 13.2. One such tool, used sparingly in the proof of Theorem 48.1 but frequently in later stages of the classification, is Glauberman's Z^*-Theorem:

Glauberman Z^*-Theorem. [Gl 3]. Let G be a finite group and t an involution in G such that t is weakly closed in $C_G(t)$. Then $t^* \in Z(G^*)$, where $G^* = G/O_{2'}(G)$.

Recall a subset S of G is weakly closed in a subgroup H of G (with respect to G) if $S^G \cap H = \{S\}$. The proof of the Z^*-Theorem uses modular character theory and is beyond the scope of this book.

The second kind of local theory used in the proof of Theorem 48.1 involves an analysis of subgroups of G of odd order using signalizer functor theory or some variant of that theory. Notice the Odd Order Theorem is one step in the proof of Theorem 48.1, since groups of odd order are of 2-rank 0. When Feit and Thompson proved the Odd Order Theorem signalizer functor theory did not exist; instead they generated their own techniques, which in time evolved into signalizer functor theory. I'll illustrate the signalizer functor approach in the generic case a little later.

The techniques used to establish Theorem 48.2 are rather different. The proof depends heavily on the fact that each pair of involutions generate a dihedral group. This observation makes possible a number of combinatorial and group theoretic arguments of the flavor of sections 45 and 46. Exercises 16.5 and

16.6 illustrate some of these arguments. Indeed Exercise 16.6 establishes a very special case of Theorem 48.2. It would be nice to have the analogue of Theorem 48.2 for odd primes, but as nothing in particular can be said about groups generated by a pair of elements of odd prime order, a different approach is required.

Let's turn next to the generic groups of odd characteristic. Observe that by Exercise 16.1 a generic group G of odd characteristic possesses an involution t such that $O_{2',E}(C_G(t)) \neq O_{2'}(C_G(t))$. I encourage you to retrace the steps in the proof of this exercise; the proof provides a good illustration of signalizer functor theory. Similar arguments reappear in the proof of Theorem 48.1 and in the analysis of the generic groups of even characteristic.

This brings us to the following fundamental property of finite groups:

B_p-**Property.** Let p be a prime, G a finite group with $O_{p'}(G) = 1$, and x an element of order p in G. Then $O_{p',E}(C_G(x)) = O_{p'}(C_G(x))E(C_G(x))$.

The verification of the B_p-Property is one of the most critical and difficult steps in the classification. Only the B_2-Property is established directly; for odd p, the B_p-Property follows only as a corollary to the Classification by inspection of the groups in $\mathcal{K}$. (Notice that by Exercise 16.4, to verify the B_p-Property it suffices to consider the case where $F^*(G)$ is simple.)

Observe next that the B_2-Property follows from the following result:

Unbalanced Group Theorem. Let G be a finite group with $F^*(G)$ simple which is unbalanced for the prime 2. Then $F^*(G)$ is a group of Lie type and odd characteristic, A_{2n+1}, $L_3(4)$, or He.

For to verify the B_2-Property it suffices to assume $F^*(G)$ is simple by Exercise 16.5. If $O_{2'}(C_G(t)) = 1$ for each involution t in G, then the B_2-Property is trivially satisfied, so we may assume G is unbalanced. Hence, by the Unbalanced Group Theorem, $F^*(G) \in \mathcal{K}$. But, by inspection of the local structure of Aut(L) for $L \in \mathcal{K}$, if $F^*(G) \in \mathcal{K}$ then G satisfies the B_2-Property.

Suppose for the moment that the Unbalanced Group Theorem, and hence also the B_2-conjecture, is established. The B_2-conjecture makes possible manipulations which prove:

Component Theorem. (Aschbacher [As 1]) Let G be a finite group with $F^*(G)$ simple satisfying the B_2-Property and possessing an involution t such that $O_{2',E}(C_G(t)) \neq O_{2'}(C_G(t))$. The G possesses a standard subgroup for the prime 2.

Actually the definition of a standard subgroup has to be relaxed a little to make the Component Theorem correct as stated above, but the spirit is accurate. Notice that at this stage we have associated to each generic group of odd characteristic the desired set of invariants: its collection of standard subgroups for the prime 2. It remains to characterize simple groups via these invariants. Thus we must consider:

Standard form problem for (L, r): Determine all finite groups G possessing a standard subgroup H for the prime r with $E(H) \cong L$.

If G is our minimal counterexample and H a standard subgroup of G, then H is a $\mathcal{K}$-group, so $E(H)/Z(E(H)) \in \mathcal{K}$. As we know the universal covering group of each member of $\mathcal{K}$, we know $E(H)$, and to prove $G \in \mathcal{K}$ it remains to treat the standard form problem for each perfect central extension of each member of $\mathcal{K}$.

How does one retrieve the structure of G from that of the standard subgroup H? Let $L = E(H)$. Then $H/C_G(H) \leq \text{Aut}(L)$, so we have good control over $H/C_G(H)$, while as $C_G(H)$ has cyclic Sylow p-groups we have good control of $C_G(H)$. Then analysis of fusion of p-elements of H gives us a conjugate H^g of H such that the intersection $H \cap H^g$ is large. Results similar to Theorem 48.2 allow us to conclude $G = \langle H, H^g \rangle$. This information can be used to define a representation of G on a subgroup geometry along the lines of the construction in section 3, or to obtain a presentation of G. For example we might represent G on a building or show it possesses a BN-pair. If so, the machinery in chapter 14 becomes available to identify G as a member of $\mathcal{K}$.

It remains to prove the Unbalanced Group Theorem. By the Odd Order Theorem $O_{2'}(C_G(t))$ is solvable for each involution t in G, so by 46.9 and Exercise 13.3, if G is not balanced for the prime 2, also $C_G(t)^* = C_G(t)/O_{2'}(C_G(t))$ is not balanced for the prime 2 for some involution t in G. Hence by 31.19 there is a component L^* of $C(t)^*$ and $X^* \leq N(L^*)$ such that $\text{Aut}_{X^*}(L^*)$ is not balanced for the prime 2. By induction on the order of G, $L^*/Z(L^*)$ is one of the groups listed in the conclusion of the Unbalanced Group Theorem. This piece of information is critical; together with some deep local theory it can be used to produce a standard subgroup for the prime 2, reducing us to a previous case.

The groups of odd characteristic have been treated; let us turn next to the groups of even characteristic. If G is a generic group of even characteristic we seek to produce a standard subgroup for some odd prime p. More precisely p is a prime in the set $\sigma(G)$ where

$$\sigma(G) = \{p : p \text{ odd}, m_{2,p}(G) \geq 3\}.$$

The actual definition of $\sigma(G)$ is a little more complicated, but again the definition above is in the right spirit. Using signalizer functor theory and other local group theoretic techniques one shows:

Theorem 48.3. (Trichotomy Theorem [As 2], [GL]) let G be a minimal counterexample to the Classification Theorem and assume G is generic of even characteristic. Then one of the following holds:

(1) G possesses a standard subgroup for some $p \in \sigma(G)$.
(2) There exists an involution t in G such that $F^*(C_G(t))$ is a 2-group of symplectic type.
(3) G is in the uniqueness case.

G is in the uniqueness case if, for each $p \in \sigma(G)$, G possesses a strongly p-embedded maximal 2-local subgroup. Recall that a p-group is of symplectic type if it possesses no noncyclic characteristic abelian subgroups, and that groups of symplectic type are described completely in chapter 8.

I've already discussed briefly how one deals with standard subgroups. In case 2 of Theorem 48.3, the structure of $F^*(C(t)) = Q$ is determined from chapter 8, as is Aut(Q) (cf. Exercise 8.5). With this information and some work one can recover $C(t)$ and then proceed as though $C(t)$ were a standard subgroup.

The arguments used to deal with the uniqueness case and the small groups of even characteristic are quite similar. They involve factoring 2-locals as the product of normalizers of certain subgroups of a Sylow 2-group of the local. The Thompson Factorization, discussed in section 32, is the prototype of such factorizations. The proof of the Thompson Normal p-Complement Theorem (39.5) and of the Solvable Signalizer Functor Theorem give some indication of how such factorizations can be used.

Remarks. See Gorenstein's series of books [Gor 2, Gor 3] for a more detailed outline of the proof of the Classification Theorem and a more complete discussion of the sporadic simple groups. In particular [Gor 2, Gor 3] contain explicit references to the articles which, taken together, supply a proof of Theorems 48.1 and 48.2 and the Unbalanced Group Theorem.

Carter [Ca] and Steinberg [St] are good places to learn about groups of Lie type.

Exercises for chapter 16

1. Let G be a finite group with $P \in \mathrm{Syl}_2(G)$ and $G = O^2(G)$. Assume $G = \Gamma_{P,2}^0(G)$, $m_2(G) > 2$, and $O_{2',E}(C_G(t)) = O_2(C_G(t))$ for each involution t in G. Assume $O_{2'}(G) = 1$. Prove G is of characteristic 2-type; that is $F^*(H) = O_2(H)$ for each 2-local subgroup H of G.

2. Let $G = A_n$ be the alternating group on $n \geq 5$ letters and $A = \text{Aut}(G)$.
   Prove:
   (1) $A = S_n$ if $n \neq 6$.
   (2) $A/G \cong E_4$ if $n = 6$.
3. Let $(G_i : 1 \leq i \leq n)$ be a family of subgroups of a group G such that
   (1) Each element of G can be written uniquely as a product

$$g_1 \ldots g_n, \text{ with } g_i \in G_i, \text{ and}$$

   (2) $G_m G_{m+1} \ldots G_n \trianglelefteq G$ for each m, $1 \leq m \leq n$.
   Then, for each permutation σ of $\{1, \ldots, n\}$, each element of G can be written
   uniquely as a product $g_{1\sigma} \ldots g_{n\sigma}$, $g_i \in G_i$.
4. Let G be a finite group and p a prime. Assume $O_{p'}(G) = 1$. Assume
   $\text{Aut}_H(L)$ satisfies the B_p-Property for each component L of G and each
   subgroup H of G with $L \leq H \leq N_G(L)$. Finally assume each component
   of G satisfies the Schreier conjecture. Prove G satisfies the B_p-Property.
5. Let H be strongly 2-embedded in G and represent G by right multiplication
   on $X = G/H$. Let I be the set of involutions of G, $t \in I \cap H$, $u \in I - H$,
   $D = H \cap H^u$, $m = |I \cap H|$, and $J = \{d \in D : d^u = d^{-1}\}$. Prove:
   (1) G is transitive on I.
   (2) H is transitive on $I \cap H$.
   (3) D is of odd order.
   (4) $uJ = uD \cap I$.
   (5) $C_G(j)$ is of odd order for each $j \in J^{\#}$.
   (6) Distinct involutions in uD are in distinct cosets of $C_G(t)$.
   (7) For each $x, y \in X$ with $x \neq y$ there are exactly m involutions in G with
       cycle (x, y).
   (8) D is transitive on $I \cap H$.
   (Hint: For (1) observe each member of I is conjugate to some $s \in I - H$;
   then use 45.2 to prove s is conjugate to t in $\langle s, t \rangle$. Use 46.4 in (2) and (3)
   and 45.2 in (4). In (5) set $Y = C_G(j)\langle u \rangle$ and observe that if $t \in C_G(j)$ then
   $H \cap Y$ is strongly embedded in Y; now appeal to (1) for a contradiction.
   Derive (6) from (5). To prove (7), let Ω be the set of triples (i, x, y) with
   $i \in I$ and (x, y) a cycle in i on G/H. Count $|\Omega|$ in two ways, using (6) to
   conclude there are at most m involutions with cycle (x, y).)
6. Let G be a finite group with noncyclic Sylow 2-group T. Assume $C_G(t)$
   is an elementary abelian 2-group for each involution t of G and T is not
   normal in G. Let $q = |T|$, $H = N_G(T)$, $X = G/H$, u an involution in
   $G - H$, and $D = H \cap H^u$. Let $F = \text{GF}(q)$ and $Y = F \cup \{\infty\}$. Regard
   $L_2(q) = G^*$ as the group of all permutations of Y of the form

$$\phi(a, b, c, d): y \mapsto (ay + b)/(cy + d) \quad a, b, c, d \in F, ad - bc \neq 0$$

as in Exercise 4.10. Let

$$H^* = \{\phi(a, b, 1, 1): a \in F^\#, b \in F\}.$$

Prove:

(1) T is elementary abelian.

(2) H is strongly 2-embedded in G.

(3) D is inverted by u, $|H| = q(q - 1)$, and D is a complement to T in H.

(4) $N_G(D) = D\langle u \rangle$ and $\{H, Hu\}$ is the fixed point set of D on X.

(5) G is 3-transitive on X with only the identity fixing 3 or more points. T is regular on $X - \{H\}$.

(6) There is an isomorphism $\pi: H \to H^*$ and a bijection $\alpha: X \to Y$ such that $H\alpha = \infty$, $(Hu)\alpha = 0$, and $(x\alpha)h\pi = (xh)\alpha$ for all $x \in X$ and $h \in H$.

(7) There exists $v \in uD$ with $(a\alpha^{-1})v = a^{-1}\alpha^{-1}$ for all $a \in F^\#$.

(8) π extends to isomorphism of G and $G^* = L_2(q)$ with $v\pi = \phi(0, 1, 1, 0)$.

(Hints: Use Exercise 16.5 in the proof of (3) and (4). In (4) show $N_G(D) = E\langle u \rangle$ where $E = O_{2'}(N_G(D))$ and $E\langle u \rangle/D$ is regular on the fixed point set Δ of D on X. Let Γ be the set of triples (i, x, y) such that $i \in uE$ and (x, y) is a cycle of i on Δ. Count Γ in two ways to get $|\Delta| = 2$. Use (4) and Phillip Hall's Theorem, 18.5, to get T regular on $X - \{H\}$. Then complete (5) using 15.11. For (6) use (5) and the observation that, as D is regular on $T^\#$, $D = \text{End}_{\text{GF}(2)D}(T)^\#$ and $\text{End}_{\text{GF}(2)D}(T) = F$.)

Appendix

Solutions to selected exercises

Chapter 3, Exercise 5. First, as α is of order n and $\langle\alpha\rangle$ is transitive on $\Delta = \{G_i: 1 \le i \le n\}$ of order n, $\langle\alpha\rangle$ is regular on Δ, so renumbering if necessary we may take $G_i\alpha = G_{i+1}$, with the subscripts read modulo n. Thus α^{i-1}: $G_1 \to G_i$ is an isomorphism.

Next, by definition of "central product" in 11.1,

$$G = \left\{\prod_{i=1}^{n} y_i: y_i \in G_i\right\},$$

with $y_i y_j = y_j y_i$ for $i \ne j$, so $\prod_i y_i$ is independent of the order of the factors. Now as $\alpha^{i-1}: G_1 \to G_i$ is an isomorphism,

(*) $$G = \left\{\prod_{i=1}^{n} x_i\alpha^{i-1}: x_i \in G_1\right\}.$$

Define $\pi: G_1 \to G$ by $x\pi = \prod_i x\alpha^{i-1}$. Then

$$x\pi\alpha = \left(\prod_i x\alpha^{i-1}\right)\alpha = \prod_i x\alpha^i = x\pi,$$

as the product is independent of the order of its factors and indices are read modulo n. Thus $G_1\pi \le C_G(\alpha)$. Similarly

$$(xy)\pi = \prod_i (xy)\alpha^{i-1} = \prod_i x\alpha^{i-1}y\alpha^{i-1} = \prod_i x\alpha^{i-1} \cdot \prod_i y\alpha^{i-1} = x\pi \cdot y\pi,$$

so π is a homomorphism. Further, if $x \in \ker(\pi)$ then $1 = \prod_i x\alpha^{i-1}$, so

$$x = \left(\prod_{i>1} x\alpha^{i-1}\right)^{-1} \in G_1 \cap G_2 G_3 \cdots G_n \le Z(G_1),$$

so $\ker(\pi) \le Z(G_1)$.

Claim $C_G(\alpha) = G_1\pi Z$, where $Z = C_{Z(G)}(\alpha)$. We just saw $G_1\pi \le C_G(\alpha)$, so $G_1\pi Z \le C_G(\alpha)$. Suppose $g \in C_G(\alpha)$. By (*), $g = \prod_i x_i\alpha^{i-1}$ for some $x_i \in G_1$. Then

$$x_1 u_{-1} = \prod_i x_i\alpha^{i-1} = g = g\alpha^{j+1} = \prod_i x_i\alpha^{i+j} = x_{-j}u_j$$

for each j, where $u_j = \prod_{i \neq -j} x_i \alpha^{i+j} \in G_2 G_3 \cdots G_n$, so

$$x_1 x_{-j}^{-1} = u_j u_{-1}^{-1} \in G_1 \cap G_2 G_3 \cdots G_n \leq Z(G_1)$$

and hence $x_j = x_1 z_j$ for some $z_j \in Z(G_1)$. Therefore $g = x_1 \pi z$, where $z = \prod_j z_j \alpha^{j-1} \in Z(G)$. As g, $x_1 \pi \in C_G(\alpha)$, also $z = g(x_1 \pi)^{-1} \in C_{Z(G)}(\alpha)$, completing the proof that $C_G(\alpha) = G_1 \pi C_{Z(G)}(\alpha)$.

Let $K = G_1 \pi$ and $Z = C_{Z(G)}(\alpha)$. As $\pi : G_1 \to K$ is a surjective homomorphism and G_1 is perfect, K is perfect and as Z is abelian $Z^{(1)} = 1$. Thus

$$C_G(\alpha)^{(1)} = (KZ)^{(1)} = K^{(1)} Z^{(1)} = K.$$

Thus it remains to show that $N_{\text{Aut}(G)}(G_1) \cap C_{\text{Aut}(G)}(K) \leq C_{\text{Aut}(G)}(G_1)$. Let $L = G_1$ and $M = C_G(L)$. As $G_2 \cdots G_n \leq M$, $G = LM$. Let $\beta \in N_{\text{Aut}(G)}(L) \cap C_{\text{Aut}(G)}(K)$. Then β acts on M. For $x \in L$, $y = x^{-1} \cdot x\pi \in M$, so as β acts on M, $y^{-1} \cdot y\beta \in M$. Now

$$y^{-1} \cdot y\beta = x^{-1} \pi \cdot x \cdot x^{-1} \beta \cdot x\pi\beta = (x \cdot x^{-1} \beta)^{x\pi} \in L,$$

as β centralizes $x\pi \in K$ and $L \trianglelefteq G$. Thus $(x \cdot x^{-1} \beta)^{x\pi} = y^{-1} \cdot y\beta \in L \cap M = Z(G)$, so $x \cdot x^{-1} \beta \in Z(L)$, so $[\beta, L, L] = 1$. Hence $[\beta, L] = 1$ by 8.9, so indeed $\beta \in C_{\text{Aut}(G)}(G_1)$.

Chapter 4, Exercise 7. (1) Let $f(x, y) = \sum_{i,j} a_{i,j} x^i y^j \in V = f[x, y]$ and $g \in G$. By definition

$$f(g\pi) = \sum_{i,j} a_{i,j} (xg)^i (yg)^j,$$

so as $xg = ax + by$ and $yg = cx + dy$ for some $a, b, c, d \in F$, we have $f(g\pi) \in V$. It is easy to check that $g\pi$ preserves addition and scalar multiplication, so $g\pi \in \text{End}_F(V)$. For $h \in G$,

$$f(gh)\pi = f(x(gh), y(gh)) = f((xg)h, (yg)h) = f(xg, yg)(h\pi) = (f(g\pi))(h\pi),$$

so $(gh)\pi = g\pi h\pi$, and hence $\pi : G \to \text{GL}(V)$ is a representation.

(2) Observe V_n has basis $B_n = \{x^i y^{n-i} : 0 \leq i \leq n\}$. Further, $(x^i y^{n-i})(g\pi) = (ax + by)^i (cx + dy)^{n-i}$ with

$$(ax + by)^k = \sum_j \binom{k}{j} a^j x^j (by)^{k-j} \in V_k,$$

so $(x^i y^{n-i})(g\pi) \in V_n$. Thus G acts on V_n of dimension $n + 1$.

(3) For $i \in \mathbf{Z}$, let $\bar{i}$ be the residue of i modulo p; that is $0 \leq \bar{i} < p$ and $i \equiv \bar{i} \bmod p$. Assume $p \leq n$ and $\bar{n} = r \neq p - 1$. Define

$$U = \{f \in V_n : a_{i,j} = 0 \text{ for all } i \text{ such that } \bar{i} > r\}.$$

Then $0 \neq U \neq V_n$ as $x^p y^{n-p} \in U$ but $x^{p-1} y^{n-p+1} \notin U$. Claim U is G-invariant, so that G is not irreducible on V_n. It suffices to show $(x^i y^{n-i})g\pi \in U$ for all i with $\bar{i} = s \leq r$. To do so, we will show that all monomials in $x^j g\pi$ and $y^j g\pi$ are of the form $x^t y^{j-t}$ with $\bar{i} \leq \bar{j}$. Hence as $\overline{n-i} = r - s$, all monomials in $(x^i y^{n-i})g\pi = x^i g\pi y^{n-i} g\pi$ are of the form $x^t y^{n-t}$ with $t \leq s + (r-s) = r$.

Let $\bar{j} = v$, so that $j = up + v$ and $x^j g\pi = (x^u g\pi)^p x^v g\pi$. As $(f + h)^p = f^p + h^p$ for all $f, g \in V$, all monomials in $(x^u g\pi)^p$ are of the form $x^{pt} y^{p(j-t)}$, so we can assume $j = v$. Then

$$x^j g\pi = (xg\pi)^j = (ax + by)^j = \sum_{k=0}^{j} \binom{j}{k}(ax)^k (by)^{j-k},$$

establishing the claim.

Finally assume $n < p$; it remains to show G is irreducible on V_n. By hypothesis, $X = \{x, y\}$ is a basis for U. Identify $g \in G$ with its matrix $M_X(g)$ and define

$$g_a = \begin{pmatrix} 1 & a \\ 0 & 1 \end{pmatrix}, \quad h_a = \begin{pmatrix} 1 & 0 \\ a & 1 \end{pmatrix}.$$

Then

$$S = \{g_a : a \in F\} \cong F \cong T = \{h_a : a \in F\}.$$

Let $M_i = \langle y^j x^{n-j} : 0 \leq j \leq i \rangle$ for $-1 \leq i \leq n$, so that $M_n = V_n$, $M_0 = \langle x^n \rangle$, and $M_{-1} = 0$.

Lemma. T acts on M_i for all i and $[M_i/M_{i-2}, T] = M_{i-1}/M_{i-2}$ for $i \geq 1$.

Proof. $xh_a = x$ and $yh_a = ax + y$, so

$$[y^j x^{n-j}, h_a] = (y^j x^{n-j})h_a - y^j x^{n-j}$$

$$= \sum_{k=0}^{j-1} \binom{j}{k} y^k a^{j-k} x^{n-k} \equiv ajy^{j-1} x^{n-j+1} \bmod M_{j-2}$$

and if $a \neq 0 \neq j$ then $aj \neq 0$.

Corollary. If $i \geq 0$ and $w \in M_i - M_{i-1}$ then $[w, T] = M_{i-1}$.

Proof. The proof is by induction on i. If $i = 0$ then $w \in \langle x^n \rangle \leq C_V(T)$, so $[w, T] = 0 = M_{-1}$. Assume the result holds at $i - 1$. By the Lemma, $[w, h] \in M_{i-1} - M_{i-2}$ for some $h \in T$, so by the induction hypothesis, $M_{i-2} = [w, h, T] \leq [w, T]$, so $[w, T] = M_{i-1}$ as $\dim(M_{i-1}/M_{i-2}) = 1$.

Now $SL(U) = H = \langle T, S \rangle$ and we will show H is irreducible on V_n. Let $0 \neq W$ be an *FH*-submodule of V_n. By the Corollary and symmetry between T and S, $y^n \in [W, S]$ or $W = \langle y^n \rangle$, so in either case $y^n \in W$. Hence by the Corollary, $M_{n-1} = [W, T]$, so $V_n = \langle y^n, M_{n-1} \rangle \leq W$.

(4) Suppose $g \in \ker(\pi_n)$. Then $x^n = x^n g\pi = (xg\pi)^n$, so $xg\pi = \lambda x$ for some nth root of unity λ. Similarly $yg\pi = \mu y$. Then $xy^{n-1} = (xy^{n-1})g\pi = \lambda\mu^{n-1}$ xy^{n-1}, so $\lambda = \mu^{1-n} = \mu$.

Chapter 4, Exercise 10. (1) and (2): First observe that if I is the identity matrix in G then $z\phi(I) = z$ for all $z \in \Gamma$, so $\phi(I) = 1$ is the identity of the monoid S of all functions from Γ into Γ.

Next, if $A = (a_{i,j})$ and $B = (b_{i,j})$ are in G then for $z \in F$ with $a_{1,2}z + a_{2,2} \neq 0$:

$$
\begin{aligned}
z\phi(A)\phi(B) &= \left(\frac{a_{1,1}z + a_{2,1}}{a_{1,2}z + a_{2,2}} \right) \phi(B) \\
&= \frac{(a_{1,1}z + a_{2,1})b_{1,1} + (a_{1,2}z + a_{2,2})b_{2,1}}{(a_{1,1}z + a_{2,1})b_{1,2} + (a_{1,2}z + a_{2,2})b_{2,2}} \\
&= \frac{(a_{1,1}b_{1,1} + a_{1,2}b_{2,1})z + a_{2,1}b_{1,1} + a_{2,2}b_{2,1}}{(a_{1,1}b_{1,2} + a_{1,2}b_{2,2})z + a_{2,1}b_{1,2} + a_{2,2}b_{2,2}} = z\phi(AB).
\end{aligned}
$$

If $a_{1,2}z + a_{2,2} = 0$ then $z\phi(A) = \infty$, so

$$
z\phi(A)\phi(B) = \infty\phi(B) = b_{1,1}/b_{1,2}.
$$

On the other hand as $a_{1,2}z + a_{2,2} = 0$ and $\det(A) \neq 0$, $a_{1,2} \neq 0$, so $z = -a_{2,2}/a_{1,2}$ and

$$
z\phi(AB) = \frac{a_{2,1}b_{1,1} - a_{1,1}a_{2,2}b_{1,1}/a_{1,2}}{a_{2,1}b_{1,2} - a_{1,1}a_{2,2}b_{1,2}/a_{1,2}} = \frac{b_{1,1}(a_{1,2}a_{2,1} - a_{1,1}a_{2,2})}{b_{1,2}(a_{1,2}a_{2,1} - a_{1,1}a_{2,2})} = b_{1,1}/b_{1,2}.
$$

Finally, $\infty\phi(A)\phi(B) = (a_{1,1}/a_{1,2})\phi(B)$ and if $a_{1,2} \neq 0$ then

$$
(a_{1,1}/a_{1,2})\phi(B) = \frac{b_{1,1}a_{1,1}/a_{1,2} + b_{2,1}}{b_{1,2}a_{1,1}/a_{1,2} + b_{2,2}} = \frac{b_{1,1}a_{1,1} + b_{2,1}a_{1,2}}{b_{1,2}a_{1,1} + b_{2,2}a_{1,2}} = \infty\phi(AB).
$$

On the other hand if $a_{1,2} = 0$ then $a_{1,1} \neq 0$ and $a_{1,1}/a_{1,2} = \infty$, so

$$
\infty\phi(A)\phi(B) = \infty\phi(B) = b_{1,1}/b_{1,2} = \frac{b_{1,1}a_{1,1} + b_{2,1}a_{1,2}}{b_{1,2}a_{1,1} + b_{2,2}a_{1,2}} = \infty\phi(AB).
$$

Thus we have verified that $\phi: G \to S$ is a monoid homomorphism. Hence as G is a group and $\phi(I) = 1$, $G^* = \phi(G)$ is a subgroup of S and $\phi: G \to G^*$ is a surjective group homomorphism. As G^* is a subgroup of S, each $\phi(A) \in G^*$ is invertible in S and hence (1) holds. Further to complete the proof of (2) it remains to show $\ker(\phi)$ is the group of scalar matrices. So let $A \in \ker(\phi)$.

Thus $z\phi(A) = z$ for all $z \in \Gamma$. In particular $\infty = \infty\phi(A) = a_{1,1}/a_{1,2}$, so $a_{1,2} = 0 \neq a_{1,1}$. Also $0 = 0\phi(A) = a_{2,1}/a_{2,2}$, so $a_{2,1} = 0 \neq a_{2,2}$. Finally

$$1 = 1\phi(A) = \frac{a_{1,1} + a_{2,1}}{a_{1,2} + a_{2,2}} = a_{1,1}/a_{2,2},$$

so $a_{1,1} = a_{2,2}$ and hence $A = a_{1,1}I$ is scalar.

(3) By construction $\alpha: \Omega \to \Gamma$ is a bijection.

Let H be the stabilizer in G of $Fx_1 \in \Omega$. By 13.5, G is 2-transitive on Ω, so G is primitive on Ω by 15.14. Hence H is maximal in G by 5.19, so $G = \langle H, t \rangle$, where $t \in G$ interchanges x_1 and x_2. Thus to show $(\omega\bar{g})\alpha = (\omega\alpha)\bar{\phi}(\bar{g})$ for each $\bar{g} \in \bar{G}$, it suffices to show $(\omega g)\alpha = (\omega\alpha)\phi(g)$ for each $g \in H$ and $g = t$.

Each $h \in H$ is the product of a scalar matrix and a matrix

$$g = \begin{pmatrix} a & 0 \\ b & 1 \end{pmatrix}, \quad a \in F^{\#}, \; b \in F,$$

so it suffices to show $(\omega g)\alpha = (\omega\alpha)\phi(g)$. But $\phi(g): z \mapsto az + b$, so g fixes Fx_1 and $\phi(g)$ fixes $\infty = (Fx_1)\alpha$. Further, for $\omega = F(\lambda x_1 + x_2)$, $\lambda \in F$,

$$(\omega g)\alpha = F((a\lambda + b)x_1 + x_2)\alpha = a\lambda + b = \lambda\phi(g) = (\omega\alpha)\phi(g).$$

Next, t interchanges x_1 and x_2, so t has cycles (Fx_1, Fx_2) and $(F(\lambda x_1 + x_2), F(\lambda^{-1}x_1 + x_2))$, $\lambda \in F^{\#}$, on Ω. Then $\phi(t): z \mapsto 1/z$, so $\phi(t)$ has cycles $(\infty, 0)$ and (λ, λ^{-1}) on Γ. Hence as $(Fx_1)\alpha = \infty$, $(Fx_2)\alpha = 0$, and $F(\lambda x_1 + x_2)\alpha = \lambda$, the proof is complete.

Chapter 5, Exercise 6. (1) Induct on m. By Jordan's Theorem, 15.17, G is 2-transitive on X, so the result holds when $m = 1$. Again by Jordan, G_x is primitive on $X - \{x\} = X'$ for $x \in Y$. So as $Y' = Y - \{x\}$ is of order $m - 1$ with $G_{x,Y'} = G_Y$ primitive on $X' - Y' = X - Y$, G_x is m-transitive on X' by induction on m. Thus G is $(m + 1)$-transitive on X by 15.12.1.

(2) Let t be a transposition or cycle of length 3 in G and set $Y = \text{Fix}(t)$. Then $t \in G_Y$ and $\langle t \rangle$ is transitive on $X - Y$. Thus G_Y is primitive on $X - Y$ as $|X - Y|$ is prime. Therefore G is $(n - 2)$-transitive on X by (1), so G contains the alternating group by 15.12.4.

(3) Let $\{x\} = M(a) \cap M(b)$, $y = xa$, $z = xb$, and $\Delta = \{x, y, z\}$. Now

$$\text{Fix}(a) \cap \text{Fix}(b) = \text{Fix}(\langle a, b \rangle) \subseteq \text{Fix}([a, b]).$$

If $v \in M(a) - \{x, y\}$ then $v, va^{-1} \in M(a) - \{x\} \subseteq \text{Fix}(b)$, so

$$v[a, b] = va^{-1}b^{-1}ab = va^{-1}ab = vb = v.$$

Similarly if $v \in M(b) - \{x, z\}$ then $v, vb^{-1} \in \text{Fix}(a)$, so

$$v[a, b] = va^{-1}b^{-1}ab = vb^{-1}ab = vb^{-1}b = v.$$

Thus $X - \Delta \subseteq \text{Fix}([a, b])$. Next $xb^{-1} \in \text{Mov}(b) - \{x\} \subseteq \text{Fix}(a)$, so

$$y[a, b] = ya^{-1}b^{-1}ab = xb^{-1}ab = xb^{-1}b = x.$$

Similarly $xa^{-1} \in \text{Fix}(b)$, so

$$x[a, b] = xa^{-1}b^{-1}ab = xa^{-1}ab = xb = z.$$

Finally $z = xb \in \text{Mov}(b) - \{x\} \subseteq \text{Fix}(a)$ and $y = xa \in \text{Fix}(b)$, so

$$z[a, b] = za^{-1}b^{-1}ab = zb^{-1}ab = xab = yb = y.$$

Thus (y, x, z) is a cycle of length 3 in $[a, b]$, completing the proof of (3).

(4) Assume G does not contain the alternating group on X and $|Y| > n/2$; we must derive a contradiction. As $|Y| > n/2$, $\Gamma = X - Y$ has order less than $|Y|$, so by minimality of $|Y|$, $G_\Gamma \neq 1$. Pick $a \in G_\Gamma^\#$, $y \in M(a)$, and set $\Delta = Y - \{y\}$. Then $|\Delta| < |Y|$, so by minimality of $|Y|$, $G_\Delta \neq 1$. Pick $b \in G_\Delta^\#$. As $G_Y = 1$, $y \in M(b)$, while

$$M(a) \cap M(b) \subseteq (X - \Gamma) \cap (X - \Delta) = X - (\Gamma \cup \Delta) = \{y\},$$

so $M(a) \cap M(b) = \{y\}$. Thus $[a, b]$ is a 3-cycle by (3), so (2) supplies a contradiction, establishing (4).

(5) Pick Y as in (4) and let $H = \text{Sym}(X)_Y$. Then $|H| = m!$, where $m = |X - Y|$ and by (4), $m \geq [(n + 1)/2]$. Now $H \cap G = G_Y = 1$, so

$$|\text{Sym}(X)| \geq |HG| = |H||G| = m!|G|,$$

establishing (5).

Chapter 6, Exercise 2. (1) Let G be the semidirect product of H by A. We must show there exists an A-invariant Hall π-subgroup of H. Let Δ be the set of Hall π-subgroups of H. By Hall's Theorem, 18.5, there is $K \in \Delta$ and H is transitive on Δ, so by a Frattini argument, $G = HN_G(K)$. By the Schur–Zassenhaus Theorem 18.1, there exists a complement B to $N_H(K)$ in $N_G(K)$ and $B^g = A$ for some $g \in G$. Hence $J = K^g$ is an A-invariant Hall π-subgroup of H.

(2) We must show $C_H(A)$ is transitive on the set $\text{Fix}(A)$ of fixed point of A on Δ. But by Schur–Zaussenhaus, $N_H(K)$ is transitive on the set $A^G \cap N_G(K)$ of complements to $N_H(K)$ in $N_G(K)$, so by 5.21, $N_H(A)$ is transitive on $\text{Fix}(A)$.

(3) We must show each A-invariant π-subgroup X of H is contained in a member of $\text{Fix}(A)$. The proof is by induction on $|G|$, so assume G is a minimal counter example. Let M be a minimal normal subgroup of G contained in H and $G^* = G/M$. By minimality of G there exists an A-invariant Hall π-subgroup Y^* of H^* containing X^*. Now $|Y|_\pi = |H|_\pi$, so Y contains a Hall π-subgroup of H, and if $H \neq Y$ then X is contained in an A-invariant Hall

π-subgroup of Y by minimality of G. Thus $H = Y$, so H/M is a π-group and hence $H = JM$, so $H^* = J^*$.

By 9.4, M is a p-group for some prime p. If $p \in \pi$ then H is a π-group, so H is an A-invariant Hall π-subgroup of H containing X. Thus $p \notin \pi$, so X is a Hall π-subgroup of MX. Let $Z = J \cap XM$. As $H^* = J^*$, $Z^* = X^*$, so Z and X are A-invariant Hall π-subgroups of XM, and hence by (2) there is $g \in C_H(A)$ with $Z^g = X$. Thus J^g is an A-invariant Hall π-subgroup of H containing X.

(4) Part (4) does not depend on p and hence remains valid.

Chapter 6, Exercise 3. As V is the permutation module for $G = \mathrm{Alt}(I)$ on I and $F = \mathrm{GF}(2)$, we can identify V with the power set of I, and for $u, v \in V$, $u + v$ is the symmetric difference of u and v; that is,

$$u + v = u \cup v - (u \cap v).$$

Define the *weight* of v to be the order $|v|$ of the subset v of I.

(1) Let $0 \neq W$ be an FG-submodule of V and $w \in W^\#$ of weight m. By 15.12.3, G is $(n-2)$-transitive on I, so as $n > 2$, G is transitive on m-subsets of I and hence on vectors of weight m. Thus for each v of weight m, $v \in wG \subseteq W$. Assume $w \neq I$, the generator of Z. Then there is v of weight m such that $w \cap v$ is of order $m - 1$, so $w + v$ is of weight 2. So as $w + v \in W$, we may take $m = 2$. But as w is of weight 2, $\langle wG \rangle = U$. Namely, if $0 \neq u \in U$, pick $i \in u$, and observe $u = \sum_{i \neq j \in u} \{i, j\}$. Thus $W = U$ or V.

(2) and (3): First observe that $G = O^2(G)$. If $n > 4$ this follows because G is simple by 15.16. If $n = 3$ or 5 it is easy to check $G = O^2(G)$ directly.

Observe next that if n is odd then $I \notin U$, so $V = U \oplus Z$ and hence $\bar{U} \cong U$ is of dimension $n - 1$. On the other hand, if n is even then $0 < Z < U < V$ with $\dim(\bar{U}) = n - 2$.

By Exercise 4.6, $Z = C_V(G)$ and $U = [V, G]$. Further, if $X \leq G$ is of odd order, then by coprime action 18.7, $C_{\bar{V}}(X) = \overline{C_V(X)}$, so as $G = O^2(G)$, $C_{\bar{V}}(G) = \overline{C_V(G)} = \bar{Z} = 0$. By 8.5.3, $[\bar{V}, G] = \overline{[V, G]} = \bar{U}$.

As $C_{\bar{U}}(G) = 0$, by 17.11 there is a largest FG-module W such that $[W, G] \leq \bar{U} \leq W$ and $C_W(G) = 0$. In particular as $\bar{V}$ is such a module, $\bar{V} \leq W$. Also by definition, $H^1(G, \bar{U}) = W/\bar{U}$, so it remains to show $W = \bar{V}$, or equivalently $H^1(G, \bar{U}) = 0$ or F for n odd or even, respectively. We prove this by induction on n.

If $n = 3$ then $G \cong \mathbf{Z}_3$ is of order prime to 2, so by 17.10, $H^1(G, \bar{U}) = 0$, as desired. Thus we may take $n > 3$. Let $H = G_1$ and $J = I - \{1\}$, so that $H = \mathrm{Alt}(J) \cong A_{n-1}$. For $j \in J$, let $v_j = \{1, j\}$. Then $\{v_j : j \in J\}$ is a basis for U with $v_j h = v_{jh}$ for $h \in H$, so U is the permutation module for H of degree $n - 1$. In particular, by Exercise 4.6.1, $v = \sum_{j \in J} v_j$ generates $C_U(H)$, and by induction on n, $\langle \bar{v} \rangle = C_{\bar{U}}(H)$. Notice if n is odd then $v = J$ so $\langle \bar{J} \rangle = C_{\bar{U}}(H)$, while if n is even then $v = I$, so $C_{\bar{U}}(H) = 0$.

Assume n is odd, so that $U \cong \bar{U}$ as an FG-module. Suppose $w \in C_W(H) - \bar{U}$ and let $M = \langle wG \rangle$. By Exercise 4.6, M is a homomorphic image of V as an FG-module. As $C_M(G) \le C_W(G) = 0$, $0 \ne [M, G] \le [W, G] = \bar{U}$, so a G is irreducible on $\bar{U}$, $\bar{U} = [M, G] \le M$. Then as $C_M(G) = 0$ and by (1), $\bar{U}$ is the only nontrivial image N of V with $C_N(G) = 0$, we conclude $M = \bar{U}$, contradicting $w \in W - \bar{U}$. Thus we have shown that $C_W(H) = C_{\bar{U}}(H) = \langle \bar{J} \rangle$, so also $C_{W/\langle \bar{J} \rangle}(H) = 0$ as $H = O^2(H)$. Further $\bar{U}/\langle \bar{J} \rangle$ is the image of the permutation module $\bar{U} \cong U$ for H modulo $C_{\bar{U}}(H)$, so by induction on n, $W/\langle \bar{J} \rangle = \bar{U}/\langle \bar{J} \rangle$, so $W = \bar{U}$, completing the proof in this case.

This leaves the case n even. Thus $n - 1$ is odd, so as U is the permutation module for H, by induction on n, $W = \bar{U} \oplus C_W(H)$. Let $K = G_2$. Then $C_W(H \cap K) = C_W(H) \oplus C_{\bar{U}}(H_2)$ is of dimension $d+1$, where $d = \dim(C_W(H))$. Thus $C_W(H)$ and $C_W(K)$ are hyperplanes of $C_W(H \cap K)$. Further

$$0 = C_W(G) = C_W(\langle H, K \rangle) = C_W(H) \cap C_H(K),$$

so $\dim(C_W(H \cap K)) \le 2$ and $d \le 1$. Therefore

$$\dim(W) = d + \dim(\bar{U}) \le n - 1 = \dim(\bar{V}),$$

so again $W = \bar{V}$, completing the proof.

Chapter 7, Exercise 5. Let $u = ax^2 + bxy + cy^2$ and $v = rx^2 + sxy + ty^2$ be in W. Then

$$Q(x + y) = (b + s)^2 - 4(a + r)(c + t)$$
$$= (b^2 - 4ac) + (s^2 - 4rt) + 2bs - 4(at + rc)$$
$$= Q(u) + Q(v) + f(u, v).$$

Evidently f is bilinear and symmetric and $Q(\lambda u) = \lambda^2 Q(u)$ for $\lambda \in F$. To prove (W, Q) is nondegenerate it suffices to observe that $X = \{x^2, xy, y^2\}$ is a basis for W with x^2, y^2 singular, $Q(xy) = 1$, $f(x^2, y^2) = -4$, and xy is orthogonal to x^2 and y^2.

(2) Let

$$g = \begin{pmatrix} a & b \\ c & d \end{pmatrix} \in G.$$

To prove (2), it suffices to check $Q(wg\alpha) = \lambda(g\alpha)Q(w)$ and $f(wg\alpha, zg\alpha) = \lambda(g\alpha)f(w, z)$ for $w, z \in X$, where $\lambda(g\alpha) = \det(g)^2 = ad - bc$. For example

$$Q(x^2 g\alpha) = Q((ax + by)^2) = Q(a^2 x^2 + 2abxy + b^2 y^2)$$
$$= 4a^2 b^2 - 4a^2 b^2 = 0 = \lambda(g\alpha)Q(x^2),$$

$$Q(xy(g\alpha)) = Q((ax + by)(cx + dy)) = Q(acx^2 + (ad + bc)xy + bdy^2)$$
$$= (ad + bc)^2 - 4abcd = (ad - bc)^2 = \lambda(g\alpha)Q(xy),$$

and

$$f(x^2 g\alpha, y^2 g\alpha) = f((ax+by)^2, (cx+dy)^2)$$
$$= f(ax^2 + 2abxy + b^2 y^2, c^2 x^2 + 2cdxy + d^2 y^2)$$
$$= 8abcd - 4(a^2 d^2 + c^2 b^2)$$
$$= -4(ad-bc)^2 = \lambda(g\alpha) f(x^2, y^2).$$

(3) Observe that u is singular iff $b^2 = 4ac$, so the set S of singular points in W is $S_0 \cup \{Fx^2\}$, where

$$S_0 = \{F(e^2 x^2 + 2exy + y^2): e \in F\}.$$

The subgroup

$$T = \left\{ \begin{pmatrix} 1 & 0 \\ c & 1 \end{pmatrix} : c \in F \right\}$$

of G is transitive on S_0, so G is 2-transitive on S and hence $\Delta(W) = G\alpha \Delta_{Fx^2, Fy^2}$. Now if $h \in \Delta_{Fx^2, Fy^2}$, then as $Fxy = \langle x^2, y^2 \rangle^{\perp}$, $M_X(h)$ is diagonal with $x^2 h = \mu x^2$, $y^2 h = \lambda \mu^{-1} y^2$, and $xyh = \sqrt{\lambda} xy$, for some $\mu \in F^{\#}$, where $\lambda = \lambda(h)$. On the other hand $g\alpha \in \Delta_{Fx^2, Fy^2}$, where

$$g = \begin{pmatrix} 1 & 0 \\ 0 & \sqrt{\lambda}\mu^{-1} \end{pmatrix} \in G,$$

and $x^2 g\alpha = x^2$, $y^2 g\alpha = \lambda\mu^{-2} y^2$, and $xyg\alpha = \sqrt{\lambda}\mu^{-1} xy$, so $h = g\alpha \cdot \mu I \in G\alpha \cdot S$. Therefore (3) holds.

(4) If U is a nondefinite 3-dimensional orthogonal space over F then $U = H \perp D$, where H is a hyperbolic line and $D = Fd$ is definite. Multiplying the quadratic form Q_U on U by a suitable scalar, we may assume $Q_U(d) = 1$. Thus $(U, Q_U) \cong (W, Q)$.

(5) Part (4) implies (5), since if F is finite or algebraically closed, then no 3-dimensional orthogonal space over F is definite by lemmas 21.3 and 20.10.1.

(6) Let $\Delta = \Delta(W, Q)$. By Exercise 4.7.4,

$$\ker(\alpha) = \{aI: a \in F \text{ and } a^2 = 1\} = \langle -I \rangle.$$

Thus $G^{(1)}\alpha \cong L_2(F)$, so as $\Delta = SG\alpha$ and $S \leq Z(\Delta)$, $\Delta^{(1)} = G^{(1)}\alpha \cong L_2(F)$. As $\Delta/\Delta^{(1)}$ is abelian, $r^g \in r\Delta^{(1)}$ for each $r \in R$ and $g \in \Delta$, so $rr^g \in \Delta^{(1)}$. But by 43.12.1, $L_2(F)$ is simple unless $|F| = 3$, so except possibly in that case, $\Delta^{(1)} = \langle rr^h: h \in \Delta^{(1)} \rangle$. Finally, if $|F| = 3$ then

$$r = \begin{pmatrix} -1 & 0 \\ 0 & 1 \end{pmatrix} \in R$$

with center Fxy, and r inverts

$$g = \begin{pmatrix} 1 & 0 \\ 1 & 1 \end{pmatrix} \in G$$

of order 3, so again $G^{(1)}\alpha$ is generated by conjugates of $rr^g = g^{-1}$ of order 3.

Chapter 8, Exercise 11. Let V be a normal elementary abelian subgroup of G of maximal rank, $H = C_G(V)$, and $G^* = G/H$. By 23.16, $V = \Omega_1(H)$. As $m(G) > 2, m(V) > 2$ by 23.17, so we may assume $m(V) = 3$. Thus $G^* \leq GL(V)$, so $m(G^*) \leq m(GL_3(p)) = 2$.

Let $E_{p^n} \cong A \leq G$ with $n > 3$. As $V = \Omega_1(H)$, $A \cap H = A \cap V$ is of rank at most 2. Also $m(A^*) \leq m(G^*) = 2$, so

$$4 \leq m(A) = m(A^*) + m(A \cap H) \leq 2 + 2 = 4$$

and hence all inequalities are equalities, so $m(A) = 4$ and $m(A^*) = m(A \cap H) = 2$.

Let $B = A \cap V$. Then A centralizes the hyperplane B of V and $m(A^*) = 2$, so A^* is the full group of transvections with axis B. Therefore $[A, V] = B = C_V(a)$ for each $a \in A - B$. Suppose $D \cong E_{p^4}$ with $D \leq AV$. Then $|AV : A| = |AV : D| = p$, so if $A \neq D$ then $m(A \cap D) = 3$, and hence there is $a \in A \cap D - B$. Then $D \leq C_{AV}(a) = AC_V(a) = AB = A$, a contradiction. Thus A is the unique E_{p^4}-subgroup of AV, so A char AV.

Let $K = N_G(A)$; we may assume $K \neq G$, so there is $g \in N_G(K) - K$. As A^* is of index p in G^*, $A^* \trianglelefteq G^*$, so $(A^g)^* = A^*$ and hence $A \neq A^g \leq AH \cap K = AX$, where $X = N_H(A)$. Now

$$[A, AX] = [A, X] \leq A \cap H = B \leq Z(AH),$$

so AA^g is of class at most 2. Hence by 23.11, AA^g is of exponent 3. Thus $AA^g = AA^g \cap X = A(AA^g \cap X)$ with $AA^g \cap X \leq \Omega_1(H) = V$. Therefore $AA^g \leq AV$, so as A is the unique E_{p^4}-subgroup of AV, $A = A^g$, contradicting $g \notin K$.

Chapter 9, Exercise 6. (1) Fix $v_i \in V_i$, $i = 1, 2$, and define

$$f_{v_1, v_2} : V_1 \times V_2 \to F,$$

$$(u_1, u_2) \mapsto f_1(v_1, u_1) f_2(v_2, u_2).$$

As f_1 and f_2 are bilinear, so is f_{v_1, v_2}, so by the universal property of the tensor product, f_{v_1, v_2} extends to a linear map $\hat{f}_{v_1, v_2} : V \to F$ with $\hat{f}_{v_1, v_2}(u_1 \otimes u_2) = f_{v_1, v_2}(u_1, u_2)$. Similarly

$$V_1 \times V_2 \to V^*,$$

$$(v_1, v_2) \mapsto \hat{f}_{v_1, v_2}$$

is bilinear, so there is $\hat{f} \in \operatorname{Hom}(V, V^*)$ with $\hat{f}(v_1 \otimes v_2) = \hat{f}_{v_1, v_2}$. Therefore $f: V \times V \to F$ defined by $f(x, y) = \hat{f}(x)(y)$ is bilinear and

$$f(v_1 \otimes v_2, u_1 \otimes u_2) = \hat{f}(v_1 \otimes v_2)(u_1 \otimes u_2)$$

$$= \hat{f}_{v_1, v_2}(u_1 \otimes u_2) = f_{v_1, v_2}(u_1, u_2) = f_1(v_1, u_1) f_2(v_2, u_2).$$

As the fundamental tensors generate $V = V_1 \otimes V_2$, f is unique subject to this property. Finally

$$f(u_1 \otimes u_2, v_1 \otimes v_2) = f(u_1, v_1) f(u_2, v_2) = (-f(v_1, u_1))(-f(v_2, u_2))$$

$$= f(v_1, u_1) f(v_2, u_2) = f(v_1 \otimes v_2, u_1 \otimes u_2),$$

so as the fundamental tensors generate $V_1 \otimes V_2$, f is symmetric.

(2) First $f(v_1 \otimes v_2, v_1 \otimes v_2) = f_1(v_1, v_1) f_2(v_2, v_2) = 0$ as f_i is symplectic. But if $\operatorname{char}(F) \neq 2$, recall from Chapter 7 that the unique quadratic form Q associated to f satisfies $Q(x) = f(x, x)/2$. Thus $Q(v_1 \otimes v_2) = f(v_1 \otimes v_2, v_1 \otimes v_2)/2 = 0$, as desired. On the other hand:

Lemma. If $\operatorname{char}(F) = 2$, X is a basis for an F-space U, f is a symplectic form on U, and $a_x \in F$ for each $x \in X$, then there is a unique quadratic form Q on U associated to f such that $Q(x) = a_x$ for each $x \in X$.

Let $X_i = \{v_i, u_i\}$ be a basis for V_i; then $X = X_1 \otimes X_2$ is a basis for V, so by the lemma there is a unique quadratic form Q on V associated to f with $Q(x_1 \otimes x_2) = 0$ for each $x_i \in X_i$. Now for $w \in V_1$, $w = av_1 + bu_1$ and

$$Q(w \otimes v_2) = Q(a(v_1 \otimes v_2) + b(u_1 \otimes v_2))$$

$$= a^2 Q(v_1 \otimes v_2) + b^2 Q(u_1 \otimes v_2) + abf(v_1 \otimes v_2, u_1 \otimes v_2) = 0.$$

Similarly $Q(w \otimes u_2) = 0$, so for $z = cv_2 + du_2 \in V_2$,

$$Q(w \otimes z) = Q(c(w \otimes v_2) + d(w \otimes u_2))$$

$$= c^2 Q(w \otimes v_2) + d^2 Q(w \otimes u_2) + cdf(w \otimes v_2, w \otimes u_2) = 0$$

completing the proof of (2).

(3) Pick X_i to be a hyperbolic basis for V_i. Then $x_1 \otimes x_2$ is orthogonal to $x_1 \otimes y_2$ and $y_1 \otimes x_2$ for all $x_i, y_i \in X_i$, while

$$f(v_1 \otimes v_2, u_1 \otimes u_2) = 1 = f(v_1 \otimes u_2, u_1 \otimes v_2),$$

so X is a hyperbolic basis for V.

(4) Let $g = (g_1, g_2) \in \Delta$. Then

$$f(v_1 \otimes v_2)g\pi, (u_1 \otimes u_2)g\pi) = f(v_1 g_1 \otimes v_2 g_2, u_1 g_1 \otimes u_2 g_2)$$

$$= f_1(v_1 g_1, u_1 g_1) f_2(v_2 g_2, u_2 g_2)$$

$$= \lambda(g_1) f_1(v_1, u_1) \lambda(g_2) f_2(v_2, u_2)$$

$$= \lambda(g_1) \lambda(g_2) f(v_1 \otimes v_2, u_1 \otimes u_2),$$

so as the fundamental tensors generate V, $g\pi \in \Delta(V, f)$ with $\lambda(g\pi) = \lambda(g_1)\lambda(g_2)$. Similarly

$$Q((v_1 \otimes v_2)(g\pi)) = Q(v_1g_1 \otimes v_2g_2) = 0 = Q(v_1 \otimes v_2)$$

as $Q(u_1 \otimes u_2) = 0$ for all $u_i \in V_i$, so by the lemma, $g\pi$ also preserves Q. Of course $g\pi \in O(V, Q)$ iff $1 = \lambda(g)$ iff $\lambda(g_2) = \lambda(g_1)^{-1}$. Also $g \in \ker(\pi)$ iff $x_1 \otimes x_2 = (x_1 \otimes x_2)g\pi = x_1g_1 \otimes x_2g_2$ for all $x_i \in X_i$, so as X is a basis for V, it follows that $g \in \ker(\pi)$ iff $g_i = \mu_i I$ for $u_i \in F^\#$ with $\mu_2 = \mu_1^{-1}$.

(5) The map

$$T_\alpha: V_1 \otimes V_2 \to V,$$

$$(u, v\alpha) \mapsto v \otimes u\alpha$$

is bilinear, so there exists a unique $t = t_\alpha \in \text{End}(V)$ with $t_\alpha(u \otimes v\alpha) = v \otimes u\alpha$ for all $u, v \in V_1$. Evidently $t^2 = 1$. Further, as α is an isometry,

$$f((v_1 \otimes v\alpha)t, (u_1 \otimes u\alpha)t) = f(v \otimes v_1\alpha, u \otimes u_1\alpha) = f_1(v, u)f_2(v_1\alpha, u_1\alpha)$$

$$= f_1(v_1, u_1)f_2(v\alpha, u\alpha) = f(v_1 \otimes v\alpha, u_1 \otimes u\alpha),$$

so t is an isometry. Next, we may choose notation so that $v_1\alpha = v_2$ and $u_1\alpha = u_2$, so

$$(v_1 \otimes v_2)t = (v_1 \otimes v_1\alpha)t = (v_1 \otimes v_1\alpha) = v_1 \otimes v_2.$$

Similarly t fixes $u_1 \otimes u_2$ and interchanges $v_1 \otimes u_2$ and $u_1 \otimes v_2$, so t is a reflection or transvection.

Let $g = (g_1, 1) \in \Delta_1$, so that $g\pi \in \Delta_1\pi$. Then

$$(u \otimes v\alpha)g^t = (v \otimes u\alpha)gt = (vg_1 \otimes u\alpha)t = u \otimes vg_1\alpha = u \otimes (v\alpha g_1\alpha^*),$$

so as $\Delta_1\alpha^* = \Delta_2$, $g^t \in \Delta_2\pi$. Thus $(\Delta_1\pi)^t = \Delta_2\pi$ and similarly $(G_1\pi)^t = G_2\pi$.

(6) and (7): For Fv a point in V_1 let $Fv \otimes V_2 = \{v \otimes u : u \in V_2\}$ and set $\mathcal{L}_1 = \{Fv \otimes V_2 : v \in V_1^\#\}$. For Fu a point in V_2 define $V_1 \otimes Fu$ and $\mathcal{L}_2$ similarly. Observe $\mathcal{L} = \mathcal{L}_1 \cup \mathcal{L}_2$ is a set of totally singular lines in V, as $Q(v_1 \otimes v_2) = 0$ for all $v_i \in V_i$.

Claim the singular vectors are the fundmental tensors $v_1 \otimes v_2$ and $\mathcal{L}$ is the set of *all* totally singular lines. For if $x \in V^\#$ is totally singular then $0 \neq x^\perp \cap Fv \otimes V_2$, so $x \in (v \otimes u)^\perp$ for some $u \in V_2^\#$. Now as V is 4-dimensional hyperbolic space and $y = v \otimes u$ is totally singular, we have $y^\perp = Fy \perp H$, where H is a hyperbolic line, and if H_1 and H_2 are the two totally singular points in H, then $Fy + H_1$ and $Fy + H_2$ are the two totally singular lines through Fy and contain all totally singular points in $y^\perp$. Thus $x \in Fy + H_i$ for some i, and as $Fv \otimes V_2$ and $V_1 \otimes Fu$ are totally singular lines through y, we may take $x \in Fv \otimes V_2$. This shows each totally singular point is a fundamental tensor, and each totally singular line l is one of the two totally singular lines $Fv \otimes V_2$

and $V_1 \otimes Fu$ through a point $F(v \otimes u)$ on l, so $\mathcal{L}$ is the set of all totally singular lines.

Observe $\mathcal{L}_1$ and $\mathcal{L}_2$ are the two classes of maximal totally singular subspaces of V described in 22.13, so by 22.13, the subgroup Γ of $\Sigma = \Delta(V, Q)$ acting on $\mathcal{L}_1$ and $\mathcal{L}_2$ is of index two in Σ, with $\Sigma = \Gamma \langle t \rangle$.

Next by Exercise 7.1.1, $G_i = SL(V_i)$. Further if $X_i = \{v_i, u_i\}$ is our hyperbolic basis, $\mu \in F^\#$ and $g_{\mu,i} \in GL(V_i)$ with $v_i g_{\mu,i} = v_i$ and $u_i g_{\mu,i} = \mu u_i$, then $g_{\mu,i} \in \Delta_i$ with $\lambda(g_{\mu,i}) = \mu$, so

$$\Delta_i = SL(V_i)\langle g_{\mu,i} \colon \mu \in F^\# \rangle = GL(V_i).$$

Now $\Delta_1 \pi \le \Gamma_0$, the stabilizer in Γ of $V_1 \otimes Fv_2$ and $V_1 \otimes Fu_2$. For $x \in V_1 \otimes Fv_2$, define $f_x \in (V_1 \otimes Fu_2)^*$ by $f_x(y) = f(x, y)$. Then the map $x \mapsto f_x$ is a Γ_0-isomorphism of $V_1 \otimes Fv_2$ with $(V_1 \otimes Fu_2)^*$, so as $C_\Gamma(V_1 \otimes Fv_2) \cap C_\Gamma(V_1 \otimes Fu_2) = 1$, Γ_0 is faithful on $V_1 \otimes Fv_2$, and thus, as $\Delta_1\pi$ acts faithfully as $GL(V_1 \otimes Fv_2)$ on $V_1 \otimes Fv_2$, it follows that $\Gamma_0 = \Delta_1\pi$. Further $\Delta_2\pi$ is 2-transitive on the points of Δ_2 and hence also on $\mathcal{L}_2$, so $\Gamma = \Gamma_0\Delta_2\pi = \Delta\pi$. This completes the proof of (6).

Finally $\Omega = \Omega(V, Q)$ is the derived group of $O(V, Q)$, and $O(V, Q) = (O(V, Q) \cap \Gamma)\langle t \rangle$ with

$$\Gamma_0 = O(V, Q) \cap \Gamma = \{(g_1, g_2)\pi \colon \lambda(g_1) = \lambda(g_2)^{-1}\}$$

by (4) and (6). Thus $\Gamma_0 = (G_1 G_2)\pi \cdot T\pi$, where $T = \{g_\mu\pi \colon \mu \in F^\#\}$ and $g_\mu = (g_{\mu,1}, g_{\mu^{-1},2})$. Further $g_{\mu^2}\pi = h_1 h_2 \pi \in (G_1 G_2)\pi$, where $v_1 h_1 = \mu^{-1}v_1, u_1 h_1 = \mu u_1, v_2 h_2 = \mu v_2$, and $u_2 h_2 = \mu^{-1}u_2$, and $g_\mu^t = g_{\mu^{-1}}$, so $[T\pi, t] \le (G_1 G_2)\pi$ and hence $\Omega = \Omega(V, Q) \le (G_1 G_2)\pi$. Conversely, if $|F| > 3$ then $G_i\pi \cong SL_2(F)$ is perfect by 43.12.1, so $\Omega = (G_1 G_2)\pi$. If $|F| = 3$ it can be checked directly that $(G_1 G_2)\pi \le \Omega$.

Chapter 9, Exercise 9. By Exercise 9.1, $\alpha = f\phi \in \operatorname{Hom}_{FG}(V, V^{\theta*})$ and as $f \ne 0$, $\alpha \ne 0$. Thus as G is irreducible on V, α is an isomorphism by Schur's Lemma 12.4. Let $\sigma = \theta*$ regarded as an automorphism of $GL_n(F)$, the group of invertible n by n matrices over F, and regard G as a subgroup of $GL_n(F)$. As $V \cong V^\sigma$, $g^\sigma = gB^*$ for all $g \in G$ and some $B \in GL_n(F)$ from the discussion in section 13. Thus

$$g^{\sigma^2} = (gB^*)^\sigma = g^\sigma (B^\sigma)^* = gB^*(B^\sigma)^* = g(BB^\sigma)^*$$

so $V \cong V^{\sigma^2}$. However, θ commutes with the transpose inverse map $*$ on $GL_n(F)$, so $V^{\sigma^2} = ((V^*)^*)^{\theta^2} \cong V^{\theta^2}$ as $(V^*)^* \cong V$.

Similarly if $V \cong V^\theta$ then $V^* \cong V^{\theta*} \cong V$, so by Exercise 9.1.4, G preserves a nondegenerate bilinear form on V. That is (1) holds in this case, so we may assume $V \not\cong V^\theta$. Then as $V \cong V^{\theta^2}$, m is even, since otherwise θ is a power of θ^2.

Let K be the fixed field of θ^2. Then $\langle\theta^2\rangle = \mathrm{Gal}(F/K)$, so by 26.3, $V = F \otimes_K U$ for some irreducible KG-submodule U of V. As θ is an automorphism of K of order 2, Exercise 9.1.4 says G preserves a nondegenerate hermitian symmetric form on U.

Chapter 9, Exercise 10. Let $G_1 = G_2 = G$. As σ is 1-dimensional, σ is irreducible and by hypothesis π is irreducible, so by 27.15, $\sigma \otimes \pi$ is an irreducible representation of $G_1 \times G_2$. Recall that if we identify G with the diagonal subgroup $\{(g, g): g \in G\}$ of $G_1 \times G_2$ via the isomorphism $g \mapsto (g, g)$, then the tensor product representation $\sigma \otimes \pi$ of G is the restriction of the tensor product representation $\sigma \otimes \pi$ of $G_1 \times G_2$ to the diagonal subgroup G, so it remains to show the diagonal subgroup G is irreducible on $U \otimes V$, where U and V are the representation modules for σ and π, respectively.

But if $u \in U^\#$ then $U = \mathbf{C}u$ and the map $\varphi: au \otimes v \mapsto av$ is an isomorphism of $U \otimes V$ with V such that $x(g, g)(\sigma \otimes \pi)\varphi = \lambda(g)(x\varphi(g\pi))$ for $g \in G$ and $x \in U \otimes V$, where $u(g\sigma) = \lambda(g)u$ and $\lambda(g) \in C^\#$. Namely $x = au \otimes v$,

$$x(g, g)(\sigma \otimes \pi)\varphi = (aug\sigma, vg\pi)\varphi = (a\lambda(g)u, vg\pi)\varphi = a\lambda(g)vg\pi$$

and $x\varphi(g\pi) = avg\pi$. Thus φ induces a bijection $\varphi: W \mapsto W\varphi$ of the $\mathbf{C}G$-subspaces of $U \otimes V$ and V, so as G is irreducible on V, it is also irreducible on $U \otimes V$.

Chapter 10, Exercise 6. (1) First, $w = 1 \cdot w$ with $l(1) = 0$, so $w \leq w$. Thus $\leq$ is reflexive. Second, if $u \leq w$ then $w = xu$ with $l(w) = l(x) + l(u)$, so as $l(x) \geq 0$ with equality iff $x = 1$, we have $l(w) \geq l(u)$ with equality iff $x = 1$ and $u = w$. Then if also $w \leq u$, by symmetry, $l(u) \geq l(w)$, so $u = w$. That is, $\leq$ is antisymmetric. Finally, assume $u \leq w \leq v$. Then $w = xu$ and $v = yw$ with $l(w) = l(x) + l(u)$ and $l(v) = l(y) + l(w)$, so $v = yw = yxu$ with

$$l(v) = l(y) + l(w) = l(y) + l(x) + l(u).$$

Further $l(yx) \leq l(y) + l(x)$, while as $v = yxu$,

$$l(yx) \geq l(v) - l(u) = l(y) + l(w) - (l(w) - l(x)) = l(y) + l(x),$$

so $l(yx) = l(y) + l(x)$ and hence $u \leq v$, proving $\leq$ is transitive.

(2) Let w be maximal with respect to $\leq$ and suppose $w \neq w_0$. Then by Exercise 10.3.2, $l(w) < |P|$, so by 30.12, there is $\alpha \in P$ with $\alpha w \in P$. Then by 30.10, $l(r_\alpha w) > l(w)$, so $r_\alpha w > w$, contradicting the maximality of w.

(3) First, (b) and (c) are equivalent by 30.10.

Next, if $l(rw) \leq l(w)$, then by 30.10, $l(w) = l(rw) + 1$, so as $l(r) = 1$, we have $l(w) = l(rw) + l(r)$ and $w = r \cdot rw$, so $rw \leq w$. That is, (b) implies (a).

Assume $rw \leq w$. Then $w = x \cdot rw$ with $l(w) = l(rw) + l(x)$, so $x = r$ and $l(rw) = l(w) - 1 = n - 1$. Thus $rw = r_2 \cdots r_n$ with $r_i \in R$, so $w = r(rw) = rr_2 \cdots r_n$. That is, (a) implies (d). Finally, if (d) holds then visibly $l(rw) = l(w) - 1$, so (d) implies (b).

(4) As $u \leq w$, $w = xu$ with $l(w) = l(x) + l(u)$. As $l(wr) \leq l(w)$, $l(wr) = l(w) - 1$ and similarly $l(ur) = l(u) - 1$. Now $wr = xur$ with

$$l(wr) = l(w) - 1 = l(x) + l(u) - 1 = l(x) + l(ur),$$

so $ur \leq wr$.

Chapter 11, Exercise 5. For $1 \leq u \leq v \leq n$, let $U_{u,v} = U_u \otimes U_{u+1} \otimes \cdots \otimes U_v$. For $k = 2j$ or $2j - 1$, let $j(k) = j$ and $l(k) = j + 2$ or $j + 1$, respectively. Let $U^1 = U_{1,j-1}$, $U^2 = U_{j,l(k)-1}$, and $U^3 = U_{l(k),n}$. Thus $V = U^1 \otimes U^2 \otimes U^3$ and $s_k = I \otimes \sigma_k \otimes \xi_k$ where $\xi_k = \gamma \otimes \cdots \otimes \gamma$ and $\sigma_k = (1 \otimes \beta + \alpha \otimes \gamma)/\sqrt{2}$ or $(\alpha + \beta)/\sqrt{2}$ for k even or odd, respectively. Let

$$a = \frac{\alpha + \beta}{\sqrt{2}}, \quad b = \frac{\gamma\beta}{\sqrt{2}}, \quad c = \frac{a\alpha}{\sqrt{2}}, \quad d = \frac{\beta a}{\sqrt{2}}, \quad e = \frac{\gamma a}{\sqrt{2}}.$$

Check that $\gamma\beta = -\beta\gamma$ and $\gamma\alpha = -\alpha\gamma$. Hence $\gamma a = -a\gamma$. Similarly, $\alpha\beta = -\beta\alpha$. Check also that α, β, γ, and a are involutions, $b^2 = -1/2 = e^2$, $ac + ca = a$, $de + ed = e$, $c^2 + 1/2 = c$, and $d^2 + 1/2 = d$.

(1) As $\gamma^2 = 1$, $\sigma_k^2 = (\gamma \otimes \cdots \otimes \gamma)^2 = \gamma^2 \otimes \cdots \otimes \gamma^2 = 1$. Similarly, if k is odd then $\sigma_k^2 = a^2 = 1$, while if k is even then

$$\sigma_k^2 = \left(\frac{1 \otimes \beta + \alpha \otimes \gamma}{\sqrt{2}}\right)^2 = \frac{1 \otimes \beta^2 + \alpha \otimes (\beta\gamma + \gamma\beta) + \alpha^2 \otimes \gamma^2}{2} = 1,$$

as $\beta^2 = \alpha^2 = \gamma^2 = 1$ and $\beta\gamma = -\gamma\beta$. Therefore

$$s_k^2 = (I \otimes \sigma_k \otimes \xi_k)^2 = I \times \sigma_k^2 \otimes \xi_k^2 = 1.$$

Next let $V^1 = U_{1,j-1}$, $V^2 = U_{j,j+1}$, and $V^3 = U_{j+2,n}$. If $k = 2j - 1$ then $s_k = I \otimes (a \otimes \gamma) \otimes \xi_k'$, where $\xi_k' = \gamma \otimes \cdots \otimes \gamma$ and $s_{k+1} = I \otimes \sigma_{k+1} \otimes \xi_{k+1}$. Thus $s_k s_{k+1} = I \otimes (a \otimes \gamma)\sigma_{k+1} \otimes \xi_k' \xi_{k+1}$ with $\xi_k' \xi_{k+1} = \gamma^2 \otimes \cdots \otimes \gamma^2 = I$ as $\gamma^2 = 1$. Thus to show $(s_k s_{k+1})^3 = -I$ we must show $x^3 = -I$, where

$$x = (a \otimes \gamma)\sigma_{k+1} = (a \otimes \gamma)((1 \otimes \beta + \alpha \otimes \gamma)/\sqrt{2}) = a \otimes b + c \otimes 1.$$

Equivalently we must show $x^2 + 1 = x$. But

$$x^2 = a^2 \otimes b^2 + (ac + ca) \otimes b + c^2 \otimes 1 = -1/2 + a \otimes b + c^2 \otimes 1,$$

as $a^2 = 1$, $b^2 = -1/2$, and $ac + ca = a$, so as $c^2 + 1/2 = c$, indeed

$$x^2 + 1 = a \otimes b + c^2 \otimes 1 + (1 \otimes 1)/2 = a \otimes b + c \otimes 1 = x,$$

completing the proof that $(s_{2j-1} s_{2j})^3 = -I$.

Similarly let $k = 2j$. Then $s_k = I \otimes \sigma_k \otimes \xi_k$ and $s_{k+1} = I \otimes (1 \otimes a) \otimes \xi_{k+1}$, so $s_k s_{k+1} = I \otimes \sigma_k (1 \otimes a) \otimes \xi_k \xi_{k+1}$, and it remains to show $y^2 + 1 = y$, where

$$y = \sigma_k(1 \otimes a) = \frac{(1 \otimes \beta + \alpha \otimes \gamma)(1 \otimes a)}{\sqrt{2}} = 1 \otimes d + \alpha \otimes e.$$

But

$$y^2 = 1 \otimes d^2 + \alpha \otimes (de + ed) + \alpha^2 \otimes e^2 = 1 \otimes d^2 - 1/2 + \alpha \otimes e,$$

as $de + ed = e$, $\alpha^2 = 1$, and $e^2 = -1/2$. Thus as $d^2 + 1/2 = d$,

$$y^2 + 1 = \alpha \otimes e + 1 \otimes d^2 + (1 \otimes 1/2) = \alpha \otimes e + 1 \otimes d = y,$$

completing the proof of (1).

(2) Let $k < i - 1$; we claim either

    (a) $l(k) \le j(i)$, or

    (b) $k = 2j$ and $i = 2(j + 1)$.

For if $j = j(k)$ then $2j - 1 \le k < i - 1 \le 2j(i) - 1$, so $j \le j(i) - 1$ and hence $l(k) = j + \epsilon \le j(i)$ unless $\epsilon = 2$ (so that $k = 2j$) and $j(i) = j + 1$. In the latter case, as $j(i) = j + 1$ and $k = 2j < i - 1$, we have $i = 2(j + 1)$, establishing the claim.

In case (a) let $V_1 = U_{1,j-1}$, $V_2 = U_{j,l(k)-1}$, $V_3 = U_{l(k),j(i)-1}$, $V_4 = U_{j(i),l(i)-1}$, and $V_5 = U_{l(i),n}$. Then $V = V_1 \otimes \cdots \otimes V_5$ and $s_r = s_{r,1} \otimes \cdots \otimes s_{r,5}$, where s_r induces $s_{r,t}$ on V_t. Further, $s_{i,u} = s_{k,1} = I$ for $u \le 3$, $s_{k,2} = \sigma_k$, $s_{i,4} = \sigma_i$, and $s_{i,5}$ and $s_{k,v}$, $v > 2$, are of the form $\gamma \otimes \cdots \otimes \gamma$. In particular $s_{k,w} s_{i,w} = s_{i,w} s_{k,w}$ for $w \ne 4$, while $s_{k,4} s_{i,4} = -s_{i,4} s_{k,4}$, since $a\gamma = -\gamma a$, $\beta\gamma = -\gamma\beta$, and $\alpha\gamma = -\gamma\alpha$. Therefore $(s_k s_i)^2 = -I$.

In case (b), let $W_1 = U_{1,j-1}$, $W_2 = U_{j,j+2}$, and $W_3 = U_{j+3,n}$. Again $V = W_1 \otimes W_2 \otimes W_3$ and $s_r = s_{r,1} \otimes s_{r,2} \otimes s_{r,3}$ with $s_{r,1} = I$, $s_{r,3}$ of the form $\gamma \otimes \cdots \otimes \gamma$, and

$$s_{k,2} = (1 \otimes \beta \otimes \gamma + \alpha \otimes \gamma \otimes \gamma)/\sqrt{2}, \quad s_{i,2} = (1 \otimes 1 \otimes \beta + 1 \otimes \alpha \otimes \gamma)/\sqrt{2}.$$

It remains to show $(s_i s_k)^2 = -I$, so we must show $s_{k,2} s_{i,2} = -s_{i,2} s_{k,2}$. This follows because $\beta\alpha = -\alpha\beta$, $\alpha\gamma = -\gamma\alpha$, and $\beta\gamma = -\gamma\beta$.

(3) As $(s_k s_i)^2 = -I$ for $i > k + 1$, we have $-I \in G^{(1)}$. Let $\bar{G} = G/\langle -I \rangle$. Then $\bar{G} = \langle \bar{s}_k : 1 \le k < 2n \rangle$, and by (1) and (2), $|\bar{s}_i \bar{s}_j| = m_{i,j}$, where $m_{i,i} = 1$, $m_{i,i+1} = m_{i+1,i} = 3$, and $m_{i,j} = 2$ for $|i - j| > 1$. Thus by 30.19, $\bar{G} \cong S_{2n}$. Let $n > 2$. As $\bar{G} \cong S_{2n}$ with $2n > 4$, $\overline{G^{(1)}} = \bar{G}^{(1)} \cong A_{2n}$ is a nonabelian simple group by 15.16. Thus as $-I \in G^{(1)}$, $G^{(1)}$ is quasisimple or $G^{(1)} = \langle -I \rangle \times G^{(2)}$ with $G^{(2)} \cong A_{2n}$. In the latter case the projection of $s_1 s_3$ on $G^{(2)}$ is an involution, which is impossible, as $(s_1 s_3)^2 = -I$.

Chapter 11, Exercise 10. We first observe Z_x is subnormal in G_y and hence also in $G_{y,z}$ for each $z \in \Gamma_y$, since $Z_x \le G_{y,z}$ by definition of Z_x. Let $Q_y = G_{y,\Gamma_y}$.

Then $Z_x \leq Q_y \leq G_x \leq N_G(Z_x)$, so $Z_x \trianglelefteq Q_y \trianglelefteq G_y$, establishing the observation.

Pick a prime q such that Z_x is not a q-group, and for $H \leq G$ let $\theta(H) = \theta_q(H) = O_q(H)E(H)$. We next show $\theta(G_y) \leq G_x \geq \theta(G_{y,z})$. For as Z_x is subnormal in G_y and $G_{y,z}$, we have $E(G_y)E(G_{y,z}) \leq N_{G_y}(Z_x)$ by 31.4, while by hypothesis $N_{G_y}(Z_x) \leq G_x$. Similarly, as Z_x is not a q-group, $1 \neq O^q(Z_x)$, so $N_{G_y}(O^q(Z_x)) \leq G_x$ and by Exercise 11.9, $O_q(G_y)O_q(G_{y,z}) \leq N_{G_y}(O^q(Z_x))$. This completes the proof of the second claim.

Third we show $\theta(G_y) = \theta(G_{x,y}) = \theta(Q_y)$. By claim 2, $\theta(G_y) \leq G_{x,y}$, so as $\theta(G_y) \trianglelefteq G_y$, $\theta(G_y) \leq \theta(G_{x,y})$ (cf. 31.3). By symmetry between x, y, z and z, x, y and claim 2, $\theta(G_{x,y}) \leq G_z$, so $\theta(G_{x,y}) \leq Q_y \leq N_G(\theta(G_{x,y}))$ and hence $\theta(G_{x,y}) \leq \theta(Q_y)$. Finally, as $\theta(Q_y) \trianglelefteq G_y$, $\theta(Q_y) \leq \theta(G_y)$, establishing claim 3.

Now if Z_y is not a q-group then we have symmetry between x and y, so by claim 3, $\theta(G_y) = \theta(G_{x,y}) = \theta(G_x)$. Then if $\theta(G_x) \neq 1$,

$$G_y \leq N_{G_y}(\theta(G_y)) = N_{G_y}(\theta(G_x)) \leq G_x$$

and by symmetry, $G_x \leq G_y$, contradicting our hypothesis that $G_x \neq G_y$. Thus $\theta(G_x) = 1$, so $O_q(G_x) = E(G_x) = 1$. But $Z_x \neq 1$, so $F^*(G_x) \neq 1$ and hence $O_p(G_x) \neq 1$ for some prime p. However we have shown $O_p(G_x) = 1$ unless Z_x or Z_y is a p-group, so interchanging the roles of x and y if necessary, we may assume Z_x is a p-group.

Next $Q_x \trianglelefteq G_{x,y}$, so $\theta(Q_x) \leq \theta(G_{x,y}) = \theta(Q_y)$. Hence

$$\theta(Q_x) \leq \bigcap_{y \in \Gamma_x} \theta(Q_y) \leq \theta(Z_x) = 1,$$

as Z_x is a p-group. That is $\theta(Q_x) = 1$. But $Z_y \leq Q_x \leq G_{x,y} \leq N_G(Z_y)$, so $\theta(Z_y) \leq \theta(Q_x) = 1$. Therefore $F^*(Z_y) = O_p(Z_y)$. Thus as $Z_y \neq 1$, $O_p(Z_y) \neq 1$, so Z_y is not a q-group for any prime $q \neq p$, and hence by an earlier reduction, $\theta(G_x) = \theta(G_{x,y}) = \theta(G_y) = 1$ for each $q \neq p$. Therefore $F^*(G_x)$, $F^*(G_y)$, and $F^*(G_{x,y})$ are p-groups.

Chapter 11, Exercise 11. Let $y \in Y$, $\mathcal{F} = \{G_x, G_y\}$, and $\Gamma = \Gamma(G, \mathcal{F})$ be the coset geometry defined in section 3. If $G_x = G_y$ then $Y = yG_x = yG_y = \{y\}$, contradicting the hypotheses that Y is nontrivial. Thus $G_x \neq G_y$.

As G is primitive on X, G_x is maximal in G by 5.19. Thus if $1 \neq H \trianglelefteq G_x$ then by maximality of G_x, $N_G(H) = G_x$ or G. But in the latter case, $H \leq \ker_{G_x}(G)$ by 5.7, while as G is faithful and transitive on X, $\ker_{G_x}(G) = 1$, contradicting $H \neq 1$. Thus $G_x = N_G(H)$, so in particular $N_{G_y}(H) \leq G_x$. Thus, in the language of Exercise 11.10, if $Z_x \neq 1$, then by Exercise 11.10, $F^*(G_x)$ is a p-group for some prime p. Therefore we may assume $Z_x = 1$. Thus G_x is faithful on

$$\Omega = \bigcup_{y \in \Gamma_x} \Gamma_y.$$

Let $m = |Y|$. Then $m = |\Gamma_x| = |\Gamma_z|$ for each $z \in \Gamma_x$, so $|\Omega| \le m^2$ and then, as G_x is faithful on Ω, $|G_x| \le m^2!$. Thus we may take $f(m) = m^2!$.

Chapter 12, Exercise 5. (1) The possible cycle structures for elements of G are:

$$1, \ (1,2), \ (1,2,3), \ (1,2,3,4), \ (1,2,3,4,5), \ (1,2)(3,4,5), \ (1,2)(3,4)$$

giving a set $S = \{g_i : 1 \le i \le 7\}$ of elements such that $|g_i| = i$ for $i < 7$ and $|g_7| = 2$ by 15.2.4. Further S is a set of representatives for the conjugacy classes of G by 15.3.2.

(2) The representation of G on X is of degree 5 and 2-transitive by 15.12.2. Arguing as in Exercise 5.1, G has a transitive representation of degree 6 on the coset space G/H, where $H = N_A(P)$ and $P = \langle g_5 \rangle \in \mathrm{Syl}_5(A)$. Namely, counting the number of 5-cycles, G has 24 elements of order 5 and hence 6 Sylow 5-subgroups, so by Sylow's Theorem, $|G : H| = 6$. Now the only possible cycle structure for the element g_5 of order 5 on the 6-set G/H is one cycle of length 5 and one fixed point, so P is transitive on $G/H - \{H\}$ and hence G is 2-transitive on G/H by 15.12.1.

Let ψ be the permutation character of G on X. By Exercise 4.5.1, $\psi(g)$ is the number of fixed points of $g \in G$ on X, so

$$\psi(g_i) = 5, \ 3, \ 2, \ 1, \ 0, \ 0, \ 1$$

for $i = 1, \ldots, 7$, respectively.

Similarly let φ be the permutation character of G on G/H. Let $h = (1,2,4,3)$. By 15.3.1, $g_5^h = (2,4,1,3,5) = g_5^2$, so $h \in H$ and $H = P\langle h \rangle$ as

$$|P\langle h \rangle| = |P||h| = 20 = |G|/|G : H| = |H|.$$

Thus the elements in H are of conjugate to g_1, g_4, g_5, and g_7, so for all other g_i, $\varphi(g_i) = |\mathrm{Fix}_{G/H}(g_i)| = 0$. By Exercise 2.7, if $Q \in \mathrm{Syl}_q(H)$ then $|\mathrm{Fix}(Q)| = |N_G(Q) : N_H(Q)|$, so as $H = N_G(P)$, $\varphi(g_5) = 1$ while as $\langle h \rangle \in \mathrm{Syl}_2(H)$ with $N_G(\langle h \rangle) \in \mathrm{Syl}_2(G)$ of order 8, $\varphi(g_4) = |\varphi(h)| = 2$. Thus $g_4 = (a,b,c,d)(e)(f)$ on G/H, so $g_7 = g_4^2 = (a,c)(b,d)$ and hence $\varphi(g_7) = 2$.

(3) By Exercise 12.1, G has $|G : G^{(1)}|$ linear characters. Let $A = \mathrm{Alt}(X)$; by 15.16, A is simple, and by 15.5, $|G : A| = 2$, so $A = G^{(1)}$ and G has two linear characters, the principal character χ_1 and the *sign character* χ_2, where

by Exercise 12.1, $\ker(\chi_2) = A$. That is $\chi_2(a) = 1$ for $a \in A$ and $\chi_2(g) = -1$ (the primitive 2nd root of 1) for $g \in G - A$.

(4) By Exercise 12.1, χ_1 and χ_2 are irreducible. By Exercise 12.6.3, $\chi_3 = \psi - \chi_1$ and $\chi_4 = \varphi - \chi_1$ are irreducible characters. Then by Exercises 9.3 and 9.10, $\chi_{i+2} = \chi_2 \chi_i$, $i = 3, 4$, are also irreducible characters.

This gives 6 irreducible characters of G, exhibited in the first six rows of the table below. As G has 7 conjugacy classes, G has 7 irreducible characters by 34.3.1. Thus it remains to determine χ_7. Let $n_i = \chi_i(1)$; by 35.5.3,

$$120 = |G| = \sum_{i=1}^{7} n_i^2,$$

so $n_7 = 6$, giving the first column of the table. Then we use the orthogonality relation 35.5.2 to calculate $\chi_7(g_i)$ for $i > 1$ to complete the table.

Character Table of S_5

	g_1	g_2	g_3	g_4	g_5	g_6	g_7
χ_1	1	1	1	1	1	1	1
χ_2	1	-1	1	-1	1	-1	1
χ_3	4	2	1	0	-1	-1	0
χ_4	5	-1	-1	1	0	-1	1
χ_5	4	-2	1	0	-1	1	0
χ_6	5	1	-1	-1	0	1	1
χ_7	6	0	0	0	1	0	-2

Chapter 13, Exercise 1. (1) Let $M = C_G(g)$. Then $|HxM| = |M : M_{Hx}|$, where M_{Hx} is the stabilizer of Hx in M. Thus $M_{Hx} = M \cap H^x = C_{H^x}(g)$, so $(|HxM|, p) = 1$ iff $C_{H^x}(g)$ contains a Sylow p-subgroup of $C_G(g)$ iff $C_H(g^{x^{-1}})$ contains a Sylow p-subgroup of $C_G(g^{x^{-1}})$ iff $g^{x^{-1}}$ is extremal in H.

(2) Let $A = H/K$, $\alpha: H \to A$ the natural map, and V the transfer of G into A via α. Recall by 37.2 that V is a group homomorphism, so as A is abelian $G^{(1)} \le \ker(V)$ and hence it suffices to show $gV \ne 1$.

Choose a set X of coset representatives for H in G as in 37.3. As $|g| = p$ the length n_i of the ith cycle of g is 1 or p. If $n_i = p$ then $1 = g^{n_i}$, so $g^{n_i x_i^{-1}} \in g^{n_i} K$; that is

$$g^{n_i x_i^{-1}} \alpha = (g^{n_i}) \alpha = (g\alpha)^{n_i} = 1.$$

On the other hand $n_i = 1$ iff g fixes Hx_i, in which case $g^{x_i^{-1}} \in H$. Now $M = C_G(g)$ permutes the fixed points of g on G/H and $g^{(x_i m)^{-1}} = g^{x_i^{-1}}$ for each $m \in M$, so if $\{Hy_j M : 1 \le j \le s\}$ are the orbits of M on $\text{Fix}_{G/H}(g) = \{Hx_i : 1 \le i \le t\}$ then

$$\prod_{i=1}^{t} (g^{x_i^{-1}})\alpha = \prod_{j=1}^{s} ((g^{y_j^{-1}})\alpha)^{k_j},$$

where $k_j = |Hy_j M|$ is the length of the jth orbit. If p divides k_j then $(g^{y_j^{-1}})^{k_j} = 1 = (g\alpha)^{k_j}$. On the other hand, if p does not divide k_j, then by (1), $g^{y_j^{-1}}$ is extremal in H, so by hypothesis, $g^{y_i^{-1}} \in gK$ and hence $(g^{y_j^{-1}})\alpha = g\alpha$, so again $(g^{y_j^{-1}}\alpha)^{k_j} = (g\alpha)^{k_j}$. Thus using 37.3.3,

$$gV = \prod_{i=1}^{r} ((g^{n_i})^{x_i^{-1}})\alpha = \prod_{i=1}^{r} (g\alpha)^{n_i} = (g\alpha)^n$$

where $n = \sum_{i=1}^{r} n_i = |G : H|$. Finally, by hypothesis, $(|G : H|, p) = 1$ and $g \in H - K$, so $g\alpha \ne 1$, and hence $|g\alpha| = p$. Therefore $gV = (g\alpha)^n \ne 1$, completing the proof.

Chapter 14, Exercise 9. (1) As $u \le w$, $w = xu$ for some $x \in W$ with $n = l(w) = l(x) + l(u)$. Then $x = r_1 \cdots r_m$ and $u = r_{m+1} \cdots r_n$ with $r_i \in S$. By 41.7, $(C_i : 0 \le i \le n)$ is a gallery from C to Cw in $\mathcal{B}$ where $C_i = Cr_{n-i+1} \cdots r_n$. By construction, $C_{l(u)} = C_{n-m} = Cr_{m+1} \cdots r_n = Cu$.

(2) Let $u = sw$ and $n = l(w)$. Thus $w = su$ and by hypothesis, $l(u) < l(w)$, so $l(u) = n - 1$ and $u \le w$. Hence by (1), Cu is adjacent to Cw, so $E = Cu \cap Cw$ is a wall of the chambers Cu and Cw.

Next B^w fixes Cw and hence each subsimplex of Cw. Similarly B^u fixes each subsimplex of Cu, so $B^u B^w$ fixes E. Thus $B \cap B^u B^w \subseteq N_B(E) = A$.

Let X be the set of galleries in $\mathcal{B}$ from C to E of length $n - 1$. By 41.7 and 42.3, $d(C, Cw) = n$ and $d(C, Cu) = n - 1$, so $d(C, E) = n - 1$. Thus $X \subseteq \Sigma$ by 42.3. As A acts on C and E, A acts on X. Further, Cu and Cw are the two chambers in Σ through E, so as $d(C, Cw) = n$, Cu is the terminal member of each member of X. Thus $A \le G_{Cu} = B^u = B^{sw}$.

(3) By hypothesis,

$$l(s \cdot sw^{-1}) = l(w^{-1}) = l(w) \le l(ws) = l(sw^{-1}),$$

so by (2), $B \cap B^{w^{-1}} B^{sw^{-1}} \subseteq B^{w^{-1}}$. Then conjugating by w, $B^w \cap BB^s \subseteq B$.

(4) As $u \le w$, $w = xu$ with $n = l(w) = l(x) + l(u)$. Induct on $l(x)$. If $l(x) = 0$ then $x = 1$ and the lemma is trivial. Thus $x = sy$ with $s \in S$ and $l(y) = l(x) - 1$, and $u \le yu$, so by induction $B \cap B^{yu} \le B^u$. Also $l(sw) = l(yu) < l(w)$, so by (2), $B \cap B^w \le B^{yu}$. Thus $B \cap B^w \le B \cap B^{yu} \le B^u$.

(5) Define

$$\tilde{H} = \bigcap_{w \in W} B^w.$$

Then $\tilde{H}$ is the pointwise stablilizer in G of Σ, so (a) and (b) are equivalent. By Exercise 10.6.2, $w \leq w_0$ for all $w \in W$, so by (4), $B \cap B^{w_0} \leq B^w$. Thus $\tilde{H} = B \cap B^{w_0}$, so (b) and (c) are equivalent.

(6) Let $\tilde{T} = (G, B, \tilde{N}, \tilde{S})$. Recall $H = B \cap N \trianglelefteq N$ with $W = N/H$ and for $w = nH \in W$, $B^w = B^n$. Thus $H \leq B^w$ for all $w \in W$, so $H \leq \tilde{H}$.

As T satisfies BN1, $G = \langle B, N \rangle$. Then as $N \leq \tilde{H}N = \tilde{N} \leq G, G = \langle B, \tilde{N} \rangle$. Also

$$B \cap \tilde{N} = B \cap \tilde{H}N = \tilde{H}(B \cap N) = \tilde{H}H = \tilde{H}.$$

By construction, $\tilde{H}$ is the pointwise stabilizer in G of Σ, so as N acts on Σ, N acts on $\tilde{H}$ and hence $\tilde{H} \trianglelefteq \tilde{H}N = \tilde{N}$. Thus $\tilde{T}$ satisfies BN1. Also $\tilde{H} \cap N$ is the kernel of the action of N on Σ, while by 43.5 and 41.8.2, $W = N/H$ is regular on Σ, so $\tilde{H} \cap N = H$. Let $\tilde{W} = \tilde{N}/\tilde{H}$. Then

$$\tilde{W} = \tilde{N}/\tilde{H} = N\tilde{H}/\tilde{H} \cong N/\tilde{H} \cap N = N/H = W,$$

so BN2 is satisfied with $\tilde{S} = \{\tilde{s} : s \in S\}$, where $\tilde{w} = n\tilde{H}$ for $w = nH \in S$. Notice that as $\tilde{H} \leq B$, we have $\tilde{w}B = wB$ and $B\tilde{w} = Bw$, so also $B^{\tilde{w}} = B^w$. Thus, as T satisfies BN3 and BN4, so does $\tilde{T}$.

Therefore $\tilde{T}$ is a Tits system. By construction,

$$\tilde{H} = \bigcap_{w \in W} B^w = \bigcap_{\tilde{w} \in \tilde{W}} B^{\tilde{w}},$$

so $\tilde{T}$ is saturated.

Chapter 14, Exercise 10.

(1) As $\alpha w > 0, l(sw) > l(w)$ by Exercise 10.6.3, so $l(w^{-1}s) = l(sw) > l(w) = l(w^{-1})$. Also by Exercise 10.6.2, $w^{-1}s \leq w_0$, and by Exercise 10.3.2, $l(w_0s) \leq l(w_0)$. Therefore by Exercise 10.6.4, $w^{-1} = (w^{-1}s)s \leq w_0s$. Then by Exercise 14.9.4,

$$B_\alpha = B \cap B^{w_0s} \leq B^{w^{-1}}.$$

(2) As $\alpha w < 0, l(sw) < l(w)$ by Exercise 10.6.3, so by Exercise 14.9.2, $B \cap B^{sw}B^w \subseteq B^{sw}$. Conjugating this containment by $w^{-1}s$, we conclude $B^{w^{-1}s} \cap BB^s \subseteq B$, so $B^{w^{-1}s} \cap B^s \leq B$.

Next by hypotheses T is saturated, so by Exercise 14.9.5, $B \cap B^{w_0} = H$. Thus

$$(B_\alpha \cap B^{w^{-1}})^s = (B \cap B^{w_0s} \cap B^{w^{-1}})^s = B^s \cap B^{w_0} \cap B^{w^{-1}s} \leq B \cap B^{w_0} = H$$

so $B_\alpha \cap B^{w^{-1}} = H^s = H$.

(3) As $l(w) \le l(ws)$, $l(w^{-1}) \le l(sw^{-1})$, so by 43.3.1, $sBw^{-1} \subseteq Bsw^{-1}B$, and hence $B \subseteq sBsw^{-1}Bw = B^s B^w$. Thus for $b \in B$, there exist $x \in B^s$ and $y \in B^w$ with $b = xy$. Then $y = x^{-1}b \in B^w \cap B^s B \subseteq B$ by Exercise 14.9.3, so $y \in B \cap B^w$ and $x = by^{-1} \in B \cap B^s$.

(4) Notice $\alpha s = -\alpha < 0$, so by (2), $B_\alpha \cap B^s = H$. Also, setting $w = w_0 s$, $l(w) \le l(w_0) = l(ws)$, so by (3), $B = (B \cap B^s)(B \cap B^w) = (B \cap B^s)B_\alpha$.

(5) If $B_\alpha = H$ then by (4), $B = H(B \cap B^s)$. Then as s is an involution, $B^s = H^s(B \cap B^s)^s = H(B \cap B^s) = B$, contrary to BN4.

(6) If $\alpha w > 0$ then $B_\alpha \le B^{w^{-1}}$ by (1). On the other hand, if $\alpha w < 0$, then by (2), $H = B_\alpha \cap B^{w^{-1}}$, so $B_\alpha \not\le B^{w^{-1}}$ by (5).

(7) Define $\varphi: \Delta \to \Phi$ by $\varphi: B_\alpha^w \mapsto \alpha w$. Let $\beta \in \pi$ and $w \in W$; we first show

(*) $B_\alpha^w = B_\beta$ iff $\beta = \alpha w$.

First, if $B_\alpha^w = B_\beta$ then as $B_\beta = B \cap B^{w_0 r_\beta}$, $B_\alpha = B^{w^{-1}} \cap B^{w_0 r_\beta w^{-1}}$. Thus by (6), $\alpha w > 0 < \alpha w r_\beta w_0$. Therefore $\alpha w r_\beta < 0$ by Exercise 10.3, so $\alpha w = \beta$ by 30.7. Similarly if $\alpha w = \beta$ then $\alpha w r_\beta w_0 > 0 < \alpha w$, so by (6), $B_\alpha \le B^{w^{-1}} \cap B^{w_0 r_\beta w^{-1}}$ and hence $B_\alpha^w \le B \cap B^{w_0 r_\beta} = B_\beta$. As $\beta w^{-1} = \alpha$, by symmetry $B_\beta^{w^{-1}} \le B_\alpha$, so $B_\beta = B_\alpha^w$. Thus we have established (*).

By (*), $B_\alpha^w = B_\beta^u$ iff $B_\alpha^{wu^{-1}} = B_\beta$ iff $\alpha w u^{-1} = \beta$ iff $\alpha w = \beta u$. Thus φ is a well defined injection. By 30.9.3, φ is a surjection. As $B_\alpha^{wu}\varphi = \alpha w u = (B_\alpha^w)\varphi u$, φ is W-equivariant.

(8) First notice

$$B_{sw}^s = (B \cap B^{w_0 w^{-1} s})^s = B^s \cap B^{w_0 w^{-1}}.$$

Next as $\alpha w < 0$, $\alpha w w_0 > 0$ by Exercise 10.3. Hence by (1), $B_\alpha \le B^{w_0 w^{-1}}$, so $B_\alpha \le B \cap B^{w_0 w^{-1}} = B_w$.

Again as $\alpha w w_0 > 0$, $l(w w_0) \le l(sw w_0)$ by Exercise 10.6.3, so $l(w_0 w^{-1} s) \ge l(w_0 w^{-1})$. Then by Exercise 14.9.3, $B^s \cap B^{w_0 w^{-1}} \le B$, so $B_{sw}^s = B^s \cap B^{w_0 w^{-1}} \le B \cap B^{w_0 w^{-1}} = B_w$. Thus $B_\alpha B_{sw}^s \subseteq B_w$. On the other hand by (4), $B = B_\alpha(B \cap B^s)$, so as $B_\alpha \le B^{w_0 w^{-1}}$,

$$B_w = B \cap B^{w_0 w^{-1}} = B_\alpha(B \cap B^{w_0 w^{-1}} \cap B^s) = B_\alpha(B \cap B_{sw}^s) \subseteq B_\alpha B_{sw}^s.$$

That is, $B_\alpha B_{sw}^s = B_w$. Finally $\alpha s = -\alpha < 0$, so by (2), $B_\alpha \cap B^s = H$. Therefore $B_\alpha \cap B_{sw}^s = B_\alpha \cap B^s \cap B^{w_0 w^{-1}} = H$.

(9) We apply (8) with $w = w_{i-1} w_0$, $\alpha = \alpha_i$, and $s = r_i$ for each $1 \le i < n$. By Exercise 10.6.3, $\alpha_i w_{i-1} w_0 < 0$ since $l(r_i w_{i-1} w_0) = l(w_i w_0) < l(w_{i-1} w_0)$. Thus by (8),

(*) $B_{w_{i-1} w_0} = B_{\alpha_i} B_{w_i w_0}^{r_i}$

and

(**)
$$B_{\alpha_i} \cap B_{w_i w_0}^{r_i} = H.$$

Claim

(!)
$$B_{w_{i-1} w_0}^{w_{i-1}} = \prod_{j=i}^{k} B_{\alpha_j w_{j-1}} \cdot B_{w_k w_0}^{w_k}$$

for $1 \le i < n$ and $i \le k < n$. We induct on k. By (*),

$$B_{w_{i-1} w_0}^{w_{i-1}} = B_{\alpha_i}^{w_{i-1}} B_{w_i w_0}^{r_i w_{i-1}} = B_{\alpha_i w_{i-1}} B_{w_i w_0}^{w_i},$$

so (!) holds when $k = i$. Assume for $k - 1$. By (*),

$$B_{w_{k-1} w_0}^{w_{k-1}} = B_{\alpha_k}^{w_{k-1}} B_{w_k w_0}^{r_k w_{k-1}} = B_{\alpha_k w_{k-1}} B_{w_k w_0}^{w_k},$$

so by the induction assumption

$$B_{w_{i-1} w_0}^{w_{i-1}} = \prod_{j=i}^{k-1} B_{\alpha_j w_{j-1}} \cdot B_{w_{k-1} w_0}^{w_{k-1}} = \prod_{j=i}^{k} B_{\alpha_j w_{j-1}} \cdot B_{w_k w_0}^{w_k},$$

establishing (!). Notice that $w_{n-1} w_0 = r_n$, so

$$B_{w_{n-1} w_0}^{w_{n-1}} = B_{r_n}^{w_{n-1}} = B_{\alpha_n}^{w_{n-1}} = B_{\alpha_n w_{n-1}}.$$

Substituting this equality into (!) with $k = n - 1$, we get

(!!)
$$B_{w_{i-1} w_0}^{w_{i-1}} = \prod_{j=i}^{n-1} B_{\alpha_j w_{j-1}} \cdot B_{w_{n-1} w_0}^{w_{n-1}} = \prod_{j=i}^{n} B_{\alpha_j w_{j-1}}$$

Further $B_{w_0} = B \cap B^{w_0 w_0} = B$, so applying (!!) at $i = 1$ we conclude

$$B = \prod_{j=1}^{n} B_{\alpha_j w_{j-1}}.$$

Next by (**) and (!!),

$$H = H^{w_{i-1}} = \left(B_{\alpha_i} \cap B_{w_i w_0}^{r_i} \right)^{w_{i-1}} = B_{\alpha_i w_{i-1}} \cap B_{w_i w_0}^{w_i}$$

$$= B_{\alpha_i w_{i-1}} \cap \prod_{j=i+1}^{n} B_{\alpha_j w_{j-1}}.$$

It remains to show $\psi : i \mapsto \alpha_i w_{i-1}$ is a bijection of $I = \{1, \ldots, n\}$ with Φ^+. We saw at the beginning of the proof that $\alpha_i w_{i-1} w_0 < 0$, so $\alpha_i w_{i-1} > 0$ and hence $I\psi \subseteq \Phi^+$. By Exercise 10.3, $|\Phi^+| = n$, so it remains to show ψ is an injection. But if $\alpha_i w_{i-1} = \alpha_k w_{k-1}$ for some $k > i$, then

$$\alpha_i = \alpha_k w_{k-1} w_{i-1}^{-1} = \alpha_k r_{k-1} \cdots r_i,$$

so $0 > \alpha_i r_i = \alpha_k r_{k-1} \cdots r_{i+1}$, contrary to Exercise 10.6.3, since

$$l(r_k \cdots r_{i+1}) > l(r_{k-1} \cdots r_{i+1}).$$

Chapter 15, Exercise 4. (1) To show $H = C_G(a)$ is balanced we must show

$$O_{r'}(C_H(X)) \leq O_{r'}(H) \quad \text{for each } X \text{ of order } r \text{ in } X.$$

But as H is solvable, this is immediate from 31.15.

(2) We must show θ is balanced. But for $a, b \in A^{\#}$

$$\theta(b) \cap C_G(a) = O_{r'}(C_G(b)) \cap C_G(a) = O_{r'}\big(C_{C_G(a)}(X)\big) \leq O_{r'}(C_G(a)) = \theta(a),$$

where $X = \langle a, b \rangle$, as $C_G(a)$ is balanced by the hypotheses of part (2).

(3) Let θ be the signalizer functor defined in (2); θ *is* a signalizer functor by (1) and (2). By the Solvable Signalizer Functor Theorem, θ is complete; that is, there exists an r'-subgroup Y of G such that $\theta(a) = C_Y(a)$ for each $a \in A^{\#}$.

For B a noncylic subgroup of A define

$$Y(B) = \langle \theta(b) : b \in B^{\#} \rangle.$$

Now for $g \in G$ with $a, a^g \in A^{\#}$,

$$\theta(a)^g = O_{r'}(C_G(a))^g = O_{r'}(C_G(a^g)) = \theta(a^g),$$

so $N_G(B) \leq N_G(Y(B))$. Further, by construction $Y(B) \leq Y$, while by 44.8.1, $Y \leq Y(B)$, so $Y = Y(B)$. Thus $N_G(B) \leq N_G(Y)$, so $\Gamma_{2,A}(G) \leq N_G(Y)$. But by hypothesis, $G = \Gamma_{2,A}(G)$, so $Y \trianglelefteq G$, and hence as Y is an r'-group, $Y \leq O_{r'}(G)$. Therefore $O_{r'}(C_G(a)) \leq Y \leq O_{r'}(G)$.

In particular if $O_{r'}(G) = 1$ then $O_{r'}(C_G(a)) = 1$. Hence, as $C_G(a)$ is solvable, $F^*(C_G(a)) = F(C_G(a)) = O_r(C_G(a))$.

Chapter 16, Exercise 5. (1) By definition of strong embedding, H is proper of even order, so there is an involution $i \in H$. Indeed, 46.4.3 is one of the equivalent conditions for strong embedding, so

(*) $\qquad\qquad |H \cap H^g|$ is odd for $g \in G - H$.

Let $j \in I$; it suffices to show $j \in i^G$. Suppose $j^G \subseteq H$. Then for $g \in G$, $j^g \in H \cap H^g$, so $|H \cap H^g|$ is even and hence $g \in H$ by (*). As this holds for each $g \in G$, $G = H$, contradicting H proper. This contradiction shows there is a conjugate s of j in $G - H$, and it suffices to show $s \in i^G$. But if not, then by 45.2, $|is|$ is even, so by 45.2.3 there is an involution $z \in \langle is \rangle$ centralizing i and s. Now as 46.4.1 is one of the equivalent conditions defining strong embedding, $C_G(t) \leq H$ for each $t \in I \cap H$, so $z \in C_G(i) \leq H$, and then also $s \in C_G(z) \leq H$, contrary to the choice of s.

(2) Let $i, t \in I \cap H$. By (1), $i^g = t$ for some $g \in G$, so $t \in H \cap H^g$ and hence, as above, $g \in H$ by (*). That is, H is transitive on $I \cap H$.

(3) As $u \notin H$, $D = H \cap H^u$ is of odd order by (*).

(4) Let $j \in J$. Then $j \in D$ and $j^u = j^{-1}$ by definition of J. Thus $(uj)^2 = ujuj = j^u j = j^{-1}j = 1$, so $uj \in I$. As $j \in D$, $uj \in uD \cap I$, so $uJ \subseteq uD \cap I$. On the other hand, suppose $d \in D$ with $ud \in I$. Then $1 = (ud)^2 = udud = d^u d$, so $d^u = d^{-1}$ and hence $d \in J$; that is, $uD \cap I \subseteq uJ$.

(5) Set $Y = C_G(j)\langle u \rangle$. As $J \subseteq D$ and $|D|$ is odd by (3), $|j|$ is odd, so as $j \neq 1$, $|j| > 2$. Then as u inverts j, we have $u \notin C_G(j)$. On the other hand $u \in N_G(\langle j \rangle)$, so $Y \leq N_G(\langle j \rangle)$ and hence $C_G(j) \trianglelefteq Y$.

Suppose $t \in C_G(j)$ and let $K = H \cap Y$. By construction, $u \in Y - H$, so K is proper in Y and $t \in Y$ so K has even order. Further for $y \in Y - K$, $y \notin H$, so $|K \cap K^y|$ is odd by (*), and hence K is strongly embedded in Y. Thus applying (1) to Y, the involutions u and t are conjugate in Y. This is impossible, as $t \in C_G(j) \trianglelefteq Y$ while $u \in Y - C_G(j)$.

(6) Let r, s be distinct involutions in uD. Then $rD = uD$ and $1 \neq rs \in D$, so

$$s = r \cdot rs \in rD \cap I = uD \cap I = J$$

by (4). But if $rC_G(t) = sC_G(t)$, also $rs \in C_G(t)$, contrary to (5). That is, $rC_G(t) \neq sC_G(t)$.

(7) and (8): Define $n = |G : H|$ and

$$\Omega = \{(i, x, y): i \in I \text{ and } (x, y) \text{ is a cycle in } i \text{ on } G/H\}.$$

As 46.4.4 is one of the equivalent conditions for strong embedding, i fixes a unique point of G/H, so i has $(n-1)/2$ cycles of length 2. Thus $|\Omega| = |I|(n-1)$ and $|I| = n|I \cap H| = nm$, so $|\Omega| = n(n-1)m$.

On the other hand, if $\Delta = X \times X$ and $\delta(x, y)$ is the number of involutions with cycle (x, y), then

$$|\Omega| = \sum_{(x,y) \in \Delta} \delta(x, y).$$

Now, up to conjugation in G, we have $x = H$, and if $\delta(x, y) \neq 0$ then $y = Hu$ for some $u \in I - H$. Let $\delta = \delta(x, y)$, $u_1, \ldots, u_\delta$ be the involutions with cycle (x, y), and $d_i = u_1 u_i$ for $1 \leq i \leq \delta$. Then $d_i \in D \leq H$, and if $i \neq j$ then by (6),

$$d_i^{-1} d_j = u_i u_1 u_1 u_j C_G(t) = u_i u_j C_G(t) \neq C_G(t),$$

so $d_i C_G(t) \neq d_j C_G(t)$. Thus

$$\delta \leq |H : C_H(t)| = |t^G| = |I \cap H| = m$$

by (2). Hence

$$n(n - 1)m = |\Omega| = \sum_{(x,y) \in \Delta} \delta(x, y) \leq |\Delta|m = n(n - 1)m,$$

so all inequalities are equalities and therefore $\delta(x, y) = m$ for all $x \neq y$ in G/H.

This establishes (7). Further, as $d_i C_G(t) \neq d_j C_G(t)$ for $i \neq j$, we have $|D : C_D(t)| \geq \delta = m = |H : C_H(t)|$, so $H = DC_H(t)$ and hence (2) implies (8).

Chapter 16, Exercise 6. (1) Let z be an involution in $Z(T)$. By hypothesis, $C_G(z)$ is an elementary abelian 2-group, so as $T \leq C_G(z)$, it follows that $C_G(z) = T$. Thus (1) holds. Further as T is elementary abelian, each $t \in T^\#$ is an involution in the center of T, so by symmetry between t and z, $T = C_G(t)$. Therefore T is a TI-set in G.

(2) As T is a TI-set in G, $N_G(S) \leq H$ for each $1 \neq S \leq T$, and as $T \in \mathrm{Syl}_2(G)$, this is one of the equivalent conditions listed in 46.4 (with $k = 1$) for H to be strongly embedded in G.

(3) By (2), we may apply Exercise 16.5. In particular, by part (3) of that exercise, D is of odd order. Thus $|C_D(u)|$ is odd, while as u is an involution, $C_D(u)$ is a 2-group, so $C_D(u) = 1$. Hence u inverts D.

Adopt the notation of Exercise 16.5. Then $I \cap H = T^\#$, so $m = |I \cap H| = q - 1$, as T is elementary abelian of order q. By Exercise 16.5.2, H is transitive on $I \cap H$, so $q - 1 = |H : C_H(t)| = |H : T|$, and hence $|H| = |T|(q - 1) = q(q - 1)$. As u inverts D, $J = uD$ is of order $|D|$, while J is the set of involutions with cycle (H, Hu) on X, so by Exercise 16.5.7, $|J| = m$. Thus $|D| = |J| = m = q - 1$. Therefore, as $D \leq H$ with $T \cap D = 1$ as D is of odd order, we have

$$|TD| = |T||D| = q(q - 1) = |H|,$$

so $H = TD$ and D is a complement to T in H.

(4) First, as T is noncyclic, $q > 2$, so $|D| = q - 1 > 1$. Further, as u inverts D, D is abelian, and T is abelian, while $H/T \cong D$, as D is a complement to T in H by (3). Thus H is solvable and D is a Hall $2'$-subgroup of H, so $D^G \cap H = D^H$ by Phillip Hall's Theorem 18.5.

Let $M = N_G(D)$ and $u \in S \in \mathrm{Syl}_2(M)$. As T is elementary abelian, so is S, so if $S \neq \langle u \rangle$ then S is noncyclic and hence $D = \langle C_D(s) : s \in S^\# \rangle$ by Exercise 8.1. This is impossible, as $D \neq 1$ while $C_D(s) = 1$, since $C_G(s)$ is a 2-group and $|D|$ is odd.

Thus $S = \langle u \rangle$, so by 39.2, $M = \langle u \rangle E$, where $E = O(M)$. Let $\Delta = \mathrm{Fix}(D)$ be the fixed points of D on X. As $D^G \cap H = D^H$, M is transitive on Δ by 5.21. Thus $k = |\Delta| = |M : M \cap H|$. But

$$M \cap H = N_H(D) = N_{DT}(D) = DN_T(D) = D,$$

as $H = TD$ and $N_D(T) = 1$, as $C_D(t) = 1$ for each $t \in T^\#$. Thus M/D is regular on Δ.

Let Γ be the set of triples (i, x, y) such that $i \in uE$ and (x, y) is a cycle of i on Δ. As $|E|$ is odd, $C_E(u) = 1$, so $uE = I \cap M$ is of order $|E|$, and as M/D is regular on Δ, each $i \in uE$ is regular on Δ. Hence there are k members of Γ whose first entry is i, so

$$|\Gamma| = |E|k = |D||E:D|k = |D|k^2/2 = (q-1)k^2/2.$$

On the other hand, there are $k(k-1)$ choices for (x, y), and by Exercise 16.5.7 there are $m = q-1$ involutions i with cycle (x, y), each in $iG_{x,y} = iD \subseteq uE$, so $|\Gamma| = k(k-1)(q-1)$. Thus $k = 2$, so $D = E$ and (4) is established.

(5) Let $x = H \in X$, $y = xu$, and $z \in X - \{x\} = X'$. By Exercise 16.5.7, there is $v \in I$ with cycle (x, z), so by symmetry between u and v, $\mathrm{Fix}(D_0) = \{x, z\}$ and $D_0 = H \cap H^v$ is a Hall $2'$-subgroup of H. Therefore there is $h \in H$ with $D^h = D_0$ by Phillip Hall's Theorem 18.5, as observed during the proof of (4). Now by (4)

$$\{x, yh\} = \mathrm{Fix}(D)h = \mathrm{Fix}(D^h) = \mathrm{Fix}(D_0) = \{x, z\},$$

so $yh = z$. Finally, as $H = DT$, we have $h = dt$ for some $d \in D$ and $t \in T$, so $z = yh = ydt = yt$, and therefore T is transitive on X'. But by 46.4, $\{x\} = \mathrm{Fix}(t)$ for each $t \in T^\#$, so T is regular on $|X'|$.

As T is regular on X', the action of D on X' is equivalent to its action on T by conjugation by 15.11. Thus as D is regular on $T^\#$, D is regular on $X - \{x, y\}$, so G is 3-transitive on X and only the identity fixes 3 or more points of X.

(6) Let $q = 2^e$, and regard T as an e-dimensional vector space over $\mathrm{GF}(2)$ and D as a subgroup of $\mathrm{GL}(T)$ as in 12.1. Let $E = \mathrm{End}_D(T)$. As D is regular on $T^\#$, D is irreducible, so by Schur's Lemma 12.4, E is a finite division algebra over $\mathrm{GF}(2)$ of dimension at most e. Therefore by 26.1, E is a finite field and $f = |E : \mathrm{GF}(2)| \le e$. But $D^\# \le E^\#$, so

$$q - 1 = |D| \le |E| = 2^f - 1 \le 2^e - 1 = q - 1,$$

so $E = F$. Thus T can be regarded as the additive group of F, and D as the multiplicative group $F^\#$, and the action of D by conjugation is $a : b \mapsto ab$ for $a \in D$ and $b \in T$. Thus D and T are determined up to isomorphism and the representation of D on T is determined up to quasiequivalence, so the semidirect product H is determined up to isomorphism by 10.3. Hence there exists an isomorphism $\pi : H \to H^*$ with $D\pi = D^* = \{\phi(a, 0, 0, 1) : a \in F^\#\}$, and the map

$$\beta : H/D \to H^*/D^*,$$

$$Dt \mapsto D^*(t\pi)$$

is a permutation isomorphism of the representations of H on H/D and H^* on H^*/D^*. Finally the map

$$\gamma: H^*/D^* \to F,$$

$$D^*\phi(1, b, 0, 1) \mapsto b$$

is an equivalence of the representations of H^* on H^*/D^* and F, so composing these two isomorphisms, we get the isomorphism α of (6).

(7) As D is transitive on $X - \{H, Hu\}$, there is $d \in D$ with $(1\alpha^{-1})ud = 1\alpha^{-1}$. Let $v = ud$. Note that as u inverts D and D is abelian, v inverts D. For $a \in F^{\#}$, let $\psi(a) = \phi(a, 0, 0, 1)$. Thus $\psi(a)^{-1} = \psi(a^{-1})$ and $\psi(a)\pi^{-1} \in D$, so $\psi(a)\pi^{-1}$ is inverted by v. Therefore

$$(a\alpha^{-1})v = ((1\psi(a))\alpha^{-1})v = (1\alpha^{-1})\psi(a)\pi^{-1}v = (1\alpha^{-1})v(\psi(a)\pi^{-1})^v$$

$$= (1\alpha^{-1})(\psi(a)^{-1})\pi^{-1} = (1\psi(a^{-1}))\alpha^{-1} = a^{-1}\alpha^{-1}$$

(8) As $\alpha: X \to Y$ is an isomorphism of sets, $\alpha^*: \operatorname{Sym}(X) \to \operatorname{Sym}(Y)$ is an isomorphism of groups, where $\alpha^*: g \mapsto \alpha^{-1}g\alpha$. Thus $\alpha^*: L \to L^*$ is an isomorphism, where $L = \langle H, v \rangle$, $L^* = \langle H^*, v^* \rangle$, and $v^* = v\alpha^*$. However, by construction, v^* fixes 1 and has cycles $(0, \infty)$, (a, a^{-1}), for $a \in F - \{0, 1\}$. That is, $v^* = \phi(0, 1, 1, 0) \in G^*$. Further α^* extends π as $x\alpha h\pi = (xh)\alpha$ for all $x \in X$ and $h \in H$. Thus it remains to show that $G = \langle H, v \rangle$ and $G^* = \langle H^*, v^* \rangle$. But as G is 3-transitive on X, G is primitive on X by 15.14, so H is maximal in G by 5.19, and hence indeed $G = \langle H, v \rangle$. Similarly $G^* = \langle H^*, v^* \rangle$.

References

Al. Alperin, J., Sylow intersections and fusion, *J. Alg.* **6** (1967), 222–41.

AL. Alperin, J. and Lyons, R., On conjugacy classes of p-elements, *J. Alg.* **19** (1971), 536–7.

Ar. Artin, E., *Geometric Algebra*, Interscience, New York, 1957.

As 1. Aschbacher, M., On finite groups of component type, *Illinois J. Math.* **19** (1975), 78–115.

As 2 Aschbacher, M., Finite groups of rank 3, I, II, *Invent. Math.* **63** (1981), 357–402; **71** (1983), 51–162.

Be 1. Bender, H., On groups with abelian Sylow 2-subgroups, *Math. Z.* **117** (1970), 164–76.

Be 2. Bender, H., Goldschmidt's 2-signalizer functor theorem, *Israel J. Math.* **22** (1975), 208–13.

BT. Borel, A. and Tits, J., Eléments unipotents et sousgroupes paraboliques de groupes réductifs, *Invent. Math.* **12** (1971), 97–104.

Bo. Bourbaki, N., *Groupes et algebras de Lie*, 4, 5, 6, Hermann, Paris, 1968.

BF. Brauer, R. and Fowler, K., On groups of even order, *Ann. Math.* **62** (1965), 565–83.

CPSS. Cameron, P., Praeger, S., Saxl, J. and Seitz, G., The Sims conjecture, to appear.

Ca. Carter, R., *Simple Groups of Lie Type*, Wiley-Interscience, New York, 1972.

Ch 1. Chevalley, C., *The Algebraic Theory of Spinors*, Columbia University Press, Morningside Heights, 1954.

Ch 2. Chevalley, C., *Théorie des groups de Lie*, Tome II, Hermann, Paris, 1951.

Co. Collins, M., Some problems in the theory of finite insolvable groups, Thesis, Oxford University, 1969.

Di. Dieudonné, J., *La géometrie des groupes classiques*, Springer-Verlag, Berlin, 1971.

FT. Feit, W. and Thompson, J., Solvability of groups of odd order, *Pacific J. Math.* **13** (1963), 775–1029.

FS. Fong, P. and Seitz, G., Groups with a (B, N)-pair of rank 2, I, II, *Invent. Math.* **21** (1973), 1–57; **24** (1974), 191–239.

Gl 1. Glauberman, G., On solvable signalizer functors in finite groups, *Proc. London Math. Soc.* **23** (1976), 1–27.

Gl 2. Glauberman, G., Failure of factorization in p-solvable groups, *Quart. J. Math. Oxford* **24** (1973), 71–7.

Gl 3. Glauberman, G., Central elements in core-free groups, *J. Alg.* **4** (1966), 403–20.

Gol 1. Goldschmidt, D., 2-signalizer functors on finite groups, *J. Alg.* **21** (1972), 321–40.

Gol 2. Goldschmidt, D., Solvable signalizer functors on finite groups, *J. Alg.* **21** (1972), 137–48.

Gor 1. Gorenstein, D., On the centralizers of involutions in finite groups, *J. Alg.* **11** (1969), 243–77.

Gor 2. Gorenstein, D., *The Classification of Finite Simple Groups*, I, Plenum Press, New York, 1983.

Gor 3. Gorenstein, D., *Finite Simple Groups; An Introduction to their Classification*, Plenum Press, New York, 1982.

Gor 4. Gorenstein, D., *Finite Groups*, Harper and Row, New York, 1968.

GL. Gorenstein, D. and Lyons, R., The local structure of finite groups of characteristic 2 type, *Memoirs AMS* **276** (1983), 1–731.

GW. Gorenstein, D. and Walter, J., Balance and generation in finite groups, *J. Alg.* **33** (1975), 224–87.

HH. Hall, P. and Higman, G., The *p*-length of a *p*-solvable group and reduction theorems for Burnside's problem, *Proc. London Math. Soc.* **7** (1956), 1–41.

HKS. Hering, C., Kantor, W. and Seitz, G., Finite groups having a split (*B*, *N*)-pair of rank 1, *J. Alg.* **20** (1972), 435–75.

He. Herstein, I., *Topics in Algebra*, Zerox, Lexington, 1975.

Hi. Higman, D., Finite permutation groups of rank 3, *Math. Z.* **86** (1964), 145–56.

La. Lang, S., *Algebra*, Addison-Wesley, Reading, 1971.

Mc. McBride, P., Nonsolvable signalizer functors on finite groups, *J. Alg.* **78** (1982), 215–38.

Ri. Richen, F., Modular representations of split (*B*, *N*)-pairs, *Trans. AMS* **140** (1969), 435–60.

Sh. Shult, E., On groups admitting fixed point free operator groups, *Illinois J. Math.* **9** (1965), 701–20.

St. Steinberg, R., Lectures on Chevalley groups, Yale University, 1968.

Su. Suzuki, M., *Group Theory*, Springer-Verlag, New York, 1982.

Th 1. Thompson, J., Normal *p*-complements for finite groups, *Math. Z.* **72** (1960), 332–54.

Th 2. Thompson, J., Normal *p*-complements for finite groups, *J. Alg.* **1** (1964), 43–6.

Th 3. Thompson, J., Bounds for the orders of maximal subgroups, *J. Alg.* **14** (1970), 135–8.

Ti. Tits, J., *Buildings of Spherical Type and Finite (B, N)-Pairs*, Springer-Verlag, Berlin, 1974.

Wi 1. Wielandt, H., Eine Verallgemeinerung der invarianten Untergruppen, *Math. Z.* **45** (1939), 209–44.

Wi 2. Wielandt, H., *Finite Permutation Groups*, Academic Press, New York.

List of symbols

Symbol	Page	Symbol	Page
$\Delta(V, f), \Delta(V, Q)$	78	$J(X, f)$	78
A^θ	78	N_F^K	84
$\mathrm{sgn}(Q), \mathrm{sgn}(V)$	86, 87	$\mathrm{Sp}_n(F), \mathrm{Sp}(V)$	88
$\mathrm{SO}(V), \mathrm{O}(V), \Omega(V)$	88, 89	$\mathrm{Sp}_n(q), \mathrm{SU}_n(q)$	89
$\mathrm{O}_n^\varepsilon(q), \mathrm{SO}_n^\varepsilon(q)$	89	$\Omega_n^\varepsilon(q)$	89
$\mathrm{PSp}_n(q), \mathrm{PGU}_n(q)$	89	$\mathrm{PO}_n^\varepsilon(q), \mathrm{P}\Omega_n^\varepsilon(q)$	89
G^+	97	$\Phi(H)$	105
$\mathrm{Mod}_{p^n}, D_{2^n}$	107	$\mathrm{SD}_{2^n}, Q_{2^n}$	107
$p^{1+2n}, D_8^n Q_8^m$	110, 111	$L(V_1, \ldots, V_m; V)$	117
$V \otimes U$	118	$U^K, K \otimes_F U$	119
π^K	119	π^σ, V^σ	121
$\mathrm{Grp}(Y : W)$	140	$l(h), l_R(h)$	143
$W(\Sigma)$	148	Σ^+	149
$\mathrm{Comp}(G)$	157	$E(G)$	157
$F(G)$	158	$\mathrm{O}_\infty(G)$	158
$F^*(G)$	159	$\mathrm{O}_{p',E}(G)$	159
$\mathscr{A}(G)$	162	$J(G)$	162
$\mathscr{P}(G, V)$	163	$\alpha_G(B)$	175
$\mathrm{char}(G)$	180	$\mathrm{cl}(G)$	179
α^G	188	$\mathscr{E}_k^p(G)$	246
$\mathscr{D}(X)$	245	$\Gamma_{P,k}(G), \Gamma_{P,k}^0(G)$	247
$\mathscr{E}_k^p(G)^0$	248	$\mathrm{e}(G)$	261
$m_{2,p}(G)$	261		

Index

algebraic integer 184
Alperin's Fusion Theorem 200
alternating group 55
apartment 215
automizer 5
automorphism 7

Baer–Suzuki Theorem 204
bar convention 6
basis
  dual 47
  hyperbolic 81
  orthogonal 79
  orthonormal 79
BN-pair 218
Borel subgroup 219
B_p-Property 263
Brauer–Fowler Theorem 244
building 215
  Weyl group 219
Burnside Normal p-Complement
    Theorem 202
Burnside $p^a q^b$-Theorem 187

Cartan subgroup 219
category 6
  coproduct 7
  product 7
Cauchy's Theorem 20
central product 32
  with identified centers 33
centralizer 3
chamber 209
character 49
  degree 179
  generalized 180
  induced 189
  irreducible 179
character table 183
characteristic value 127
characteristic vector 127
class function 179
classical group 88
Classification Theorem 260
Clifford algebra 95
Clifford group 96
Clifford's Theorem 41
cocycle 64

commutator 26
complex 209
  chamber 209
  chamber graph 209
  connected 209
  reflection 212
  thick 209
  thin 209
Component Theorem 263
composition factors 24
composition series 23
conjugate 3
Coprime Action Theorem 73
covering 168
Coxeter complex 211
  diagram 141
  group 142
  matrix 141
  system 142, irreducible 146
critical subgroup 108
cycle 54
cycle structure 54

dihedral group 141
direct product 4
distance 8
dual space 47
Dynkin diagram 251

edge 8
exact sequence 47
  short 47
  split 47
Exchange Condition 143
extension, central 166
  perfect 168
  universal 166
extension problem 10

FG-homomorphism 36
FG-representation 35
field, perfect 92
Fitting subgroup 158
  generalized 159
fixed point 14
flag 8
folding 211
  opposite 212

Printed in the United States
By Bookmasters